D1654871

Rodney A. Gayer
Reinhard O. Greiling
Andreas K. Vogel
(Eds.)

Rhenohercynian and Subvariscan Fold Belts

International **earth evolution sciences** Monograph Series

R. A. Gayer
The Tectonic Evolution of the Caledonide-Appalachian Orogen

A. Vogel • H. Miller • R. O. Greiling
The Renish Massif

J. Pohl
Research in Terrestrial Impact Structures

C.-Y. King • R. Scarpa
Modeling of Volcanic Processes

S. El-Gaby • R. O. Greiling
The Pan-African Belt of Northeast Africa and Adjacent Areas

C. O. Ofoegbu
Groundwater and Mineral Resources of Nigeria

C. O. Ofoegbu
The Benue Trough: Structure and Evolution

S. J. Freeth • C. O. Ofoegbu • K. M. Onuoha
Natural Hazards in West and Central Africa

R. A. Gayer • R. O. Greiling • A. K. Vogel
Rhenohercynian and Subvariscan Fold Belts

Rodney A. Gayer
Reinhard O. Greiling
Andreas K. Vogel
(Eds.)

Rhenohercynian and Subvariscan Fold Belts

Die Deutsche Bibliothek – CIP-Einheitsaufnahme

Rhenohercynian and subvariscan fold belts /
Rodney A. Gayer ... (ed.). – Braunschweig;
Wiesbaden: Vieweg, 1993
(Earth evolution sciences)
ISBN 3-528-06488-9

NE: Gayer, Rodney A. [Hrsg.]

Vieweg is a subsidiary company of the Bertelsmann Publishing Group International.

Produced by W. Langelüddecke, Braunschweig
Printed on acid-free paper
Printed in Germany

ISBN 3-528-06488-9

Contents

Editorial 1

Convection Tectonics – Global Tectonics in the Light of Mantle-Wide Convection 5
A. Vogel

Basin Evolution

Comparative Evolution of Coal Bearing Foreland Basins along the Variscan Northern Margin in Europe 47
R. A. Gayer, J. E. Cole, R. O. Greiling, C. Hecht, J. A. Jones

Palaeographic and Metamorphic Evolution of the Ligerian Belt in Europe 83
M. S. Oczlon

An Outline of Evolution of the Late Devonian Munster Basin, South-West Ireland (Extended Abstract) 131
E. A. Williams, M. Ford, H. E. Edwards, M. J. O'Sullivan

Facies Complexes of the Upper Carboniferous in North-West Germany and their Structural Implications 139
B. Jankowski, F. David, V. Selter

Silesian Sedimentation in South-West Britain: Sedimentary Responses to the Developing Variscan Orogeny 159
A. Hartley

Basin Inversion and Foreland Basin Development in the Rhenohercynian Zone of South-West England 197
L. N. Warr

Structural Geology

Magnetic Fabric Relationship between Crystalline and Variscan Sedimentary Complexes in Eastern Bohemian Massif 227
F. Hrouda

Some Aspects of Interactivity between Folding and Thrusting in the Ruhr Carboniferous 241
W. Wrede

On the Structure of the Variscan Front in the Eifel-Ardennes-Area 269
V. Wrede, G. Drozdzewski, J. Dvorak

Displacement Gradients of Thrusts, Normal Faults and Folds from the Ruhr and the South Wales Coalfields 297
P. Gillespie

Variscan Thrust Deformation in the South Wales Coalfield – a Case Study from Ffos-Las Opencast Coal Site 315
K. Frodsham, R. A. Gayer, E. James, R. Pryce

Economic Geology

Exploration for Bituminous Coal in NW Germany. Research in Progress (Extended Abstract) 351
M. von Sperber, D. Schmitz, Ä. Strack

The Origin of Kupferschiefer Mineralization in the Variscan Fold Belt of Southwestern Poland 369
S. Speczik

Post-Variscan Evolution

The Post-Variscan Development of the Rhenish Massif 387
W. Meyer

Editorial

The present volume comprises a selection of papers presented at the conference on Rhenohercynian and Subvariscan Fold Belts organized by the Editorial Group of Earth Evolution Sciences (EES) and sponsored by the Tectonic Studies Group (TSG) of the Geological Society of London.

The aim of the conference was to bring together a wide range of geologists working on the northern, external part of the Variscan orogen to study the different aspects of foreland basin development and deformation. These aspects also included the economic importance of the foreland fold-and-thrust belt and its subsequent, post-orogenic evolution.
The conference took place within the Rhenohercynian Belt at Boppard on the river Rhine in June 1989. It was a sequel to a previous seminar in 1984 which resulted in an Earth Evolution Sciences volume on "The Rhenish Massif" edited by A.Vogel et al. (1987).
Due to the sponsorship of the TSG the 1989 meeting attracted a wide range of delegates, particularly from geologists working on the Variscan of the British Isles. Therefore the collection of papers, which resulted from the conference and is presented here, provides a relatively wide regional coverage of the Variscan northern margin from Ireland in the west through Wales and SW England to central Europe including Poland in the east.

The conference was opened by a workshop on
Global Tectonics in the Light of Mantle Convection.
This volume is preceded by a frontier article on recent progress in understanding tectonic forces from this aspect.
For proper understanding of the history of orogenic belts their evolution should be seen in the context of global geotectonics. While the formulation of the plate tectonic concept has initiated an era of great discoveries from the phenomenological point of view, our rapidly increasing knowledge about the dynamic processes in the Earth's interior begins to reveal a causative view, leading to a new breakthrough in our understanding of the geological evolution of our planet.
The core-mantle boundary divides two dynamic systems with extremely different physical properties and composition. The initial article discusses the role of
core-mantle interaction in guiding mantle convection, thus providing an explanation for the complexity of geotectonic processes both at present and in Earth's history. It also explains the function of the geomagnetic field as a reference system in geotectonics.

The papers from the conference selected for this volume are grouped into three major topics:

Basin evolution
Structural geology
Economic geology
These sections are followed by a review of the
Post-Variscan Evolution of the Rhenish Massif.

Of the six papers within the section on Variscan **Basin Evolution**, the first on the evolution of coal-bearing foreland basins along the Variscan northern margin by R.A. Gayer and coworkers (Cardiff, Heidelberg, Cheltenham) sets the scene for the following papers. It reviews examples of coal-bearing basins from SW Britain, N France, Belgium and NW Germany, highlighting the characteristics of foreland basin development. The analysis of tectonic load and effects of the Wales-Brabant Massif on basin development serve to contrast the Ruhr basin from those lying to the west. The second paper on Silesian sedimentation in SW Britain by A.J. Hartley (Cardiff/Aberdeen) describes in some detail the Namurian to Early Cantabrian sedimentary sequences and depositional environments of two major sedimentary basins in SW England and S Wales, producing a series of palaeogeographic maps for the main stages of basin evolution. The paper discusses the relationship between basin development and Variscan deformation in terms of a progressively advancing deformation front. The following paper on Upper Carboniferous facies complexes of the Upper Carboniferous in NW Germany by B. Jankowski and coworkers (Bochum) serves as an interesting comparison between this and the SW Britain basins of the previous paper. It analyses in some detail the basin-fill into facies complexes developed in the Ruhr coal basin and demonstrates the evolution of the basin from an early marine stage, through stages of lower delta plain to upper delta plain and fluvial-alluvial plain. The paper also discusses the relationship between tectonics and basin development. The next two papers discuss the Rhenohercynian Zone in SW Ireland and in SW England. E.A. Williams and coworkers (Cork,Zürich,Keele) investigate the evolution of the Munster basin - the most westerly extension of the Rhenohercynian in Europe. This extended abstract describes the main basin-fill features of the Upper Palaeozoic, including an unusual development of the Devonian Upper Old Red Sandstone type sequence, and outlines the Variscan deformation in terms of the strain developed and the tectonic evolution during Variscan inversion. The paper by L.N. Warr (Exeter,Heidelberg) on basin inversion and foreland basin development in the Rhenohercynian of SW England describes in general terms the principal sedimentary sequences of an early extensional basin in Devonian and Early Carboniferous times and a late foreland basin of Silesian age. This serves as a template to analyse the deformation and metamorphism of the region in two principal Variscan stages, accommodating over 50% shortening. The final paper of this section describes the main features of the southern European Variscides of the Ligerian Belt by M.S. Oczlon (Heidelberg). The paper outlines the evolution of this belt from Marocco through NW Spain to Bohemia. It demonstrates through an analysis of palaeogeography, deformation and metamorphism, the development of a Wilson Cycle from an oceanic phase in the Early Silurian, through Late Silurian subduction, to Early Devonian collision tectonics as Gondwana underthrust the Ligerian Terrane.

The section involved with **Structural Geology** consists of five papers, four dealing with aspects of thrust and fold deformation in the external northern zones of the Variscides and the fifth with magnetic anisotropy in the eastern margins of the Bohemian Massif. This latter paper by F. Hrouda (Brno) presents the results of a comprehensive study of magnetic fabric in both the crystalline basement of the eastern Bohemian massif and its Lower Carboniferous cover. It is demonstrated that the fabric in both basement and cover is tectonic and related to Variscan nappe development. In a paper describing the relationship between folding and thrusting in the Ruhr coal basin, V. Wrede (Krefeld) shows that folding and thrusting are intimately related within particular stockwerks of the structure. These thrusts are not linked and do not represent imbrication from a common detachment. The third paper of the section analyses the Variscan Front in the Eifel-Ardennes area. V. Wrede and his coworkers (Krefeld and Brno) show that the Midi-Aachen thrust system is not a megathrust with 100s km displacement, as has been suggested for the French sector, but has a maximum of 4 km throw in the Aachen area and grades eastwards into the autochthonous Ruhr basin. The paper by P. Gillespie (Cardiff/Liverpool) analyses folds and faults in terms of displacement gradients, using mine data from both the S Wales and Ruhr coal basins. He shows that both folds and faults have elliptical tip-line loops of zero displacement, with horizontal displacement gradients varying from 2% to 20%. He also demonstrates the importance of such analyses for fault prediction, particularly in mines. The final paper of the section by K. Frodsham and coworkers (Cardiff and British Coal) studies the development of thrusts in the S Wales coalfield using data from almost 3000 boreholes and 3D exposure in an opencast coal mine. The thrusts are shown to develop in an unusual break-back sequence, giving a highly characteristic geometry.

The third section on **Economic Geology** has two papers. The first by M. v.Sperber and colleagues (Bochum and Saarbrücken) outlines the approach to coal exploration in the north of the Ruhr coal basin. The paper by S. Speczik (Warsaw) on the origin of Kupferschiefer mineralization in the Polish Variscides describes fluid inclusion and vitrinite-liptinite reflectivity studies to establish anomalously high geothermal gradients in the Polish Rhenohercynian Zone basement. It concludes that availability of high thermal energy related to the Variscan orogeny is an important control on the formation of Kupferschiefer mineralization.

The final paper of the volume by W. Meyer (Bonn) demonstrates the factors leading to the **post-Variscan development** of the Rhenish Massif, emphasizing uplift processes and continental volcanism.

The editors are specifically aware of the engagement of Mrs. Elka von Hoyningen-Huene-Vogel (Berlin), who contributed greatly to the success of the seminar. They also appreciate the contributions of M. Ohadi (Berlin) and C.Hecht, C.Hofmann, M.Oczlon and E.Stephan (Heidelberg). Editing of the volume was supported by the following colleagues, who reviewed one or more of the contributions:

J.Andrews/ Southampton, B.M.Besly/ Keele, C.Cornford/ Hallsannery, W.-Chr.Dullo/ Kiel, L.Fontboté/ Genéve, M.Ford/ Zürich, W.Franke/ Gießen, A.Hartley/ Aberdeen, D.Heling/ Heidelberg, G.Kelling/ Keele, M.Miliorizos/ Cardiff, H.Miller/ München, A.Moiseyev/ Hayward CA., O.Oncken/ Würzburg, J.S.Rathore/ Trondheim, H.G.Reading/ Oxford, J.Rippon/ Nottingham, D.Sanderson/ Southampton, D.Warnke/ Hayward CA., L.N.Warr/ Heidelberg, J.Watterson/ Liverpool, K.Weber/ Göttingen.

Final proofreading by A.H.N.Rice and L.N.Warr (Heidelberg) is greatfully acknowledged, as is technical assistance by I.Knapp, R.Eisenhauer, R.Koch and K.Schacherl (Heidelberg) and L.Jakob and L.Krauss (Berlin).

January 1992

Rodney A.Gayer, Cardiff
Reinhard O.Greiling, Heidelberg
Andreas K.Vogel, Berlin

Reference:

Vogel,A., Miller,H. and Greiling,R. (Eds.) 1987.
The Rhenish Massif - Structure, Evolution, Mineral Deposits and Present Geodynamics. Earth Evolution Sciences, Vieweg Braunschweig/Wiesbaden.

Convection Tectonics - Global Tectonics in the Light of Mantle-Wide Convection

A. Vogel

Mathematical Geophysics Group, Free University of Berlin, Podbielskiallee 60, D-1000 Berlin 33

Abstract

A seismological investigation on the shape of the Earth's fluid core carried out by the author at the end of the fifties revealed irregularities which were considered as a first evidence for convective flow in the Earth's deep interior. In recent years the irregular shape has been confirmed by seismic computer tomography. Combined with further results of seismological, geophysical and geochemical research a model on the structure and dynamics of the core-mantle boundary has been developed. It provides an unexpected and surprisingly clear insight into the steering process of mantle convection and thus into the driving mechanism of global geotectonics, past and present.

In view of inconsistencies inherent in the concept of plate tectonics it seems more adequate to reveal a causative view by introducing the concept of "convection tectonics" as a framework to describe and explain the complexity of geotectonic processes. The model under consideration also explains why both the geomagnetic field and the hot spot framework form an independent reference system for geotectonic processes.

1. Retrospective view on Alfred Wegener's theory of continental drift

Alfred Wegener's first article published in 1912 on his theory of continental drift can be regarded as the beginning of an epoch of great discoveries about the history of Earth. Although Wegener was not the first to recognize the matching of coast lines of continents and to draw the conclusion that a large continent may have broken apart, to form the present continents, he collected further evidence in favor of this hypothesis (Wegener 1915, 1929). He was aware of the early results of geophysical research which clearly indicated that the structure of the ocean bottom is totally different from continents and that the continental blocks according to Archimedes' principle are floating in the heavier material which form the ocean bottom. This simple and clear idea is being followed in the present paper.

In spite of overwhelming evidence, Wegener was not accepted by the authorities of his time because he could not find a satisfactory explanation for the driving forces of continental drift. Theoreticians of his time maintained that the mantle was rigid like steel, and thus unable to permit any movements of continents. These arguments were encountered by Wegener with the first seismological results which had revealed a fluid zone in the upper mantle, and the work of his Austrian colleagues Ampferer and Schwinner who proposed a flow of mantle material and convection as the causes of displacement of continents, respectively.

2. First evidence for an irregular shape of the earth's fluid core and its implication

Vogel (1960 a) had a first opportunity to carry out an investigation on irregularities in the shape of the earth's fluid core by means of earthquake waves reflected at the core-mantle boundary (Fig. 1). At that time it was already known that the Earth's interior consists of crust, mantle, outer and inner core. Traveltime residuals of both PcP- and ScS-Phases were considered as the effect of deviations from the normal depth to the core of 2898 +/- 4 km. They were assessed quantitatively by comparing observed traveltimes with theoretical curves for various depths to the core-mantle boundary (CMB). In order to eliminate the effect of inhomogeneities in the crust and upper mantle and also uncertainties of the focal parameters, traveltime differences between reflected waves and direct waves were considered (Fig. 1).

Earthquake records from seismological observatories worldwide were used to determine the depths to the core at about 500 reflection points. Regions with a tendency to depths less than normal could be separated from those with greater depths. Because of gaps and scatter of the data, it was difficult to quantify the regional deviations upward and downward. Mean values indicated elevations and depressions of a few tens of kilometers in amplitude. Fig. 2 a shows a simplified presentation of the topography of the core-mantle boundary as obtained by these investigations.

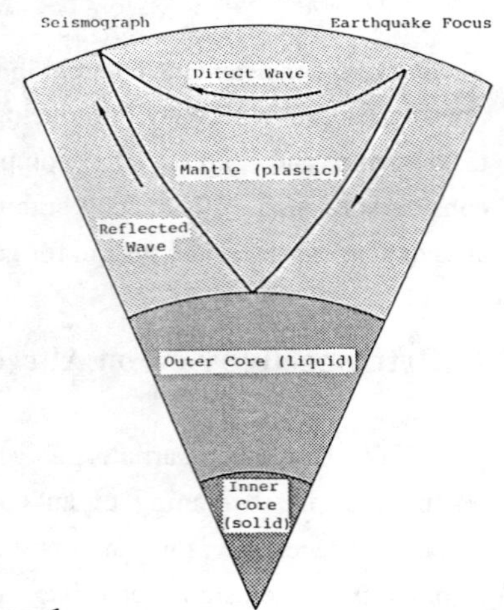

Figure 1
Ray paths of direct and reflected earthquake waves at the core-mantle boundary.

The existence of irregularities in the shape of the fluid core, or simply "bumps" of the core-mantle boundary involved some highly significant geophysical implications. The bumps were considered as an evidence for large-scale dynamic processes in the Earth's deep interior. A homogeneous fluid core which is only affected by gravitational and centrifugal forces should assume the shape of an oblate sphere or an ellipsoid. The irregular shape was interpreted as an indication of instabilities and mass displacements. Thermal convection was suggested as the reason for the core's irregular shape, whereby according to the laws of thermodynamics rising currents cause an elevation, descending currents a depression of the core-mantle boundary.

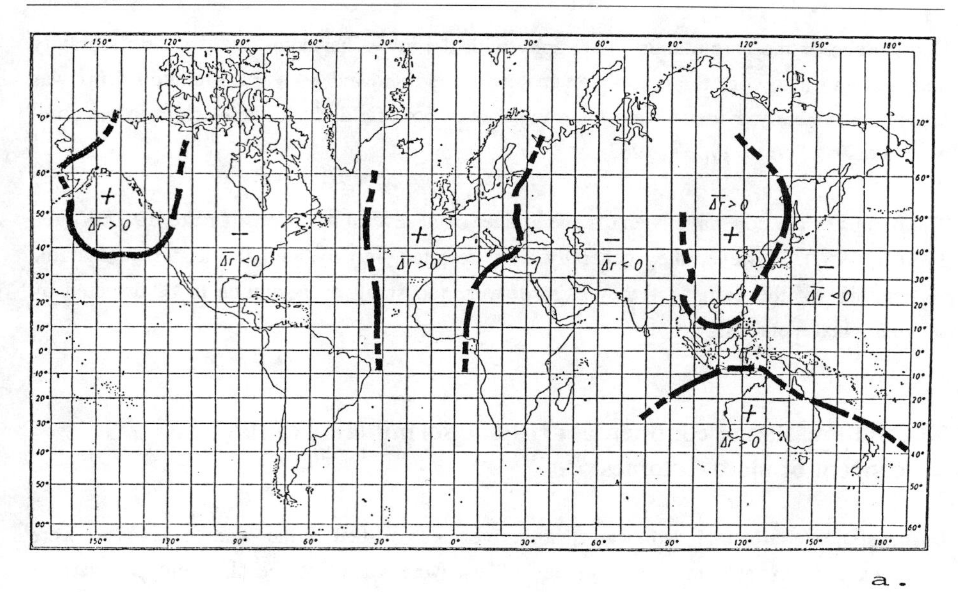

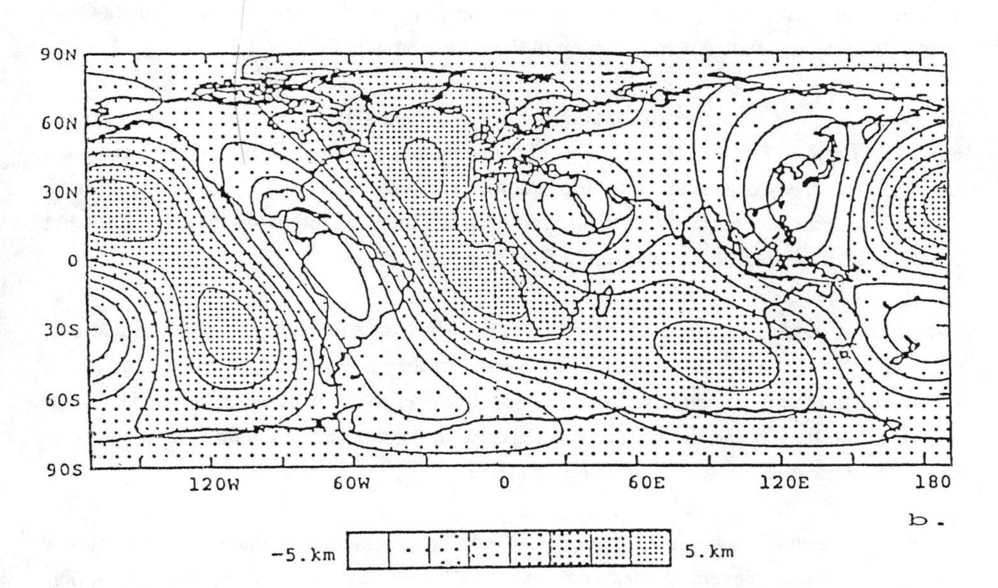

Figure 2
a. Depth to Earth's core from reflected earthquakes waves during the period 1948 - 1954. + less, - more than 2900 km (after A. Vogel, 1960).
b. Topography of core-mantle boundary obtained by inversion of PcP residuals, corrected for lower mantle heterogeneity, for spherical harmonic coefficients up to degree and order 4. + higher, - deeper than normal (after Morelli, 1989).

Comparison of the core-mantle topography with the lower harmonics of the earth's gravity potential and also with the non-dipole magnetic field showed a clear correlation. This was interpreted as early evidence that main features of both fields have their origin in the core-mantle region (Vogel 1960 a, 1963).

The bumps of the core have been discussed later on at several occasions (Lucke, 1960; Vogel, 1960 b; 1967). However they generally were rejected, mostly with the argument that the vertical size of the bumps, if they existed, were too small in amplitude to be detected by seismological methods.

3. Confirmation of continental drift and formulation of the concept of plate tectonics

In the sixties continental drift was quantitatively established. Polar wandering paths were constructed from paleomagnetic studies. They were found to be different for various continents. By proving that the magnetic field has been a dipole through most of the Earth's history, the pole wandering curves were interpreted as movements of the Earth's surface and of individual continents against a stable magnetic pole as a reference, a technique which opened the way to detailed reconstruction of their position with time.

At the same time intensive exploration of the sea floors revealed the now well-known remarkable relief (Fig. 3). The most prominent features are the mid-oceanic ridges which extend continuously over ten thousands of kilometers along the ocean bottom cut into sections by displacements along transform faults, and deep sea trenches following the coast lines and island areas of the Pacific and western Indian Ocean.

Interestingly, some plateaus, such as between the Malayan Archipelago, lie for a great part submerged below the sea level in spite of the fact that they are continental masses. Similarly, volcanic sea-mounts sometimes rise above the sea-level to form islands like the Hawaiian Islands in the Pacific and the Canary and Cape Verde Islands in the Atlantic, although their natural isostatic position should be below sea level.

The identification of magnetic anomalies arranged in stripes parallel to the mid-oceanic ridges as a record of the magnetic polarity time scale led to the discovery of sea-floor spreading, which could be reconstructed to provide further details of continental drift.

When new ocean bottom is generated continuously, it has to be consumed somewhere. With the improvement of worldwide seismograph networks, a more detailed picture of the

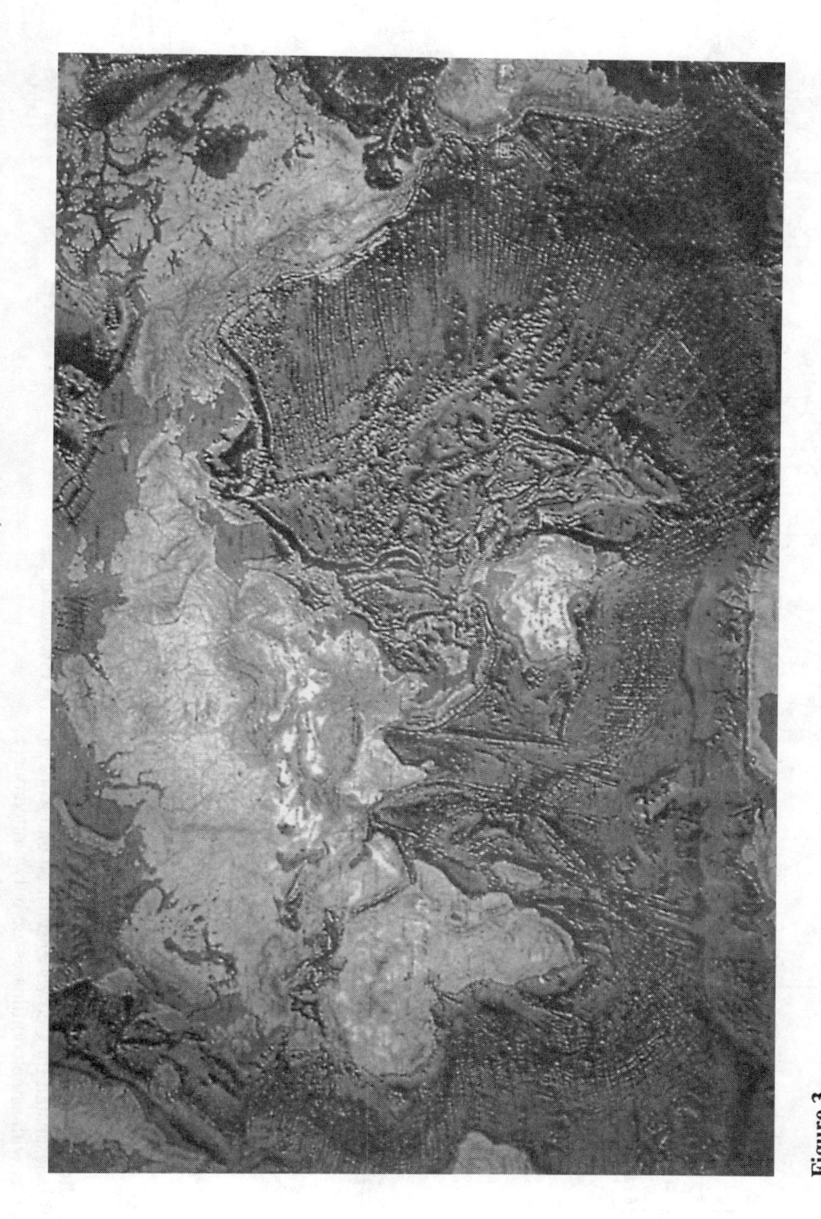

Figure 3
Relief of the Earth's solid surface.

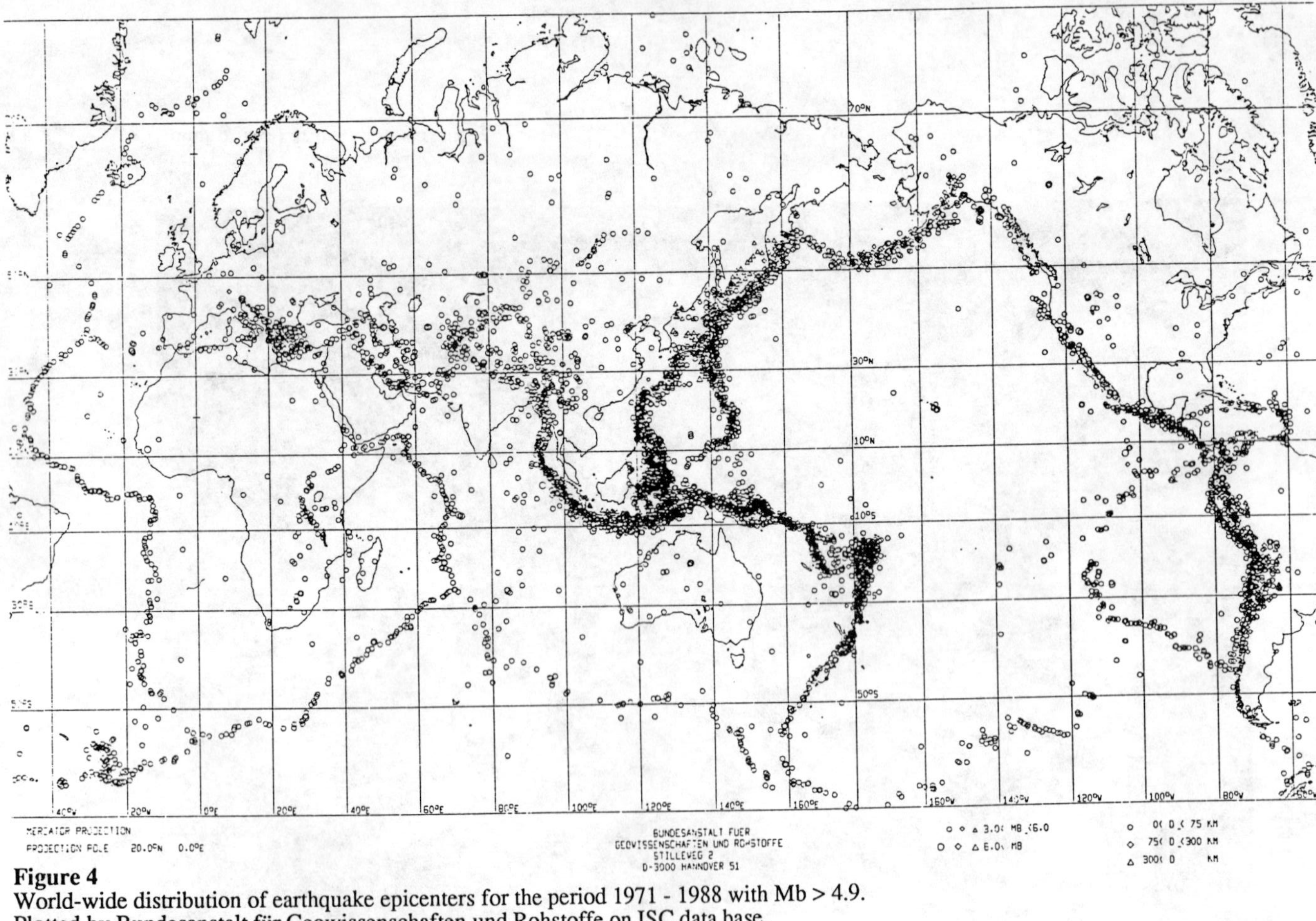

Figure 4
World-wide distribution of earthquake epicenters for the period 1971 - 1988 with Mb > 4.9.
Plotted by Bundesanstalt für Geowissenschaften und Rohstoffe on ISC data base,
whereby events with same coordinates are only plotted once.

distribution of earthquakes could be obtained. Early it was found (Turner, 1930; Wadati, 1935) that below island areas and continental margins, the depth of earthquakes increases continuously. Benioff (1954) identified this pattern as the traces of oceanic crust which are subducted into the mantle, today known as the Wadati-Benioff zones.

Some places of continuing volcanic activities, the hot spots, were found to have a fixed position with respect to each other over millions of years thus forming a reference framework for a detailed reconstruction of motions of the Earth's surface (Wilson, 1963).

The spatial distribution of earthquakes as revealed by a worldwide network of seismograph stations showed that earthquakes occur mainly in very distinct zones (Fig. 4). Major earthquake activity turned out to be confined to the main topographic features of the earth's solid surface which are at the same time the youngest geological structures, such as mid-oceanic ridges, deep-sea trenches, oceanic and continental rifts and young mountain chains (Fig. 3). It showed that the tectonic forces which have been forming the youngest geological structures are still active, as revealed by seismic as well as volcanic activity.

Investigations of seismic wave propagation confirmed the original discovery by Gutenberg (1928) of a low-velocity layer below the lithosphere which is supposed to allow plastic flow, extending to depths of 100 to 400 km.

All these findings formed the basis for the formulation of the theory of plate tectonics. More details about the issues that were crucial to establishing plate tectonics can be found in a compilation of original papers by Cox (1973). According to the original concept of plate tectonics, the outer shell of the earth consists of about a dozen rigid plates, comprising both continents and ocean bottom. Floating on the viscous material of the asthenosphere, the plates are moving with respect to each other. Continents are embedded in the lithospheric plates, whereby their drift is caused by relative plate motion (Fig. 5 a).

Due to the relative motion, the plates interact with each other at their boundaries causing tectonic forces still active as revealed by seismic and volcanic activity. According to the conventional theory of plate tectonics the thermal, mechanical and chemical properties of the lithospheric plates below a depth of 100 km or so are largely independent of the crustal type. The origin of the lithospheric plates is explained by a boundary layer evolution, according to a cooling plate model, whereby old oceans and old continents have the same subcrustal lithospheric structure. The stress and strain fields within the plates, and thus intraplate tectonic processes, are caused by forces exerted at the plate boundaries. It explains intra-continental tectonics which is evident in the geologic record and presently observable, including the diffuse nature of continental seismicity.

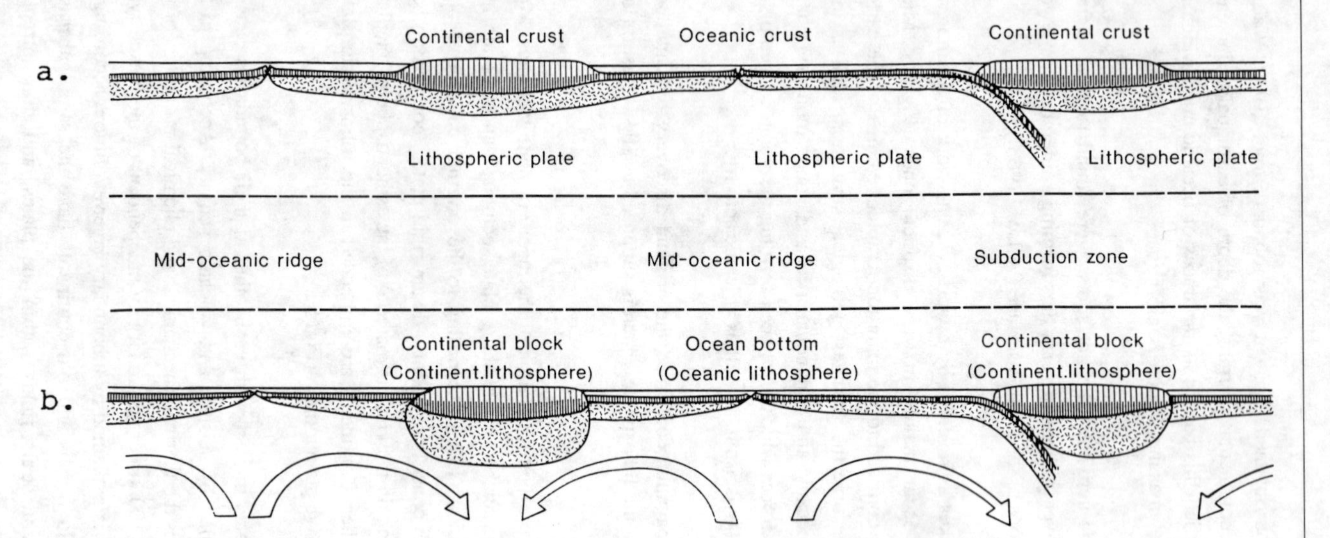

Figure 5
Idealized models of lithospheric structure.
a. According to the original concept of plate tectonics, the continental crust is embedded in lithospheric plates, comprising both continents and oceans.
b. In our present view continents have to be considered as individual blocks and the ocean floor as temporarily solidified mantle material. This is in accordance with the original ideas of A. Wegener, however with the difference that the continental blocks blocks are not drifting at the basis of their crust, but at the basis of their inherent lithosphere.

4. Confirmation of the irregular shape of the Earth's fluid core and its hypothetical relation to global tectonics

A few years ago a comprehensive study of the topography of the Earth's core by reflected earthquake waves was carried out by Morelli and Dziewonski (1987) and continued by Morelli (1989). Taking into consideration the effect of the presently known mantle structure, traveltime residuals of about 26 000 readings of PcP-waves were used to obtain the shape of the core-mantle boundary. The results of their study by application of computer tomography, as presented in Fig. 2 b, exhibit striking similarities with my early results in Fig. 2 a based on more limited data, as already demonstrated (Vogel, 1989). Below North and Central America the core boundary is clearly depressed. It is elevated in the Eastern Atlantic, in Western Europe and Western Africa. In Eastern Europe, Western Asia and Eastern Africa it is again depressed. On the map of Morelli and Dziewonski there is a slight tendency upwards in Central Asia. In my map a high appears somewhat displaced to the east, but there was however a lack of data in the central part of Asia. In the Western Pacific both maps show a depression. Southwest of Australia the core boundary rises upward, and in the Eastern Pacific we find a large elevation of the core boundary in the tomographic map. My early results show an elevated area which is limited to the northeastern part of the Pacific. Again, the southern bound of this area is of little significance because there were almost no data available for the Central and Southeastern Pacific.

The amplitude of the "bumps" according to modern tomography is +/- 6 km, when spherical harmonic expansion is applied to the data (Morelli and Dziewonski, 1987), up to +/- 15 km, when the data undergo adaptive filtration (Morelli, 1989). Filtering reveals some better coincidence with the earlier results, such as a local elevation south of Alaska.

5. Inconsistencies of plate tectonics

The processes of seafloor spreading and subduction are well established. Continental drift is well documented. However, the explanation of these processes in terms of plate tectonics meets some inconsistencies.

Large rigid plates in the sense of plate tectonics, comprising oceans and continents of common subcrustal lithosphere, do not exist. The different nature of continents and oceans becomes obvious just by comparison of their appearance. Most striking is the difference between elevation of continents and depths to ocean bottom. The age of the sea-bottom varies systematically with its distance from the spreading centers and its main tectonic features, the transform faults aligned perpendicularly to these centers. In comparison, continents consist of a diversity of rock and deformation units of different age and origin.

According to Jordan et al. (1989) it is convenient to discriminate between three types of boundary layers, when describing the structure of the lithosphere:

1. Thermal boundary layer (TBL), which is defined by the thermal thickness of the lithosphere, being equal to the depth where the geotherm reaches the mantle adiabat.
2. Mechanical boundary layer (MBL), which is defined by the effective mechanical thickness for long-term loads
3. Chemical boundary layer (CBL), which is formed by chemical differentiation. In a layered convective body with a chemical composition as complicated as the Earth, it can be expected that the mechanism of mass transport will, over time, build up compositionally distinct boundary layers of intermediate density at major chemical transitions such as the Earth's surface and core-mantle boundary.

From a kinematic point of view, the TBL defines the lithosphere or plate in the conventional terminology of plate tectonics. Traditionally, lithosphere denotes the strong outer shell, the MBL or tectosphere, which is capable of withstanding large deviatoric stresses over geologically long periods of time.

In a review of seismological and other evidence, Andersson (1979) stated that the continental roots extend no deeper than 150 - 200 km. Sclater et al. (1981) concluded that the thermal structure beneath old stable continents is indistinguishable from that beneath old oceans below 150 to 200 km and that oceans and continents are part of the same thermal system.

In the simplest model of plate tectonics, the plates themselves are the upper thermal boundary layer of the mantle convection. The cold, mechanically strong boundary layer has a high solidity and ability to transmit deviatoric stresses from the plate boundaries.

The thermal boundary layer theory of plate structure has enjoyed great success in explaining the large-scale physiography and evolution of the major ocean basins as well as many of their features. Attempts to integrate the continents into this thermal boundary layer have been less successful (Jordan et al., 1989).

A comparison of the lithospheric structure as it has become known in recent years reveals the different nature of continental and oceanic lithosphere. In Fig. 6 the main structural peculiarities have been drawn for comparison.

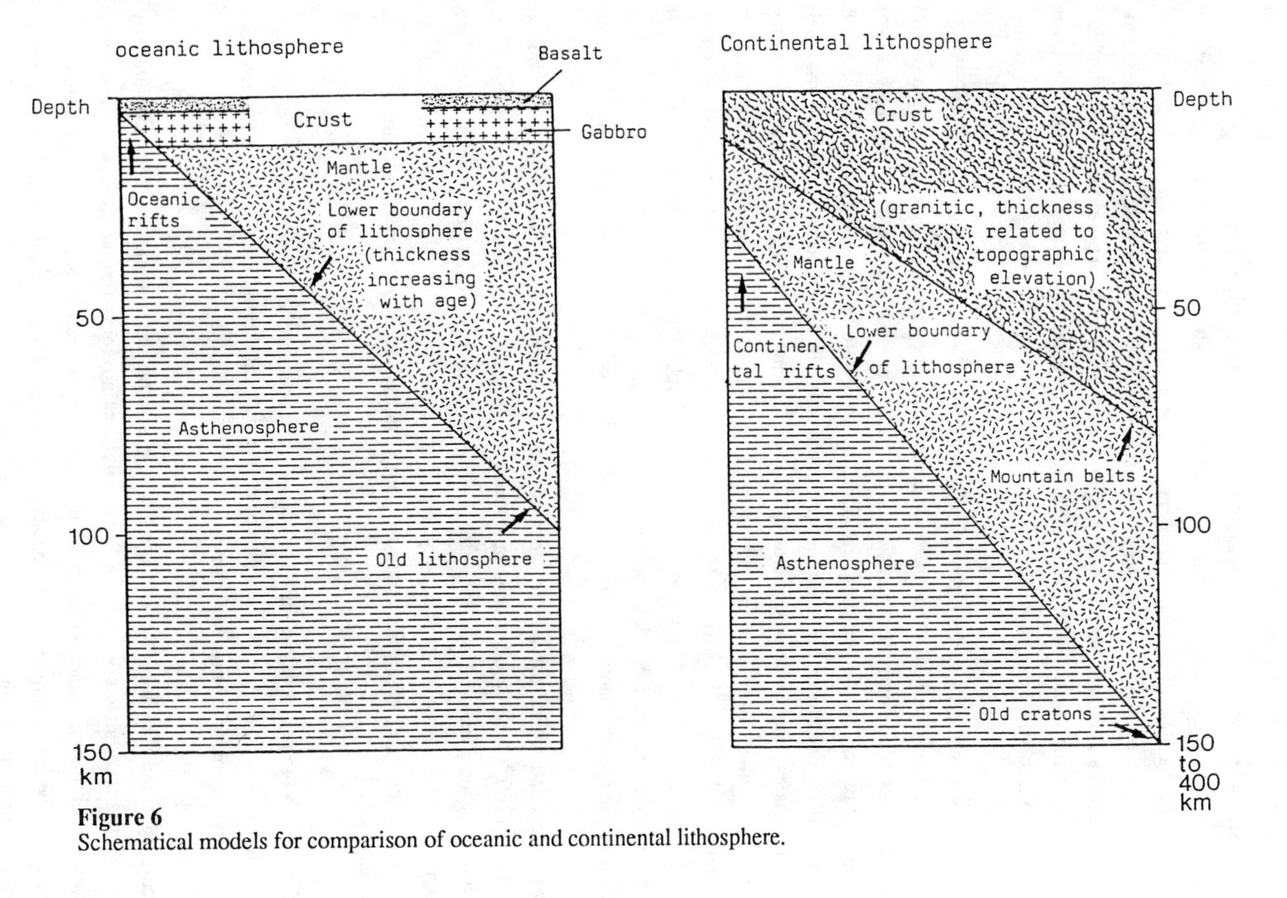

Figure 6
Schematical models for comparison of oceanic and continental lithosphere.

The ocean floor to the left consists of a crust, generally about 7 km thick and the subcrustal lithosphere increasing with age to an asymptotic thickness of slightly more than 100 km. The oceanic crust is a chemical boundary layer which is produced at the spreading centers by extraction of basaltic melt from the upper mantle material, whereby a residual layer of depleted mantle material of 25 - 30 km thickness is formed. This ophiolithic sequence of about 35 km is a chemical boundary layer that translates away from the ridge crests with little modification (Ringwood, 1969). There is sufficient evidence that oceanic lithosphere evolves according to a cooling layer model.

The continents in Fig. 6 to the right are composed of the crust, which varies considerably in thickness and the subcrustal lithosphere which in old cratons locally can reach a depth of 400 km. It has now been confirmed by a variety of seismic methods that the continental blocks are generally substantially thicker than any oceanic lithosphere of any age (Jordan et al., 1989; Condie, 1989). Continents are underlain by an extensive layer of anomalous mantle material, which translates with the continental drift. Thus, continents are individual blocks which carry their deep mantle with them as they move about the surface. The average thickness of continental lithosphere is considerably greater than allowed by the simple thermal boundary model of classical plate tectonics. Advective thickening by lateral accretion across subduction zones or by diapirism from descending slabs as well as continent to continent collision has been proposed as mechanism for the growth of buoyant continental masses (Jordan, 1978).

Petrological data indicate that in continental areas the temperature at a depth of 150 - 200 km is 1000 - 1200 °C, while oceanic temperatures are inferred to be 300 °C higher. It becomes obvious that the subcrustal lithosphere of continents consists of material which in spite of the lower temperatures has a lower density than the oceanic upper mantle at the same depth. In such a way the continental blocks are buoyantly trapped at the surface and stabilized against convective recycling.

Fig. 5 b shows in principle the resulting view on the lithospheric structure in comparison with the plate tectonic concept. Continents consist of individual lithospheric blocks and ocean floor of temporarily solidified mantle material.

According to the plate tectonics concept in Fig. 5 a, intraplate tectonics has been explained by forces acting from plate boundaries or by subdivision into mini- or microplates. In our current view intracontinental tectonics in geologic time and at present as revealed by earthquake activity can be easily explained by convective flow directly acting on the continental blocks from the underlying mantle.

6. A refined model of the core-mantle boundary

Results of geophysical research have shown that the core-mantle boundary is not implicitly a sharp interface between presumedly plastic mantle material and fluid core alloy. The core-mantle boundary (CMB) separates dynamic systems of remarkably contrasting composition and material properties. A natural consequence is the development of regionally distinct thermal and chemical boundary layers comparable to the different conditions in oceanic and continental regions at the Earth's surface.

On the basis of reduced seismic velocity gradients, Bullen (1949) found a transition layer 200 to 300 km thick at the base of the mantle which he designated the D" region. Intensive studies in recent years have shown that we have to consider a boundary layer which varies laterally both in thickness and radial structure. The most important constraints on the properties of the core-mantle transition are provided by seismology, mineral physics, geomagnetism and geodynamics. Lay (1989) reviewed these conditions in detail. Its salient results are listed below.

Seismological methods have provided critical information about the variation in velocity structure in the transition zone for the longest period of time. By inversion of large data sets of P and S wave travel times and waveforms, Dziewonski (1984) found some spatial continuity between the regions of faster than average velocity in D" and the fast regions in the overlying mantle, located roughly below the circum-Pacific belt. Shear velocities obtained by waveform modeling also show coherent large-scale variations, exhibiting a pronounced ring of high shear velocities with several percent velocity perturbations in D" beneath the circum-Pacific (Woodhouse and Dziewonski, 1987). One has to note that depressed regions in the long-wavelength topography of Morelli and Dziewonski (1987) tend to correlate with the zones of higher P- and S-velocity in D".

Obviously depressed topography is dynamically supported by downwelling cool mantle material. The shear velocity structure for isolated patches of D" has been investigated by various authors as reviewed by Lay (1989). They found precursors of the core-reflections ScS, most probably reflections from the top of the D"-layer, in areas with large-scale high-velocity perturbations, while in the low velocity area below the Pacific such reflections have not been evident. It is most likely that this discontinuity at the top of the D" layer is caused by compositional stratification which appears to vary laterally and vanishes, like under the Pacific.

Mineral physics has placed some important constraints on the most probable lower mantle mineral phase, silicate perovskite, as well as iron-oxygen and iron sulfur systems relevant to the core-alloy. Recent estimates of the lower mantle adiabats range from 2600 to 3800 K (Jeanloz and Morris, 1986). The outer core temperature has been estimated between 3800 and 4400 K (Ahrens and Hager, 1987). These estimates require a temperature increase of at least 700 K across the D" transition zone.

Both seismic velocity models and experiments in mineral physics suggest that chemical dregs are located in D" which are aggregations by the action of deep mantle convection systems, either products of downwelling subduction or chemical reactions with the core. Experiments under high pressure and temperature indicate that chemical reactions between the mantle and core may be occurring, leading to the development of a chemically distinct transition zone in D". However, given the low diffusion coefficients involved, enrichment of a chemically distinct zone more than a few hundred meters thick requires that a dynamic mechanism is sweeping away the reaction zone and exposes continuously fresh material to the core. Such a process can take place in regions where convecting mantle material moves horizontally along the core-mantle boundary, entraining the iron-enriched mantle material in areas of rising currents, thereby accumulating an iron-enriched transition zone at the base of the mantle.

Geodynamic constraints which have been inferred from the Earth's forced nutation do not permit a topography of the core-mantle boundary larger than a few kilometers. Modelling of the whole mantle convection as constrained by the seismic structure and the geoid permits a dynamic topography of the core-mantle boundary of only 3 kilometers (Hager et al., 1985) in case of a sharp boundary.

There is an obvious discrepancy between the seismologically revealed and geodynamically constrained topography of the core-mantle boundary.

7. Pattern of mantle convection

Today it is generally agreed that global tectonics is driven by convective flow in the mantle. A vigorous debate has been focused on the vertical extent of the convective layer (K. C. Condie, 1989). According to one school, mantle convection takes place in two layers which are strictly divided by the seismic discontinuity in 670 km depth, separating the upper from the lower mantle. According to another it extends over most, if not the entire depth, of the mantle. The question of stratified versus whole mantle convection was raised nearly two decades before the advent of plate tectonics (Gutenberg et al., 1951). Of decisive importance remained the question whether this seismic discontinuity is a chemical or a phase boundary. Defenders of the two-layer convection are in favor of a chemical boundary which in their opinion would not allow any convective flux of material across this boundary.

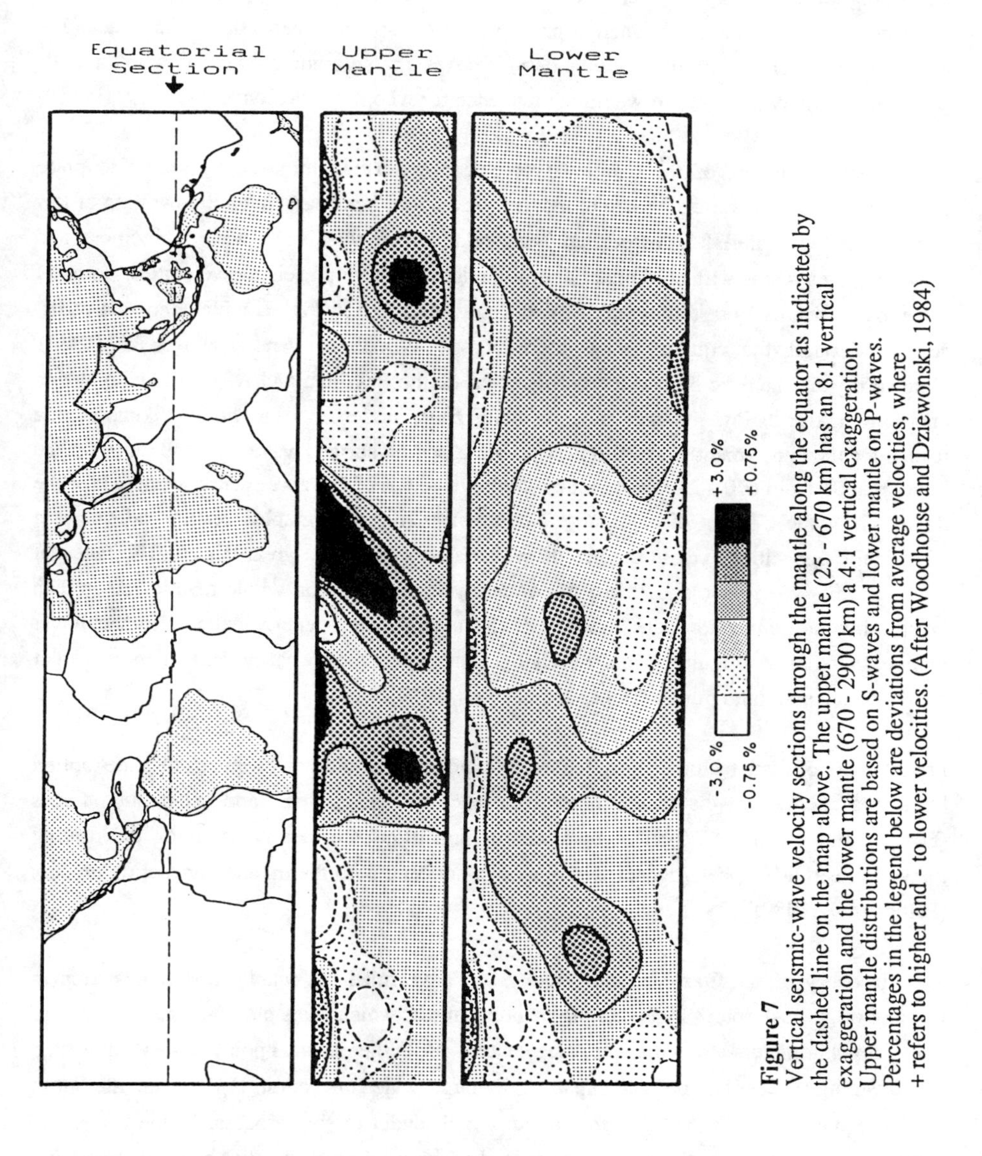

Figure 7
Vertical seismic-wave velocity sections through the mantle along the equator as indicated by the dashed line on the map above. The upper mantle (25 - 670 km) has an 8:1 vertical exaggeration and the lower mantle (670 - 2900 km) a 4:1 vertical exaggeration. Upper mantle distributions are based on S-waves and lower mantle on P-waves. Percentages in the legend below are deviations from average velocities, where + refers to higher and - to lower velocities. (After Woodhouse and Dziewonski, 1984)

According to laboratory experiments under high pressure and temperature, the 670-km discontinuity marks the depth where a phase shift from spinel to perovskite takes place. The probability of a phase equilibrium boundary removes a main argument against whole mantle convection. Today there is overwhelming evidence for whole mantle convection.

The velocity structure of the mantle as revealed by seismic tomography seems to place important constraints on the pattern of mantle convection. It appears that hot upwelling of low density mantle material is associated with relatively low seismic wave velocities, and downwelling streams with cold material of high density and high seismic velocities. Some of the existing results from tomographic studies are illustrated in Fig. 7, which shows a cross-section through the equator plane according to Woodhouse and Dziewonski (1984). Noteworthy are the deep high-velocity roots beneath the Brazilian and African Shields which appear to merge below 400 km. Low velocity zones underlie ocean ridges. Tomographic results for the upper mantle down to 670 km are based on surface wave data and those for the lower mantle from 670 - 2900 km on travel time residuals of P-waves, which accounts for the mismatch at the 670-km discontinuity. Nevertheless, the tomographic results indicate that large inhomogeneities exist in the mantle which are extending to great depths. The fact that many anomalies cross the 670 km discontinuity tends to favor whole-mantle rather than layered-mantle convection. The low and high velocity heterogeneity patterns in the lower mantle and their continuity in the overlying mantle become more evident in more recent results of seismic tomography (Hager and Clayton, 1989).

Practitioners of plate tectonics have tended to explain the driving mechanisms of plate motion by forces exerted by the plates themselves, such as "ridge-push" and "slab pull" forces (Forsyth and Uyeda, 1975). Recent model computations are clearly ascribing the driving forces to deep lateral heterogeneity of density associated with the mantle convection process (Peltier et al., 1989).

The discovery of sea-floor spreading seemed to have ruled out mantle-wide convection in favor of ridge-push forces by upwelling mantle material which takes place at a narrow zone of only a few kilometers along the spreading centers. From the physical point of view, however, the spreading process can only be explained by large-scale convection. Fig. 8 is an attempt to condense our present knowledge into a simplified model of this process. In the centers of rising and diverging mantle convection, partial melting takes place. By chemical, fractional and gravitational differentiation basic magma rises to form a magma chamber. The magma chamber is covered by a solid basalt layer of 2 - 3 km thickness. While diverging mantle convection is pulling apart the solid basalt layer, cracks open and magma rises to the surface of the sea bottom, where it erupts to form a layer of pillow lava of about two kilometers

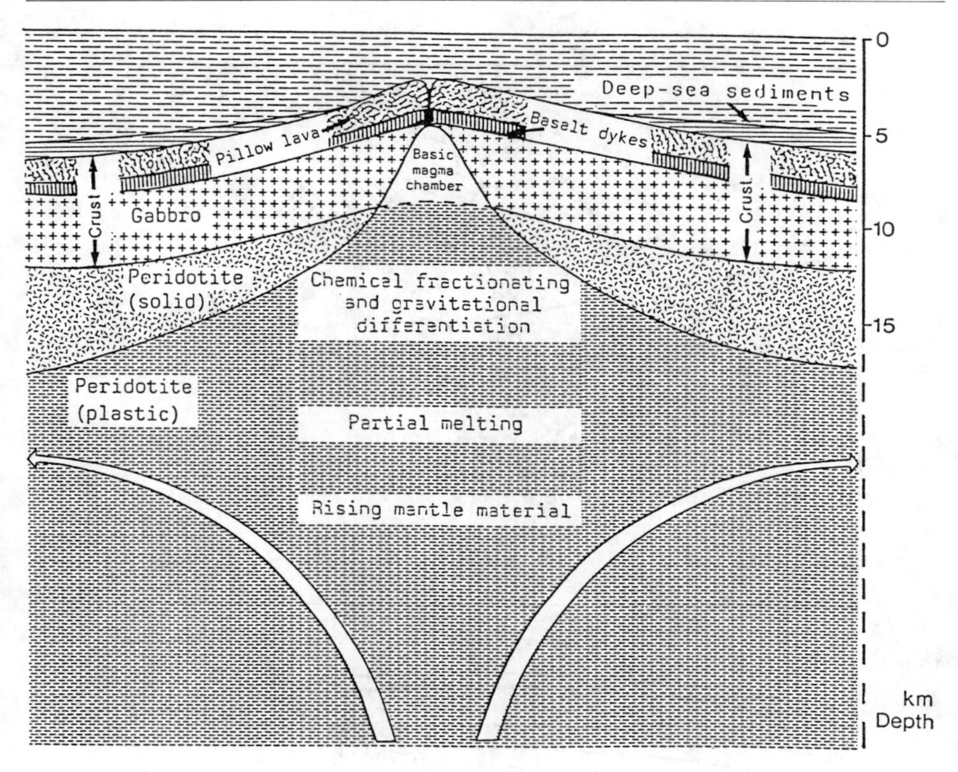

Figure 8
Seafloor spreading mechanism as evidence for mantle-wide convection.

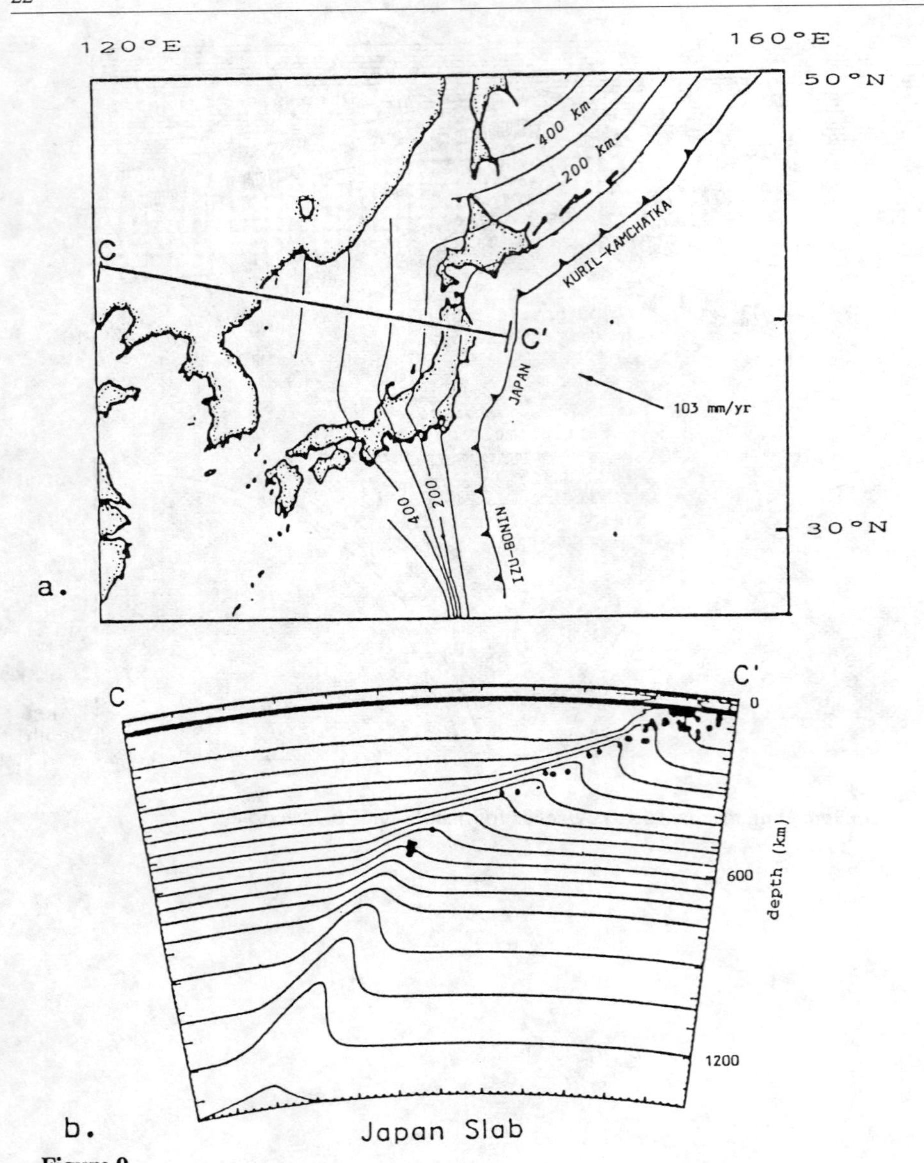

Figure 9
Slab penetration below Japan.
a. Section CC' across the Japan subduction zone with depth distribution of slab penetration contoured.
b. Dots are projections of well recorded ISC hypocenters along section CC'. P-wave velocities on the vertical cross-section are contoured at 0,25 km/s intervals. Ambient values are from a radial model. Velocity perturbations were calculated from a thermal model characterized by slab penetration to 1400 km, assuming $dv_p/dT = -0{,}5$ m/s/°K (effective thermal coefficient of P-wave velocity). (Modified after Jordan et al., 1989)

thickness. Below, the lava intrusions form lava dikes of about 1 km thickness. The melt of the magma chamber solidifies sideways to form a gabbro layer of about 5 km thickness. The ophiolithic series consisting of 7 - 8 km oceanic crust and 25 - 30 km of depleted mantle material is the chemical boundary layer that translates away from the ridge crest with little modification. While mantle convection is pulling apart the sea-floor, solidification continues to increasing depth, forming oceanic lithosphere which in its oldest parts at continental margins reaches depths of somewhat more than 100 km.

At subduction zones, where seismicity marks the trajectories of cold descending sea-floor we do have direct evidence for the interior flow pattern. Seismicity ceases at a depth close to the 670-km discontinuity. High-resolution seismic images, however, show clearly the penetration of descending slabs into the lower mantle beneath at a number of subduction zones and therefore the existence of a significant mass flux across the 670-km discontinuity. The pattern of seismicity distribution and the aseismic extension of velocity anomalies provide strong evidence that subducting slabs are part of mantle-wide convection (Fig. 9).

Latest results of seismic computer tomography indicate that the traces of cold subducted slabs are not only broadening at the 670-km discontinuity, but continuing to deeper parts of the mantle without keeping their slablike configuration (Fukao et al., 1991). Subducted material is obviously accumulated at this discontinuity. After phase changes a considerable part is obviously transported to lower parts of the mantle.

Seafloor spreading and subduction are the surface manifestation of convective circulation which comprises the whole mantle.

The geoid provides a fundamental observational constraint on models of mantle convection. It presents some apparent paradoxes, when interpreted in terms of mantle flow. Although subduction zones are always associated with relative highs of the long-wavelength geoid, similar association does not exist with ridges. Obviously, mass inhomogeneities are reflected in the geoid where mantle flow is disturbed such as in subduction zones. There is also no simple relation between geoidal undulation and density distribution as inferred from seismic tomography. The geoid predicted on the basis of seismic tomography assuming a non-convecting Earth matches the observed geoid well in pattern but is of opposite sign.

In order to interpret the geoid in terms of mantle convection, the dynamic topography both at the Earth's surface and at the core-mantle boundary has to be considered. The geoid can be described in terms of mantle flow by the calculation of fluid mechanic models, constrained by density models as inferred from seismic tomography. For the case of such a combined whole-mantle convection with density heterogeneities obtained from seismic tomography, Hager and

Clayton (1989) calculated the dynamic topography of the Earth's surface and the core-mantle boundary. The resulting theoretical geoid matches the observed geoid by 90 %, which must be regarded as further evidence for whole-mantle convection.

8. A comprehensive model of geodynamic processes

The previous discussion clearly suggests that there exists a system of mantle-wide convection. Continents are ancient blocks which are floating on the heavier mantle material. The seafloor is part of the flowing mantle material which differentiates and solidifies transitorily at the surface. The core-mantle boundary is an interface of irregular shape, however, with regionally distinct chemical boundary layers embedded.

Piecing together the fragmentary information about the structure and dynamics of the core-mantle boundary and surface manifestations as revealed by global geotectonic processes, it is attempted to construct a model of mantle convection. Fig. 10 shows a cross-section roughly through the equatorial plane, seen from the southern hemisphere. Rising and diverging currents of whole-mantle convection are the causes of the spreading centers at the mid-oceanic ridges of the Pacific, Atlantic and Indian Ocean. Continental blocks, such as America, Africa and the SW-Asian Archipelago and shelf seas are assembled in the regions of converging and descending currents. In these regions of downwelling currents, the core-mantle boundary is dynamically depressed. East and west of the Malay Archipelago and at the margins of South America, subduction takes place.

It is most probable that remains of subducted sea floor are deposited at the core-mantle boundary. The load of the cooler residues or dregs increases the depression to be expected in case of a sharp boundary. Where mantle currents are moving horizontally along the core as below the western Pacific, a sharp boundary exists. Chemical reactions take place enriching the mantle material with core alloy and moving and depositing it in regions of rising currents.

This model explains the discrepancy between the deformations of the core-mantle boundary, as predicted from the dynamically constrained model and the seismic observations. Where a sharp boundary exists, both should be identical. Where residues of mantle convection have been formed, the dynamic boundary becomes mathematically fictive. The boundaries and structures of the intermediate chemical layers appear in the seismic images without being resolved by dynamic models.

The conditions at the core-mantle boundary are rather similar to those at the Earth's surface. The dynamic boundary is only recognizable in oceanic areas where the surface of the mantle

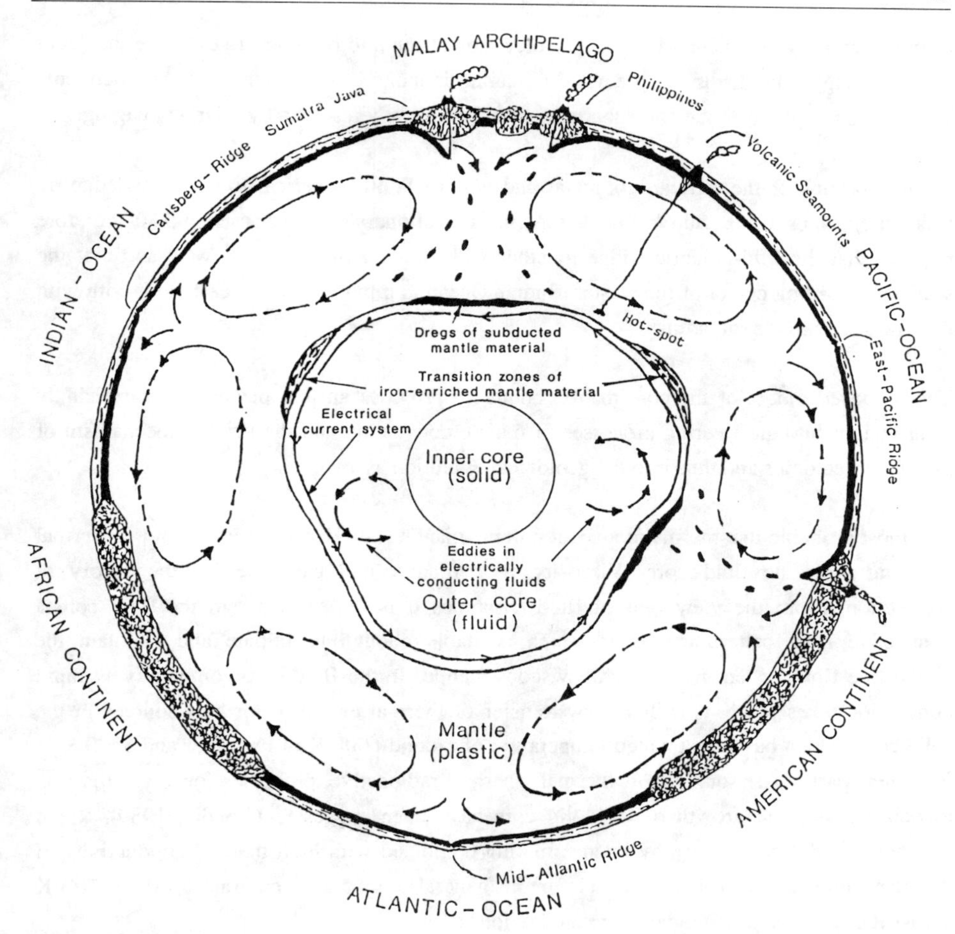

Figure 10
Tentative model of mantle convection in the equatorial plane to explain the structure and dynamics of the core-mantle boundary in relation to geotectonic processes.

is undisturbed by continental blocks. The elevation of the mid-oceanic ridges above the deep-seafloor roughly confirms the predicted dynamic boundary. The structure of the continents such as their depth to Moho and subcrustal extension is only resolved by seismic imaging.

At the margins of the African continent and eastern South America, to our knowledge no subduction takes place. Supposing that Africa is stationary, the proposed model therefore requires that the Mid-Atlantic Ridge together with South America move west and that the seafloor spreading center of the western Indian Ocean is moving roughly eastward, both with their rate of seafloor spreading.

The proposed model of the core-mantle boundary provides an unexpected and surprisingly clear insight into the steering processes of mantle convection and the driving mechanism of global geotectonics and thus into the geological evolution as a whole.

It is most probable that thermal convection in the mantle is produced by some kind of thermal coupling with the fluid core. Viscosity is a fundamental parameter in the theory of convection. From the viewpoint of fluid dynamics it is unlikely that in the fluid core a convective flow pattern can exist which is stable enough to initiate and maintain the convective flow system in the highly viscous mantle. In the fluid core comparatively rapid convection takes place. It is driven by transfer of thermal energy from heat sources in the solid core, which has an estimated temperature of around 6000 K in its center and 5500 K in its outer parts. As sources of thermal energy, radioactive processes or crystallisation associated with the growth of the solid core have been suggested (Condie, 1989). Rapid exchange of thermal energy by intense mixing combined with high thermal conductivity of the core-alloy is the probable reason for the general temperature contrast of about 700 K across the core-mantle boundary, as stated before.

Thermal conductivity across the core-mantle boundary varies with its constitution. Residues of subduction products are poor thermal conductors. Where a sharp boundary or an intermediate layer with an admixture of core alloy exists, transfer of thermal energy into the mantle is favored. To the author it seems most probable that the high temperature contrast across the core-mantle boundary associated with regionally distinct transfer of thermal energy is sufficient to maintain large-scale thermal convection in the mantle.

This does not explain the initiation of convective cycles. Nance et al. (1988) have suggested that the episodicity inherent in the alternation between continental assembly and breakup is not due to convection, but rather to the storage of heat below the poorly conducting continents. In the context of the proposed model it seems more probable that, instead, this process takes place at the core-mantle boundary. Thermal energy is stored below the insulating layers of subducted mantle material; the accumulation of heat finally leads to a breakup of the bottom layers, initiating a mantle plume and a new convective cycle which breaks apart the continents accumulated at the surface.

Below Africa seismic tomography shows persistent evidence for a hot region in the lower mantle. Possibly it indicates the initiation of a mantle plume and a new convective cycle.

The processes which basically cause the extensive active tectonics as revealed by recent crustal movements and earthquake activity can be well explained by the model drafted in Fig. 10. With respect to places of occurrence and source processes, three categories of earthquakes can be roughly distinguished. A first type are those which occur at the spreading centers and between them along transform faults. On land they are noticeable on Iceland where sea-floor spreading is projected to the surface of a volcanic island grown on a hot spot, and in the region of continental breakup in the Red Sea. Another place is California where a spreading center seems to have migrated under the American continent and where earthquake activity is mainly caused by displacements along transform faults. A second category of earthquakes are those in subduction zones which occur along the traces of downgoing slabs to a depth of 700 km.

A third category are the widely spread earthquakes in less defined, tectonically active zones on continents. These intracontinental earthquakes cannot be easily explained in terms of the plate tectonics concept. However, when we consider the model in Fig. 10, we understand that the continental blocks are directly affected by convective flow in the underlying mantle. By transfer of frictional energy from the bottom of the lithosphere, stress fields are built up which cause faulting and deformation also far from any plate boundary. They give rise to uplift and subsidence, upthrusting and downthrusting, horizontal shear motions, compression associated with folding and dilatation causing the formation of continental rifts.

All these intracontinental tectonics are accompanied by earthquakes, which differ from earthquakes in the ocean bottom in one important aspect. Because of the relatively low rock strength of the ocean bottom, earthquakes have short recurrence time, stress-drop and magnitude. Therefore, the seismotectonic pattern is well known. Old continental rock masses are characterized by high rock strength, high stress drop and long recurrence time and therefore many times high magnitude. Because of this sporadic occurrence and the short time of available earthquake records, the seismotectonic pattern of intracontinental earthquakes is virtually unknown. The unexpected occurrence of catastrophic earthquakes like in Central and East Asia and in Australia demonstrate our present lack of knowledge and understanding of the basic processes which cause intracontinental earthquakes. Intensive studies of the causative faults and extension of the recurrence rates beyond the historical record by palaeoseismic studies are beginning to improve our understanding of the source processes.

Another important source of tectonic and earthquake activity are hot spots, which will be discussed below in a different context. According to location, origin, appearance and composition of magma, three types of volcanism can be distinguished. The largest lava masses are produced by the volcanism which at the spreading centers continuously produces new oceanic lithosphere. Totally different is the volcanism of subduction zones which is characterized by explosive eruptions and accumulation of volcanic mounts, specifically along the margins of the Pacific Ocean. A third type is the "intraplate volcanism". There is strong geochemical evidence that this kind of volcanism is caused by basalt magma which rises from hot spots, magma chambers located in the lower mantle (Condie, 1989). In the oceans it forms volcanic sea-mounts and islands like Hawaii, the Canary and Cape Verde Islands. On continents it causes volcanism like in the Eifel mountains, with slag cones and craters produced by volcanic eruption of material, which is very similar to that of the volcanic islands (Schmincke, 1988).

Seafloor displacement rates along seamount chains such as the Hawaii-Emperor chain can be reconstructed if one assumes the causative hot spot to be fixed. The hot spots show only slow motions relative to each other, ten times smaller than lithospheric motion. They are part of a much more stationary system, which can be used as a frame of reference for surface motions of the lithosphere (Roeser, 1989).

In a whole-mantle convection no part of the mantle can be considered as a fixed reference. Nevertheless, the proposed model in Fig. 10 provides an explanation for the comparatively fixed position of hot spots. In case of whole-mantle convection it is most probable that remains of subducted sea-bottom follow the circulation of mantle material into places in the lower mantle of increased heat transfer from the core. Heating of the non-related mantle material leads to partial melting. By the formation of mantle diapirs, basaltic magma rises to the Earth's surface, where it produces the volcanism which is typical for hot spots. Compared to the velocity of convective flow, the pattern of convection is rather stable. Melting of the products of subduction occurs at relatively well defined places, while the sea floor at the surface moves across the hot spots with the velocity of convective flow. Volcanic island chains are formed in principle as shown in Fig. 10 for the Pacific.

On the basis of the proposed model it becomes understandable why the hot spots in spite of mantle-wide convection can form a relative stable frame of reference for geotectonic surface motions as shown in the world map in Fig. 11. The paleomagnetically and seismically constrained reconstruction of surface plate velocities by Minster and Jordan (1978) allows

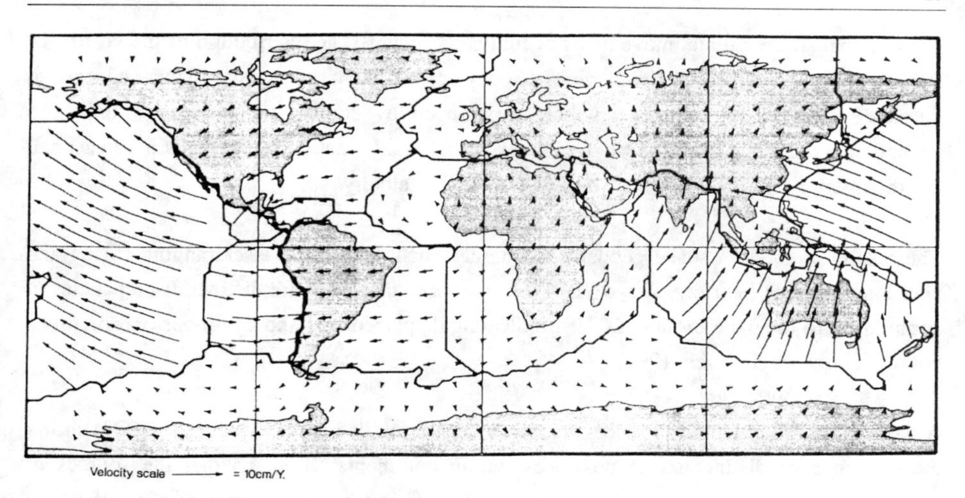

Figure 11
Surface plate velocities relative to the "hot spot" frame. Plate boundaries are shown as heavy solid lines (Minster and Jordan, 1978).

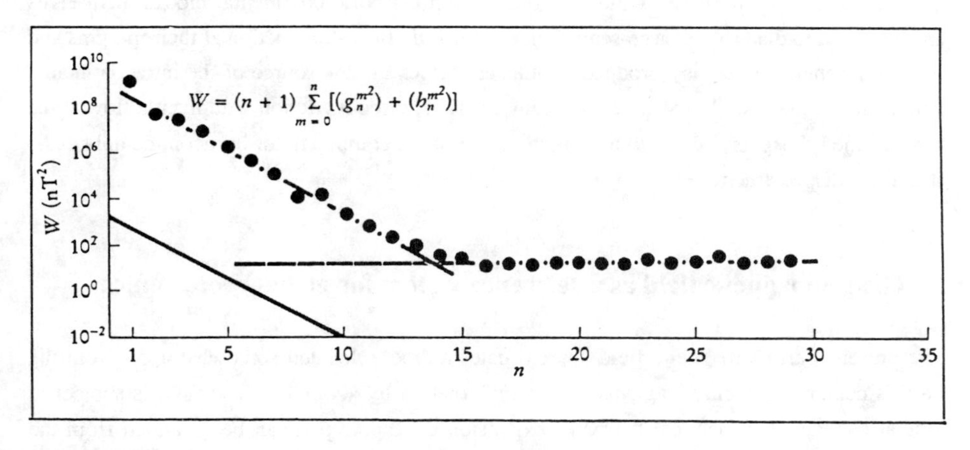

Figure 12
Spatial spectrum of the observed inner magnetic field. W is a measure for the mean energy density of the field at the Earth's surface, n the degree of spherical harmonic expansion of the inner field. The term n = 1 corresponds to the dipole (Meyer, 1986).

translational and rotational movements of individual plates. The distribution of the vectors of surface motion shows the high velocities of convective flow in ocean areas in clear contrast to the low velocities of continents which are trapped by converging convective flow. An exception is Australia which presently moves with a rapid mantle flow towards northeast and South America which moves westward with medium velocity.

The results of modern space geodetic techniques such as satellite laser ranging (SLR) and very long baseline interferometry (VLBI) clearly show that there are distinct discrepancies in comparison to the plate motion models, indicating displacements and continous deformations within continents (Drewes, 1991).

A world stress map which in recent years has been compiled by Zoback et al. (1989) shows the existence of distinct stress provinces within continents. It seems that the sources of intracontinental stress and strain have to be understood in terms of the pattern of mantle flow in the mantle beneath.

One has to consider that on the surface of convective flow in spherical shells like the Earth's mantle, besides poloidal also torroidal movements occur. These may cause the rotational displacement fields observed within continents. Furthermore, continental blocks themselves are supposed to disturb the large-scale pattern of mantle flow. Subduction at their margins and deep lithospheric roots may produce turbulences which are the source of the intracontinental stress pattern like the observed stress provinces. Thus, viscous friction with mantle flow at the basis of the lithosphere seems to be a main source of the complexity of intracontinental stress-strain and displacement fields.

9. The geomagnetic field as a reference system for global geotectonics

The present Earth's magnetic field approximates a dipole, located somewhat apart from the Earth's center and inclined against the axis of rotation by about 11°. The field is subject to secular changes. By spherical harmonic expansion the dipole part can be separated from the residual field. Both constituents are changing with time, with large-scale features of the latter appearing to drift westward by about 0,2° per year.

Since the fundamental work of Gauss (1839), who established that the sources of the Earth's main magnetic field are located in the Earth's interior, various attempts have been made to locate the sources more exactly. The evaluation of extensive global surveys together with theoretical considerations led to the conclusion that the main part of the Earth's inner field originates from the fluid core and only a small part arises from magnetic minerals in the crust, while the mantle as a whole can be regarded as source-free.

Although the source mechanism of the core field is uncertain, it can be assumed that it is caused by electrical current flow. A reduction of the surface field to an equivalent current distribution is unique, if the currents are confined to an arbitrary sphere (Vestine, 1954). Such a current system has a realistic meaning, if there is independent proof that the electrical currents really are flowing in a certain uniform depth, which subsequently is called the source layer. Actually this can be proved in a convincing way, if the surface field is expanded into spherical harmonics and the magnetic energy density W(n) is calculated according to the formula in Fig. 12, where g and h are the Gauss coefficients of n degree and m order. Fig. 12 shows the total spectrum of the inner field up to degree n = 29, as derived from measurements of the Magsat-satellite (Cain et al., 1984). One recognizes two quasi-linear terms. The horizontal branch with the lower harmonics from n = 15 upward could be identified as the crustal field. This spectrum continues theoretically to at least n = 4, however of small size compared to the core field. The linear branch of the lower harmonics up to n = 13 corresponds to a source layer in a certain depth. The theoretical slope of the spectrum from a source layer at the surface of the core in a depth of 2900 km is shown in Fig. 12 down to the left. The linear regression of the lower spectrum indicates a somewhat deeper source. As a mean value of several field models, a depth of 147 $\pm$ 16 km below the core-mantle boundary has been obtained. The term n = 1, which corresponds to the dipole field, deviates in a significant way which is interpreted as a separate source involving a mechanism in deeper parts of the core (Meyer, 1986).

Fig. 13 shows the isolines of the vertical component for the Earth's magnetic field including the harmonic terms n = 1,...,12. Based on the field of the epoch 1980, the current function in Fig. 14 for a source layer depth of 3000 km has been calculated (Meyer et al., 1985).

Paleomagnetics, the analysis of rock magnetization, has revealed astonishing details about the behavior of the geomagnetic field in the Earth's history. In the last 1000 years the magnetic pole has continuously changed its position, however always moving closely around the pole of rotation. Paleomagnetism has revealed large changes of the magnetic pole with respect to the present surface of the Earth, which have been documented by apparent polar wandering curves. Paleomagnetic studies also show strong evidence that in the last 100 million years the magnetic pole has been coinciding within 5° with the axis of rotation and that throughout the Earth's history, predominantly a dipole has existed which has been aligned extensively with the axis of rotation (Condie, 1989).

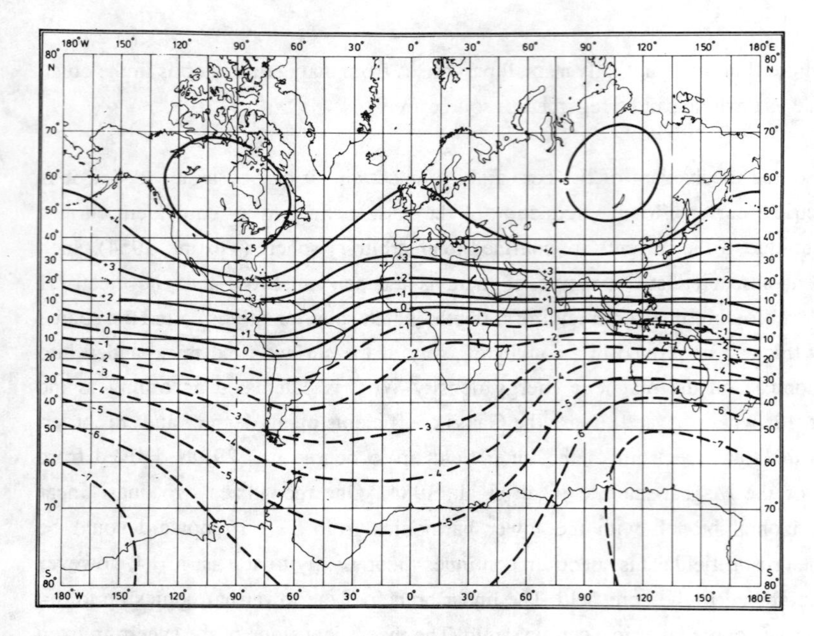

Figure 13
Isolines of the vertical component of the main magnetic field at the Earth's surface (n = 1, ..., 12; epoch 1980). The unit is 10^4 nT (Meyer et al., 1985).

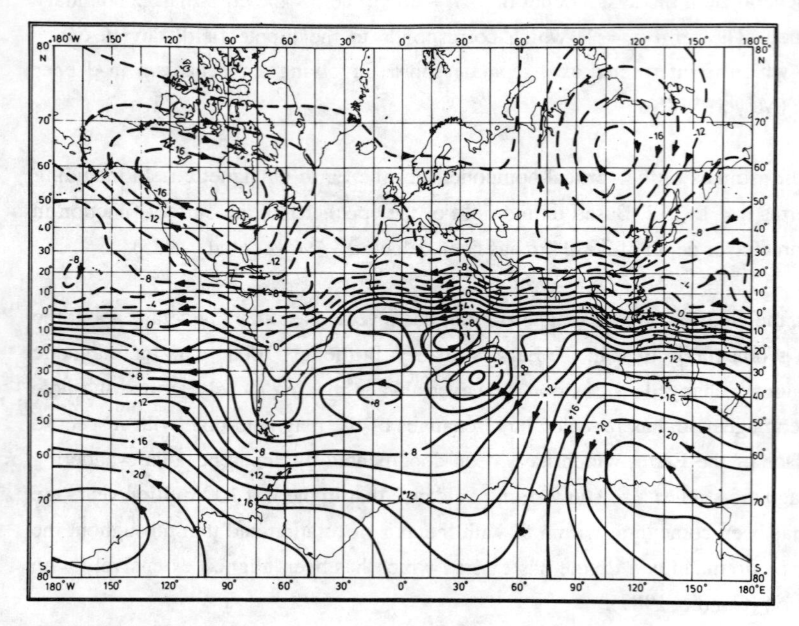

Figure 14
Isolines of the current function of the total main field (n = 1, ..., 12; epoch 1980) in the source layer depth of 3000 km. The unit is 10^8 A. The direction of the currents is marked by arrows (Meyer et al., 1985).

On the basis of these findings the magnetic poles could be regarded as relatively stable and the magnetic field considered as a reference system. The apparent polar wandering curves had to be interpreted as displacements of the Earth's surface and individual continents with respect to a fixed pole. In this way continental drift could be reconstructed rather accurately and at the same time confirmed quantitatively.

For more than 100 years the Earth's magnetic field has been continuously decreasing, which means that the field would vanish in about 1500 years, should the trend remain. Paleomagnetic studies have shown that a polarity reversal has often recurred. A polarity time scale has been constructed which has been used for dating the ocean floor and determining the rate of seafloor spreading. Paleomagnetism also provides information about the process of field reversals, as reviewed by Hoffman (1988). Today there is clear evidence that during a field reversal the dipole does not simply wander into the opposite position. Instead the field weakens while dipole positions are oscillating along certain longitudinal bands, whereafter it is rebuilt with opposite polarity (Tric et al., 1991). Sometimes, however, it returns to its original polarity through an interim time which usually lasts only a few thousand years. Thus, the magnetic field is stable and of dipole character during 98 % of the time, while it is unstable only during the rest of the time.

The last polarity change took place 730.000 years ago. The present decrease is obviously caused by the growth of magnetic field sources in the core which are opposite to the present polarity. These sources are at high latitudes below the capes of Africa and South America. In case that this trend continues, the magnetic field may change its polarity.

There is also paleomagnetic evidence that during magnetic field reversals the magnetic field remains for some time in a relatively stable intermediate position. However, the documentations of changes of the field strength show that these positions are related with rather weak field intensities. After full reversals it obviously happens, that the intensity becomes three times higher than before.

There is general consensus that the Earth's magnetic field is somehow related to convective motion of the electrically highly conducting core alloy. There is also agreement that the geomagnetic field is generated by magnetohydrodynamic processes according to the principle of a self-exciting dynamo. Most controversies have to deal with the pattern of core convection and energy supply.

The proposed model of the structure and dynamics of the core-mantle boundary as derived by the synthesis of prevailing information provides new insights and obviously overcomes present controversies. In conjunction with our present knowledge about the temporal behavior and the sources of the corefield, it provides a model of the generating processes. The high temperature contrast at the core-mantle boundary is maintained by transfer of thermal energy from the inner core through convection in the fluid core, which is independent from the convection system in the mantle. Convection of the highly conductive core fluid causes electrical currents, which are the direct sources of the magnetic field. The currents obviously are flowing in the outer part of the fluid core, which has been identified as the source layer, whereby according to Figs. 13 and 14 also the dipole field arises from this single source layer. The dominance of the dipole field and its approximate alignment with the axis of rotation suggests that fluid motion is strongly influenced by Coriolis forces.

High absorption and low velocities in an intermediate layer of a few hundred kilometers is consistent with the idea that the top of the inner core is a "musky" two-phase zone (Bloxham et al., 1990). Because of thermal convection in the fluid core, this boundary also should be of regional distinct structure with varying thermal conductivity and capability of storing thermal energy. It seems that comparable with the cycle of solar activity, however, in much larger time intervals of hundred thousands of years, exceptionally strong outbursts of hot material occur from the intermediate layer at the surface of the inner core. Turbulent fluid motion disturbs the prevailing pattern of core convection, weakens the existing dipole field and produces transient pole positions. In times of normalizing heat transfer the turbulent convection is aligned by the influence of Coriolis forces, whereby a current system is produced in the source layer, which in accordance with the presentation in Figs. 13 and 14 gives rise to a dominating dipole field, closely aligned with the axis of rotation. The new polarity of the magnetic field depends on the nature and locations of the eruptions. These most probably occur where thermal energy is stored below residues of convection at the bottom of the core fluid. Such a mechanism explains why the magnetic field during most of the Earth's history provides an independent reference framework for geotectonic processes, closely connected to the axis of rotation, while reversals occur during relatively short time intervals.

The correlation between dominant features in the lower harmonics of the gravity and non-dipole magnetic field and the irregular shape of the core (Vogel, 1960) and further investigated (Vogel, 1963) has been subject to many discussions since that time (Hide, 1969). According to the model of the core-mantle boundary in Fig. 10, it seems most probable that the source layer of the magnetic core field adapts itself to the topography of the core surface thus explaining the correlation between the lower harmonics of the gravity field and of the stable part of the magnetic field.

The west-drift of certain patterns of the non-dipole field seems to be due to the shift of the circulation pattern of the convection cells, which at the surface of the fluid core is about

0,3 mm/s. The present decrease of the magnetic field obviously is caused by regional disturbances of core convection which have been found to be characteristic for the process of polarity changes.

10. Mantle convection as generating and recycling mechanism of continental lithosphere

There is convincing geological, geophysical and geochemical evidence that the cores of our present continents were formed and stabilized against mantle convection already in the earliest time of Earth's history. The old cratons are obviously products of geochemical surface layers of convective flow, which were buoyantly trapped and accumulated in regions of cold downwelling mantle material (Jordan et al., 1989). The deep subcrustal lithospheric roots of continental blocks must be of geochemical composition which in spite of the lower temperature is less dense than the oceanic upper mantle. The continental masses must have been formed and continuously recycled by a process which has to be seen in the context of the convection model in Fig. 10.

When an ocean is opened, material derived from weathering, erosion and denudation of continental surface rocks is accumulated at river deltas or sedimented on the seafloor. The beginning of subduction initiates the recycling process, whereby the ocean floor acts like a conveyor belt for the renewal of continental masses. The sea-floor sediments are transported into subduction zones, where they are at least partially melted and by magmatism integrated again into continental masses or on an intermediate stage form island arcs. Thick sediment packages like river deltas, volcanic sea mounts and plateaus, island arcs and fragments of continental blocks are carried to continental margins where they are obducted, thus leading to a lateral growth of continents by accretion of terranes and final continental collision.

According to the model in Fig. 10, hot spot volcanism becomes an important source of continental growth caused by convectional mantle flow. Hot spot volcanism is obviously just the surface manifestation of mantle diapirs which accumulate huge masses of magma, leading to thickening of the oceanic lithosphere and subsequently formation of sub-marine plateaus like along the island chain of Hawaii. Iceland, which happens to be situated on a spreading center, gives us an idea of the immense amount of magma which can be produced by a hot spot. It is a logical consequence that hot spot volcanism which happens to occur in continental areas is associated with extensive magmatism in crust and subcrustal lithosphere. Intrusions of basaltic magma from hot spot sources which solidify in the cold continental lithosphere most probably are an important factor in the stabilization of continental blocks against disruption by mantle convection.

An indication of hot spot volcanism within continents seems to be plateau uplift accompanied by formation of rifts. The young tectonic processes in the Rhenish Massif have been brought in connection with a mantle diapir by Neugebauer et al. (1983). Our proposed convection model would imply that plateau-uplift is caused by accumulation and solidification of magma from a mantle diapir, which rises from a hot spot in the lower mantle. Lighter constituents of magma rise to the surface, where they are responsible for the young volcanism of the Eifel mountains. As a natural consequence of crustal vaulting due to magma intrusion, dilatation has taken place leading to the extensional rift structures, by which the Rhenish Massif is bisected into two halves which were shifted away from each other (Ahorner, 1987). Under similar aspects the formations of the Upper-Rhine Valley, the East African Rift and other continental rift zones have to be considered.

11. "Convection tectonics" as a motor of geological evolution

Since the introduction of plate tectonics, geologists have tended to interpret all tectonic processes in earth's history in terms of this theory. In view of our increasing knowledge about the processes in the Earth's interior on one hand and the inconsistencies inherent in the phenomenological kinematic theory of plate tectonics on the other hand, it seems more adequate to reveal a causative view by introducing the term "convection tectonics" as a unifying conceptional framework to comprise the entirety and complexity of global tectonic processes, both at present and in Earth's history.

Obviously the convection systems are not stationary but moving with time. India has been separated from southeast Africa. This could not have occured in the present position of the spreading center. It must have moved across the East-African margin before reaching its present position where it is still pushing India into the Eurasian continent. Part of the terranes at the North-American west coast has its origin in the Pacific, thousands of kilometers away from their present position (Howell, 1988). It means that the spreading center which is now moving into the North-American continent most probably originated in the western part of the Pacific.

One can draw the conclusion that convective flow originates, migrates and disintegrates, comparable with a cyclone in the atmosphere. This causes the cycle which begins with rising and diverging currents, thinning of continental lithosphere, followed by rifting, opening and widening of an ocean basin. With the beginning of subduction and obduction, orogeny is initiated. Subsequent converging currents close the ocean basin and under collision of continents orogeny is completed, followed by denudation of the mountain chain. From the phenomenological point of view this sequence of geotectonic processes is called a Wilson cycle. From the causative point of view, such a sequence of geotectonic processes has to be regarded as being caused by a "convective mantle cyclone".

Mantle convection has been forming continental lithosphere since the earliest time of Earth's history. The map in Fig. 15 (Milanovsky, 1987) shows ancient platforms, old continental cratons which already have existed at the beginning of the Paleozoic, 600 million years ago. Cratons, however, have been stabilized long before the end of the Archean, probably already 3000 Ma ago. The first cratons were accumulations of basalt and basalt-depleted ultrabasic rocks, derived from a process which takes place at mid-oceanic ridges. In the late Archean and early Proterozoic, large mantle convection has led to granitization of large platform regions, associated with further consolidation and stabilization.

The ancient blocks, however, themselves exhibit linear tectonic structures, which indicate that they have been divided and reassembled in the Archeozoic and Proterozoic of Earth's history 4000 to 600 Ma b. p. Archeozoic greenstone and charnokite granulite belts are the remnants of orogenic belts caused by early convectional cyclones.

In the Paleozoic 600 - 200 million years b. p. cratons have been drifting around, opening and closing oceans in between them. Orogenic belts like the Caledonian-Appalachan between Europe and America, the Variscan in Central Europe, the Urals-Mongolian Belt and Pacific belt are evidence of the action of convective mantle cyclones in this period of Earth's history.

Presently, we are witnessing the activity of the youngest cyclones, whereby they are passing through different phases in different parts of the world: In the Gulf of Aden and Red Sea rifting and beginning of the opening of an ocean, in the Atlantic an ocean basin that recently has been opened, in the Pacific an ocean where at the margins subduction and accretion of terranes takes place, associated with the formation of mountain belts and beginning of ocean closing, and in the collision zone between Africa, Arabia and India on one hand and Eurasia on the other hand, where the closure of oceans and collision of continental blocks has just taken place and mountain formation is being completed.

However, tectonic processes do not only take place at the boundaries of ancient cratons. Continental lithospheric blocks as a whole are subject to stress fields which are exerted by convective flow in the underlying mantle. Typical for intracontinental tectonics in Earth's history are epirogenic movements like formation of sediment basins and plateaus associated with transgressions and regressions of inland-seas, as well as extensional and sheer movements associated with the formation of continental rifts.

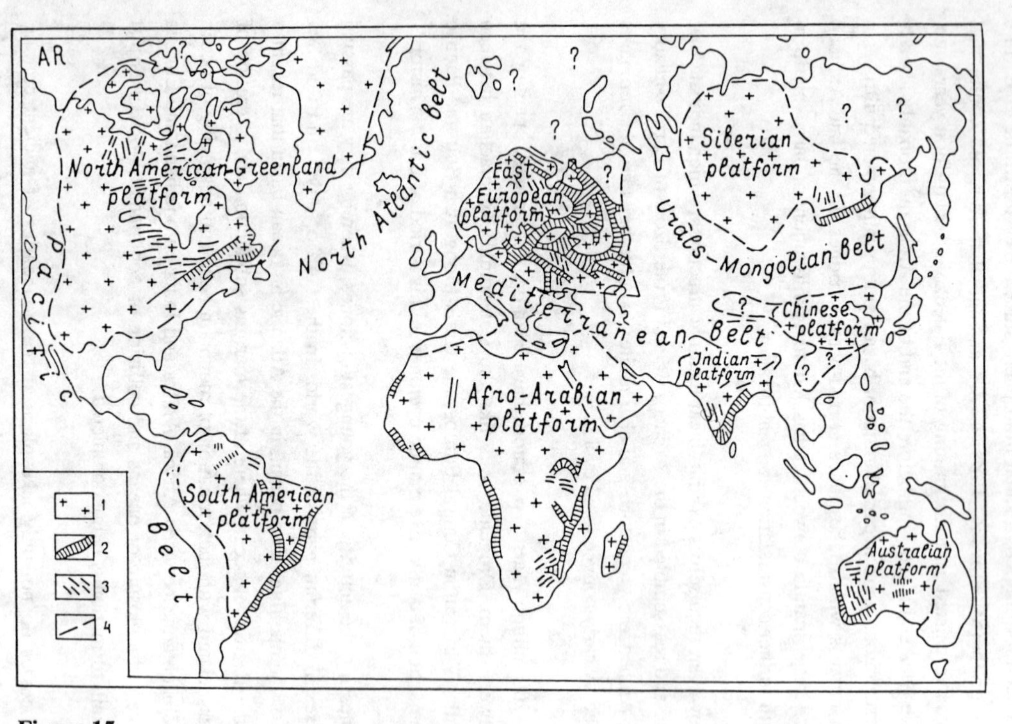

Figure 15
Archean charnokite-granulite belts and early and late Archean greenstone belts on ancient platforms and folded belts in the Paleozoic up to the beginning of the Permian.
1 = Archean gneissic basement of ancient platforms partly reworked in the early and late Proterozoic, without subdivision;
2 = charnokite-granulite belts;
3 = greenstone belts;
4 = margins of ancient platforms delineated in the late Proterozoic.
(Modified after Milanovsky, 1987).

Some of the continental rift zones, like the East-African system, are interpreted in terms of upwelling mantle plumes either initiating or vitnessing the failure of continental break-up. In view of convection tectonics there exist two further types of continental rifts. One seems to be caused by hot spot diapirs, typically accompanied by plateau uplift of the rift shoulders and magmatism. The other one does not reveal volcanism. It can be explained by toroidal components of convection which cause rotational and shear stresses in the continental lithosphere leading to lateral fault systems and shear-extensional rifts of sometimes on-echelon configuration.

Intraplate tectonics cannot be explained only by large-scale mantle convection. It seems that primary mantle flow is disturbed by the continental lithospheric blocks thus causing small-scale mantle flow of secondary order.

One example is the opening of back-arc basins as explained by Uyeda (1983). Under certain circumstances subduction disturbs and directs mantle flow in a way that it gives rise to a secondary convection system which causes the opening of a back-arc basin. As another example we may consider retroactive foreland basins which form behind continental margin arc systems. The greatest thickness of foreland basin sediment borders the fold-thrust belts reflecting enhanced subsidence caused by thrust-sheet loading and deposition of sediments (Condie, 1989). It seems that besides this isostatic effect also the dynamic effect of secondary mantle flow, induced by subduction, should be contributing to the formation of foreland basins.

In view of convection tectonics it appears self-evident, that mantle flow also is disturbed by deep roots of continental lithosphere, thus causing secondary flow and providing a source of intraplate tectonics, which must have been active since the early times of Earth's history. Thus, a variety of tectonic settings as evident in intracontinental geological evolution can be explained by viscous coupling with first and second order mantle flow.

It is the hope of the author that the causative view of convection tectonics may help to further clarify some of the issues discussed in the following articles of this volume, such as the formation and propagation of the Variscan fold and thrust belts and its coal-bearing foreland basins, palaeogeography, as well as the Post-Variscan development of the Rhenish Massif.

Finally, it is pointed out that the model of mantle convection, which is based on a combination of geological, geophysical and geochemical evidence, also provides important insight into the processes responsible for the genesis of mineral deposits. Krahn and Friedrich (1991) presented a review on the genesis of base mineralization in the Rhenish Massif. The Variscan mineralization is typical for orogenic processes driven by mantle convection. For the

Post-Variscan ore deposits, mantle diapirs are presumed to be the motor of mineralization. In view of our model of mantle convection, it seems justified to assume that these diapirs may be arising from hot spots in the deep mantle.

Acknowledgements

The author thanks Professor Alexis Moiseyev from California State University, Hayward for his extremely careful critical review of the article.

Literature

Ahorner, L., 1987: Seismicity and Seismotectonics of the Rhenish Massif, Central Europe. Extended Abstract in A. Vogel et al.: The Rhenish Massif - Geological Structure, Evolution, Deposits of Economic Minerals and Present Dynamics. Earth Evolution Sciences, Vieweg Braunschweig - Wiesbaden.

Ahrens, T. J. and B. H. Hager, 1987: Heat transport across D": Problems and paradoxes. EOS Trans. AGU, 68, 1493.

Andersson, D. L., 1979: The deep structure of continents. J. Geophys. Res. 84, 7555 - 7560.

Benioff, H., 1954: Orogenesis and deep crustal structure: Additional evidence from seismology. Geol. Soc. Am. Bull. 65, 385 - 400.

Bloxham, J., M. Baron, J. Cain, D. Fearn, D. Loper and R. Merrill, 1990: Summary Report on the Second SEDI Symposium at Sante Fe, New Mexico. Deep Earth Dialog 4.

Bullen, K. E., 1949: Compressibility-pressure hypothesis and the Earth's interior. Mon. Not. R. Astron. Soc., 5, 355.

Cain, J. C., D. R. Schmitz and L. Muth, 1984: Small-scale features observed by Magsat. J. Geophys. Res. 89, 1070 - 1076.

Condie, K. C., 1989: Plate Tectonics and crustal evolution. Third Edition, Pergamon Press, Oxford.

Cox, A., 1973: Plate Tectonics and geomagnetic reversals. W. H. Freeman, San Francisco.

Davies, G. and M. Gurnis, 1986: Interaction of mantle dregs with convection: Lateral heterogeneity at the core-mantle boundary. Geophs. Res. lett., 13, 1517.

Drewes, H., 1991: Discrepancies between geodetic and geological plate motion models: Episodic movements or deformations? Report to the XX General Assembly of IUGG, Vienna, Austria of DGFI Munich/Germany.

Dziewonski, A. M., 1984: Mapping of the lower mantle: Determination of lateral heterogeneity in P velocity up to degree and order 6. J. Geophys. Res. 89, 5929.

Forsyth, D. W. and S. Uyeda, 1975: On the importance of the driving forces of plate motion. Geophys. J. R. Astr. Soc., 43.

Fukao, Y., N. Suda, H. Kumagai and M. Obayashi, 1991: Seismic tomogram of the Earth and pattern of mantle convection. Abstract of paper presented at the XX General Assembly of IUGG in Vienna.

Gauss, C. F., 1839: Allgemeine Theorie des Erdmagnetismus. In C. F. Gauss und W. Weber (Herausgeber): Resultate aus den Beobachtungen des Magnetischen Vereins im Jahre 1838, 1 - 57, Verlag d. Weidmannschen Buchhandlg., Leipzig.

Gutenberg, B., 1928: Mechanik und Thermodynamik des Erdkörpers. In: Müller-Pouillet, Bd. V, 1 (Geophysik), Braunschweig 1928.

Gutenberg, B., H. Benioff, J. M. Burgers, and D. Griggs, 1951: Colloquium on plastic flow and deformation within the earth. Trans. AGU 32, 497 - 543.

Hager, B. H., R. W. Clayton, M. A. Richards, R. P. Comer and A. M. Dziewonski, 1985: Lower mantle heterogeneity, dynamic topography and the geoid. Nature, 313, 541 - 545.

Hager, B. H. and R. W. Clayton, 1989: Constraints on the structure of mantle convection using seismic observations, flow models, and the geoid. In W. R. Peltier (Editor): Mantle Convection - Plate Tectonics and Global Dynamics. Gordon and Breach Science Publishers, 658 - 763.

Hide, R., 1969: Interaction between the Earth's liquid core and solid mantle. Nature, 222, 1055-1056.

Hoffman, K. A., 1988: Umkehr des Erdmagnetfeldes: Aufschluß über den Geodynamo. Spektrum der Wissenschaft, 7.

Howell, D. G., 1988: Terrane. Spektrum der Wissenschaft, 7.

Jeanloz, R. and S. Morris, 1986: Temperature distribution in the crust and mantle. Ann. Rev. Earth Planet. Sci. 14, 377.

Jordan, T. H., 1978: Composition and development of the continental tectosphere. Nature 274, 544 - 548.

Jordan, T. H., A. L. Lerner-Lam and K. C. Creager, 1989: Seismic imaging of boundary layers and deep mantle convection. In: W. R. Peltier (Ed.): Mantle convection - Plate tectonics and global dynamics. Gordon and Breach, New York.

Krahn, L. und G. Friedrich, 1991: Zur Genese der Buntmetall-Vererzung im westlichen Rheinischen Schiefgebirge. Erzmetall, 44, 1, 23 - 29.

Lay, T., 1989: Structure of the Core-Mantle Transition Zone: A Chemical and Thermal Boundary Layer. EOS, 70, 4, 49 - 58.

Lucke, O., 1960: Bemerkungen zur Dissertation von A. Vogel: Über Unregelmäßigkeiten der äußeren Begrenzung des Erdkerns auf Grund von reflektierten Erdbebenwellen. Zeitschr. f. Geophysik, 1, 50 - 56.

Meyer, J., J.-H. Hufen und M. Siebert, 1985: Stromfunktion des erdmagnetischen Hauptfeldes in der Quellschicht an der Kern/Mantel-Grenze. J. Geophys., 57, 42 - 50.

Meyer, J., 1986: Fortschritte in der Erforschung des Erdmagnetismus. Geowissenschaften in unserer Zeit, 4, Nr. 5, 164 - 174.

Milanovsky, E. E., 1987: Rifting evolution in geological history. Tectonophysics, 143, 103 - 118.

Minster, J. B. and T. H. Jordan, 1978: Present-day plate motions. Jour. Geophys. Res., 83, 5331 - 5354.

Morelli, A. and A. M. Dziewonski, 1987: Topography of the core-mantle boundary and lateral homogeneity of the liquid core. Nature, 325, No. 6106, 678 - 683.

Morelli, A., 1989: FIR filters in spherical geometry: core-mantle boundary topography from PcP travel times. In A. Vogel, R. Gorenflo, B. Kummer and C. O. Ofoegbu (Eds.): Inverse Modeling in Exploration Geophysics, Vieweg Braunschweig-Wiesbaden, 439 - 451.

Nance, R. D., T. R. Worsley and J. B. Moody, 1988: The supercontinent cycle. Scientific American, 7, 72 - 79.

Neugebauer, H.-J., W.-D. Woidt and H. Wallner, 1983: Uplift, volcanism and tectonics: Evidence for mantle diapirs at the Rhenish Massif. In: K. Fuchs et al. (Eds.): Plateau Uplift, 381 - 403, Springer Heidelberg - Berlin - New York.

Peltier, W. R. (Ed.), 1989: Mantle Convection - Plate Tectonics and Global Dynamics. Gordon and Breach Science Publishers, New York.

Peltier, W. R., G. T. Jarvis, A. M. Forte and L. P. Solheim, 1989: The Radial Structure of the Mantle General Circulation. In: W. R. Peltier: Mantle Convection - Plate Tectonics and Global Dynamics, Gordon and Breach Science Publishers, 765 - 815.

Ringwood, A. E., 1969: Composition and evolution of the upper mantle. In P. J. Hart (Ed.): The Earth's crust and upper mantle. Am Geophys. Union Monograph 13, 1 - 17.

Roeser, H. A., 1989: Motions of the presently exposed oceanic crust (Extended abstract). Special Workshop, International Seminar on the Rhenohercynian and Subvariscan Fold Belts held in Boppard/Rhein, June 6 - 10, 1989.

Schmincke, H. U., 1988: Unruhige Erde - Plattentektonik, Vulkanismus und Erdbeben. Praxis Geographie, 5.

Sclater, J. G., B. Parsons, and C. Jaupart, 1981: Oceans and Continents: Similarities and differences in the mechanisms of heat loss. J. Geophys. Res. 86, 11535 - 11552.

Tric, E., C. Laj, C.Jéhanno, J.-P. Valet, C. Kissel, A. Mazaud and S. Iaccarino, 1991: High-resolution record of the upper Olduvai transition from Po Valley (Italy) sediments: support for dipolar transition geometry ?
Physics of the Earth and Planetary Interiors, 65, 319 - 336.

Turner, H. H., 1930: Deep Focus. In: International Seismological Summary for 1927, 1 - 108.

Uyeda, S., 1983: Comparative Subductology. Episodes, 2, 1 - 14.

Vestine, E. H., 1954: The Earth's Core. Symposium on the Interior of the Earth. Trans. Amer. Geophys. Union, 35, 63 - 72.

Vogel, A., 1960 a: Über Unregelmäßigkeiten der äußeren Begrenzung des Erdkerns auf Grund von reflektierten Erdbebenwellen. Gerlands Beiträge zur Geophysik, 69, 3, 150 - 174.

Vogel, A., 1960 b: Laufzeitanomalien von am Erdkern reflektierten Erdbebenwellen und deren Korrelation zum Schwerkraft- und Nicht-Dipol-Magnetfeld der Erde. Zeitschrift für Geophysik, 26, 6, 273 - 275.

Vogel, A., 1963: Secular Variations in the Lower Harmonics of the Earth's Gravity Field due to Convection Currents in the Earth's Core. Meddelande fran geodetiska institutionen vid Uppsala universitet.

Vogel, A., 1967: Travel-Time Anomalies of Elastic Waves Reflected at the Core-Mantle Boundary (Abstract). In: K. Runcorn (Ed): Mantles of the Earth and Terrestrial Planets, V, Interscience Publishers.

Vogel, A., 1989: The irregular shape of the Earth's fluid core - A comparison of early results with modern computer tomography. In: A. Vogel, R. Gorenflo, B. Kummer and C. O. Ofoegbu (Eds.): Inverse Modeling in Exploration Geophysics. Vieweg Braunschweig-Wiesbaden, 453 - 463.

Wadati, K., 1935: On the activity of deep-focus earthquakes in the Japan Islands and neighbourhoods. Geophys. Mag. 8, 305 - 325.

Wegener, A., 1912: Die Entstehung der Kontinente. Peterm. Mitt., 185 - 195, 253 - 256, 305 - 309.

Wegener, A., 1915: Die Entstehung der Kontinente und Ozeane. 1st edition, Friedr. Vieweg & Sohn, Braunschweig

Wegener, A., 1929: Die Entstehung der Kontinente und Ozeane. 4th revised edition, Friedr. Vieweg & Sohn, Braunschweig. (Reprint of the first and fourth edition with an introduction and concluding remarks by A. Vogel, Vieweg Braunschweig/Wiesbaden, 1980.)

Wilson, J. T., 1963: Evidence from islands on the spreading of the ocean floor. Nature, 197, 536 - 538.

Woodhouse, J. H. and Dziewonski, A. M., 1984: Mapping the lower mantle: Determination of lateral heterogeneity in P-velocity up to degree and order 6. J. Geophys. Res., 89, 5929.

Woodhouse, J. H. and A. M. Dziewonski, 1987: Models of the upper and lower mantle from waveforms, mantle waves and body waves. EOS Trans. AGU, 68, 356.

Zoback, M. L.; Zoback, M. D.; Adams, J.; Assumpcao, M.; Bell, S.; Bergman, E. A.; Bluemling, P.; Denham, D.; Ding, J.; Fuchs, K.; Gregersen, S.; Gupta, H. K.; Jacob, K.; Knoll, P.; Magee, M.; Mercier, J. L.; Mueller, B. C.; Paquin, C.; Rajendran, K.; Stephansson, O.; Suter, M.; Udias, A.; Xu, Z. H.; Zhizhin, M., 1989: Global patterns of intraplate stress: a status report on the world stress map project of the International Lithosphere Program. Nature 341, 291 - 298.

Basin Evolution

Comparative Evolution of Coal Bearing Foreland Basins along the Variscan Northern Margin in Europe

R.A.Gayer[1] / J.E.Cole[1] / R.O.Greiling[2] / C.Hecht[2] / J.A.Jones[3]
1 Department of Geology, University of Wales Cardiff, PO Box 914, Cardiff CF1 3YE, UK
2 Geologisch-Paläontologisches Institut, Ruprecht-Karls-Universität, Im Neuenheimer Feld 234, D-6900 Heidelberg, Germany
3 Department of Geography and Geology, Cheltenham and Gloucester College of Higher Education, Shaftesbury Hall, Cheltenham, GL50 3PP, UK

ABSTRACT

The principal features of stratigraphy, sedimentology and deformation that reflect tectonic control of basin development are summarised for each of the main coal-bearing basins along the Variscan northern margin from southern Britain through northern France and Belgium to northwestern Germany. The basins show many features in common that suggest they developed as peripheral foreland basins by downflexure of the lithosphere in response to a Variscan tectonic load to the south. These features include: northward migrating depocentres; rapid subsidence; and a switch to a southerly immature sediment supply that commonly also involves cannibalisation of earlier developed Coal Measures. Minor basins lying relatively closer to the Variscan northern margin show similar features, but a later onset of rapid subsidence.

In addition, the more westerly basins, lying over the southern edge of the cratonic Wales-Brabant Massif, demonstrate the influence on basin evolution of pre-existing basement structures. Reactivation of such structures during deposition is revealed by variations of sediment thickness, patterns of coal seam split axes, stacking of channel sandstone bodies, and palaeocurrent maps. During subsequent Variscan deformation the same structures are re-used and may control the main structural framework of the basins. These features are most clearly demonstrated by the South Wales basin, although in some cases a predominantly northward propagating linked thrust system may override the basement structures. No such control by pre-existing structures is seen in the Ruhr basin that lies to the east of the Brabant Massif termination. The Variscan deformation of the Ruhr basin shows a gradual northwestward diminution of strain.

A series of published deep seismic reflection profiles across Variscan tectonic units in the immmediate hinterland of the coal basins have been analysed to determine the size of the tectonic loads that may have induced flexural subsidence. Estimates of crustal-scale shortening of between 35-41 % suggest crustal thicknesses varying from 50 km in SW Britain, N France and Belgium to 60 km in the Rheinisches Schiefergebirge. It is argued that these loads are in approximate agreement with the subsidence curves for the respective coal-bearing foreland basins.

INTRODUCTION

Foreland basins are sedimentary basins developed along the margins of actively forming orogenic belts between the mountain chain and the external craton. Two main categories can be recognised, depending on the type of destructive plate margin with which they are associated. With B-type subduction (Bally & Snelson 1980) the foreland basin is related to the loading of the magmatic arc above the subduction oceanic lithosphere - the retro-arc foreland basins of Dickinson (1974). In the case of continental collision orogenic belts, associated with A-type subduction (Bally & Snelson 1980), the foreland basin results from fold and thrust sheet loading during the orogenic development - the peripheral foreland basins of Dickinson (1974). It is the latter type with which this article is concerned.

Peripheral foreland basins have long been recognised and described by European geologists; they are the foredeeps of Alpine and Variscan literature (e.g. Collet 1927) and the Molasse troughs along the northern margins of the Alps. The significant advances in recent understanding of foreland basins come from three sources: firstly, kinematic understanding of the development of the thrust sheet load derived from knowledge of foreland propagation of deformation in thin-skinned thrust systems (e.g. Boyer & Elliott 1982); secondly, investigations into the mechanisms of downflexure of the lithosphere resulting from this tectonic load; and thirdly, models of basin-fill architecture and of erosion in the basin catchment. The study by Stockmal et al. (1986), modelling convergent margin tectonics, gave insight into the relative importance of size of tectonic load and the downflexure imparted on lithosphere of varying thermal maturity. The models explained the long realised phenomenon that foreland basins migrate in time from hinterland towards the foreland in the external parts of mountain belts and that the early more internal basins tend to be deeper water and starved of sediment (e.g. the Eocene Flysch Noire basins of the French Alps), whilst the younger more external basins tend to be fluviatile/lacustrine dominated with a coarse clastic basin-fill (e.g. the Alpine Molasse basin). Other models (e.g. Howard et al. 1989) investigated the rheology of the lithosphere and the effects of loading on a brittle/elastic upper lithosphere and a ductile/viscous lower lithosphere. They showed that the flexural response was dependent upon the age (thermal maturity) of the lithosphere and upon the angle of dip of the load-generating fault in the upper lithosphere. Work by Sinclair et al. (1989) on modelling rates of sediment accumulation and erosion and their effects on basin subsidence has shown the importance of the basin-fill.

It is thus clear that foreland basin development is controlled by a number of factors. These include: a) the geometry, mass and kinematics of the lithospheric (thrust sheet) load; b) the flexural response of the underlying lithosphere to this load, in turn affected by i) the age and rigidity of the lithosphere, ii) the presence of suitably oriented pre-existing faults capable of reactivation; and c) the volume and rate of sediment supply (related to erosion of the tectonic load) affecting the basin-fill architecture.

The coal basins flanking the northern margin of the Variscan orogenic belt in Europe have many features characteristic of peripheral foreland basins. Several published accounts suggest a "molasse" environment for these "paralic" coal basins, with sediment derived from Variscan tectonic sourcelands to the south (e.g. Bless et al. 1980, Engel & Franke 1983). However, only in the case of the South Wales coal basin has basin development been related to downflexure of the lithosphere resulting from the Variscan tectonic load to the south (e.g. Kelling 1988, Jones 1989b, Gayer & Jones 1989).

It is the purpose of this paper to investigate the various aspects of the foreland basin model in relation to the coal-bearing basins along the northern Variscan margin and by comparing their development, gain additional understanding of the controls on foreland basin evolution.

THE VARISCAN NORTHERN MARGIN IN EUROPE

The Variscan northern margin trends WNW-ESE across southern Britain where its position is marked by a series of major thrust ramps with thrust displacement in excess of several kms (Fig. 2). These are largely obscured by younger strata resting unconformably on the eroded Variscan mountain chain, and their presence is detected by deep seismic reflection profiling (e.g. Chadwick et al. 1983, Brooks et al. 1988, Donato 1988, Le Gall 1990, Miliorizos 1991). The thrusts are exposed in the Mendip area (Williams & Chapman 1986), where it is argued that they form a linked thin-skinned thrust system. Where Upper Palaeozoic rocks crop out, e.g. in South Wales (see below, and Frodsham et al. this volume) the Variscan thrust deformation is seen to extend northwards, but with diminishing intensity, forming typical mountain front structures (Shackleton 1984, Gayer & Jones 1989, Jones 1989b, 1991).

The northern margin continues in a WNW-ESE direction across the English Channel to Pas-de-Calais, where its position is marked by the Faille du Midi thrust (Fig. 1). This thrust continues into Belgium where the strike gradually swings through W-E to SW-NE to the south of the Campine basin around the southeast termination of the Brabant Massif (e.g. Bless et al. 1983). The thrust passes laterally to the northeast into the Aachen thrust system (Wrede 1987). The Midi thrust carries the internally strained Dinantian and older strata of the Dinant Nappe over Silesian coal basins with displacements of 40 - 120 km in the French sector (Raoult & Meilliez 1987) but with only 10's km in the Belgian sector south of the Late Carboniferous Namur basin (Bless et al. 1989). The Midi-Aachen thrust system has been imaged on deep seismic reflection profiles and shown to extend southwards as a shallow dipping reflector

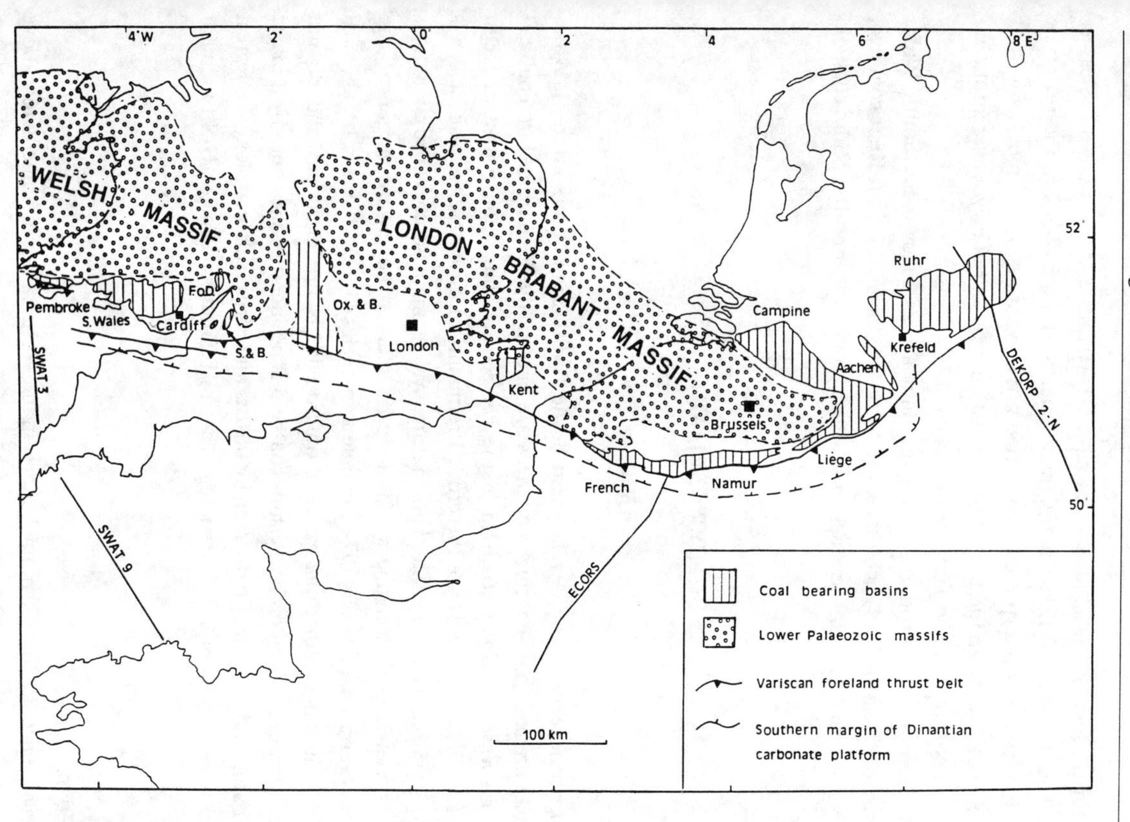

FIGURE 1
Map of the northern Variscan margin and the Wales-Brabant Massif to show the coal basins studied in this paper, and the seismic lines discussed. F.O.D. = Forest of Dean; S & B. = Somerset and Bristol; Ox. & B. = Oxfordshire and Berkshire.

interpreted as a thrust detachment (Meissner et al. 1984, Raoult & Meilliez 1987, Bless & Bouckaert 1988). The northern Variscan margin continues to trend SW-NE in the region of the Ruhr coal basin where it is interpreted as a buried thrust within a typical mountain front thrust system (Brix et al. 1988). To the northeast its position is obscured by younger strata but it is thought to extend NE towards the Tornquist Lineament.
The coal-bearing basins to be considered in this account are as follows: the South Wales coalfield, the Forest of Dean (Bristol) coalfields, the Berkshire basin, the East Kent coalfield, the French and Namur coal basins, the Aachen-Erkelenz coalfields, and the Ruhr coalfield (Fig. 1).

SOUTHERN BRITAIN

South Wales

The most widespread foreland basin development in the British area occurs in South Wales where three distinct areas of coal-bearing sequences are known: the Pembrokeshire coalfield in the west; the main South Wales coalfield in the centre; and the small Forest of Dean coalfield in the east. These coalfields are the eroded remnants of an extensive basin that has been deformed by Variscan, post-Early Stephanian orogeny. The main South Wales coalfield is preserved in an asymetric E-W trending synform with a marked northward vergence. The patterns of thickness variation (Thomas 1974), sedimentary facies (Jones 1989 a & b), and coal rank (White 1991) all suggest that the present outcrop limits are erosional and do not represent the basin margins. The Pembrokeshire coalfield is bounded to the south by a major thrust, the Ritec Fault, and is strongly deformed by thrusts and thrust-related folds. The Forest of Dean coalfield is preserved in a N-S trending synform related to a major N-S basement fault.

i) Stratigraphy/Sedimentology
The coal bearing sequence, up to 2750 m thick, starts in the Westphalian A, immediately above the basal Gastrioceras subcrenatum Marine Band. However, a major change in sedimentation occurred earlier at the base of the Namurian with the onset of clastic coarsening upwards cyclic sedimentation following a prolonged period of carbonate platform deposition during the Dinantian. It is this sudden change to a clastic dominated regime that is thought to register the onset of flexural subsidence in a foreland basin. The change is marked by a significant unconformity, greatest in the east along the N-S Usk axis between the main coalfield and the Forest of Dean. Subsidence curves, derived from stratigraphic thicknesses, are typical of foreland basins (Kelling 1988 and Jones 1989b) and show an increasing rate of subsidence at

the base of the Westphalian (Fig. 3). This might reflect a forelandward migration of the tectonic load at this time. Although isopach maps for the basin have been interpreted to imply a northward migrating depocentre from Namurian through Westphalian C/D time (Kelling 1988 and Jones 1989b), detailed examination of the data shows a more complex pattern that suggests localised variations in thickness possibly related to syn-sedimentary faulting. The Namurian is essentially a marine sequence with only minor evidence for fluvial conditions at the top of coarsening upwards cycles. During the Westphalian A fluvial conditions became more frequent and marine influence was restricted to numerous marine bands, especially at the base of the sequence. At this level orthoquartzitic sandstones are developed in channels indicating a southerly transport vector in the north but northwesterly in the south and east of the coalfield (Bluck 1961; Kelling 1974). The principal productive coals occur at the top of the Westphalian A and into the Westphalian B where the essentially fluvial, mud-dominated conditions continued. Marine bands are found throughout but occur at horizons that can be correlated across western Europe and North America (e.g. Bless et al. 1977) thought to be associated with eustatic change (e.g. the Amman (Donetzoceras vanderbecki) Marine Band at the Westphalian A/B boundary, and the Cefn Coed (Anthracoceras aegiranum) Marine Band at the Westphalian B/C boundary). The highest marine band - the Upper Cwm Gorse (Anthracoceras canmbriense) Marine Band (Calver 1969) - occurs at a significant change to a coarse fluvial lithic sandstone facies (Pennant Sandstone) in which coals and mud-rich overbank sequences are thin and infrequent. The Pennant Sandstone ranges in age from Late Westphalian C to Early Stephanian (Cleal 1987). Jones (1989b) interpreted the sandstones as bars formed in a braided channel sequence. Palaeocurrent orientations suggest a dominantly southerly source (Kelling 1974) and this sudden influx of less mature sand material into the basin suggests erosion of tectonic lands within the Variscan hinterland to the south (Kelling 1988; Jones 1989 a & b; Gayer & Jones 1989). The presence of rounded coal clasts in some Pennant Sandstone channel lags suggests that this southerly land area contained already compacted coal sequences that became cannibalised as Variscan deformation propagated towards the foreland.

Detailed analysis of the palaeocurrent data shows variations within the basin that are related to major E-W trending folds (Jones 1989 a & b), and demonstrate an influence by reactivation of inferred basement structures. Movement along basement structures has also been sugested to account for both thickness variation of Westphalian A and B sequences and for coal seam splitting across the same folds that affected later (Westphalian C) palaeocurrent patterns (Jones 1989 a & b). In addition syn-sedimentary faulting has been widely reported from the main South Wales basin, in some instances related to down-slope sliding, but in at least one instance to a major basement lineament (Cole et al. 1991).

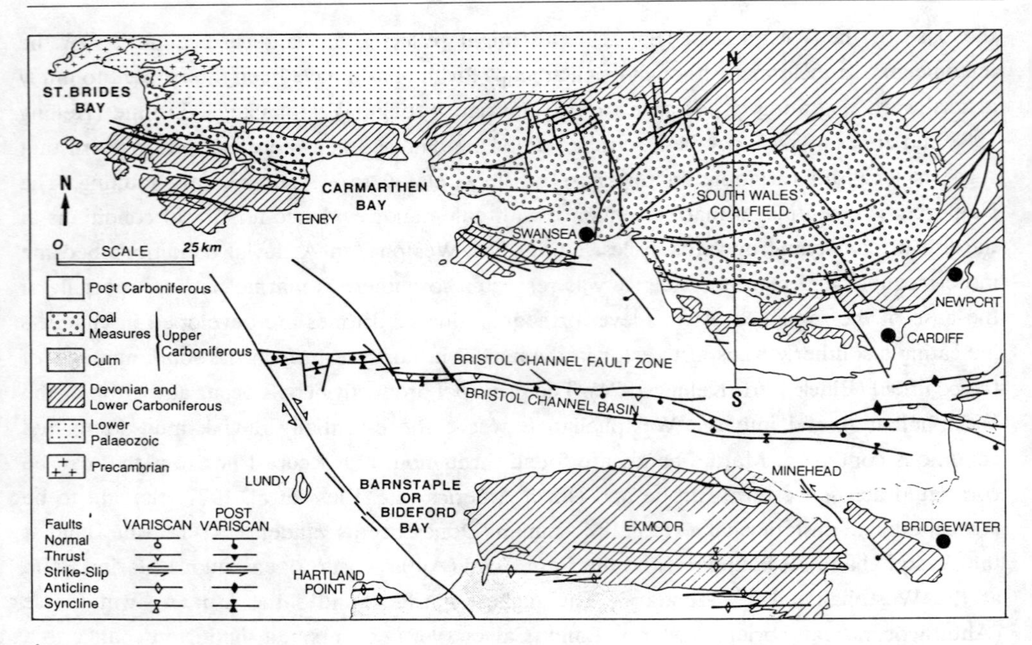

FIGURE 2

A) Map of the South Wales Coalfield in relation to the Variscan and post-Variscan structures of South Wales and the Bristol Channel. Named structures are: BF, Benton Fault; BT, Blaina Thrust; BtF, Bettws Fault; CaBT, Caswell Bay Thrust; CaCA, Castell Coch Anticline; CBT, Caer Bryn Thrust; CCA, Cardiff-Cowbridge Anticline; CF, Cothelstone Fault; CCFZ, Carreg Cennen Fault Zone; CHT, Cross Hands Thrust; CS Caerphilly Syncline; DPF, Dinas Powys Fault; GdS, Gelligaer Syncline; GF, Gardeners Faul; GS, Gorseinon Syncline; JT, Johnston Thrust; KT, Kenfig Thrust; LT, Llannon Thrust; LlT, Llanharan Thrust; MA, Margam Anticline; MGF, Moel Gilau Fault; NFZ, Neath Fault Zone; NT, Newlands Thrust; OA, Orielton Anticline; PA, Pontypridd Anticline; PCT, Pont-y-Clerc Thrust; RhT, Rhigos Thrust; RT, Ritec Thrust; SF, Sticklepath Fault; SVFZ, Swansea Valley Fault Zone; TF, Tongwynlais fault; TT, Trimsaran Thrust; UA, Usk Anticline; WA, Winsle Anticline. Modified from Gayer & Jones (1989).

B) Typical stratigrahic column for the Silesian sequence in the South Wales basin.

C) N-S cross-section showing the principal structures along the line X-Y marked in A). Modified from Jones (1991).

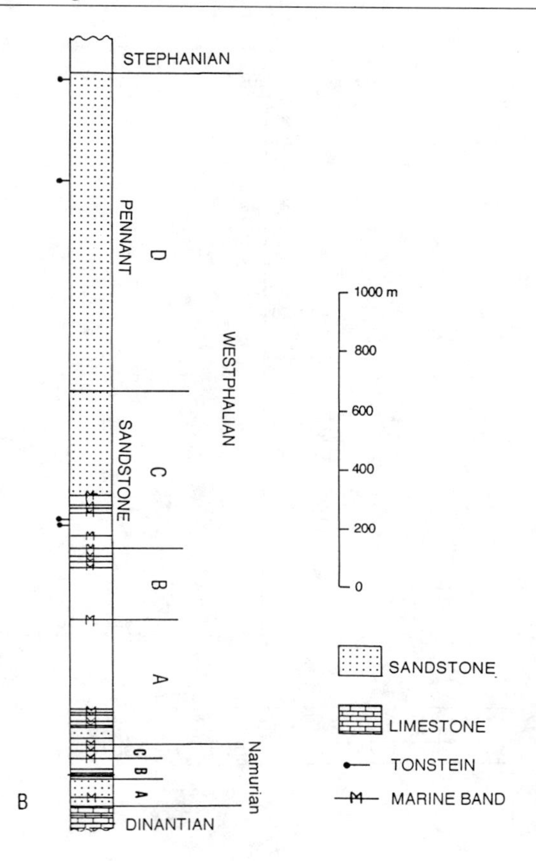
STEPHANIAN
PENNANT
SANDSTONE
D
C
B
A
WESTPHALIAN
C
B
A
Namurian
DINANTIAN
B
1000 m
800
600
400
200
0
SANDSTONE
LIMESTONE
TONSTEIN
MARINE BAND

C

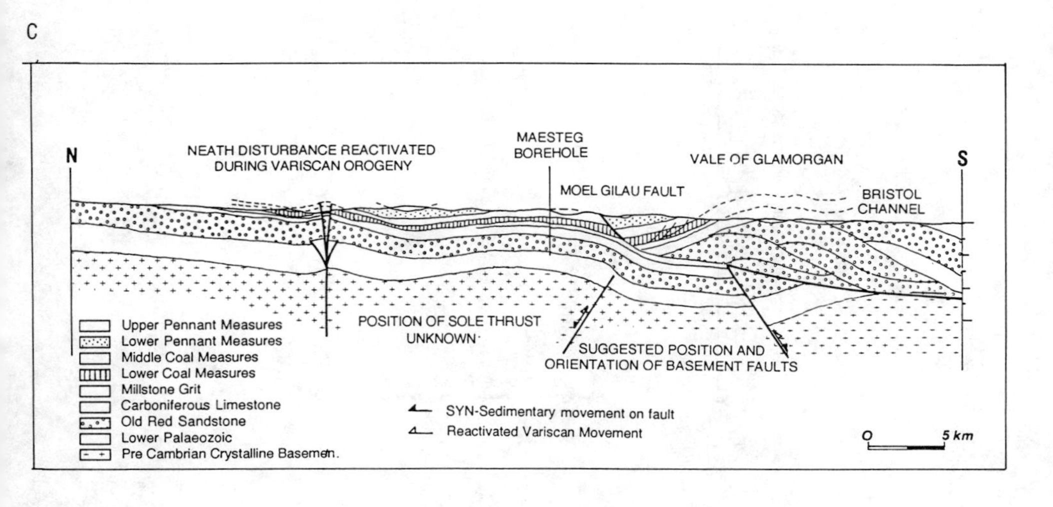
N
S
NEATH DISTURBANCE REACTIVATED DURING VARISCAN OROGENY
MAESTEG BOREHOLE
VALE OF GLAMORGAN
MOEL GILAU FAULT
BRISTOL CHANNEL
POSITION OF SOLE THRUST UNKNOWN
SUGGESTED POSITION AND ORIENTATION OF BASEMENT FAULTS
Upper Pennant Measures
Lower Pennant Measures
Middle Coal Measures
Lower Coal Measures
Millstone Grit
Carboniferous Limestone
Old Red Sandstone
Lower Palaeozoic
Pre Cambrian Crystalline Basemen.
SYN-Sedimentary movement on fault
Reactivated Variscan Movement
O
5 km

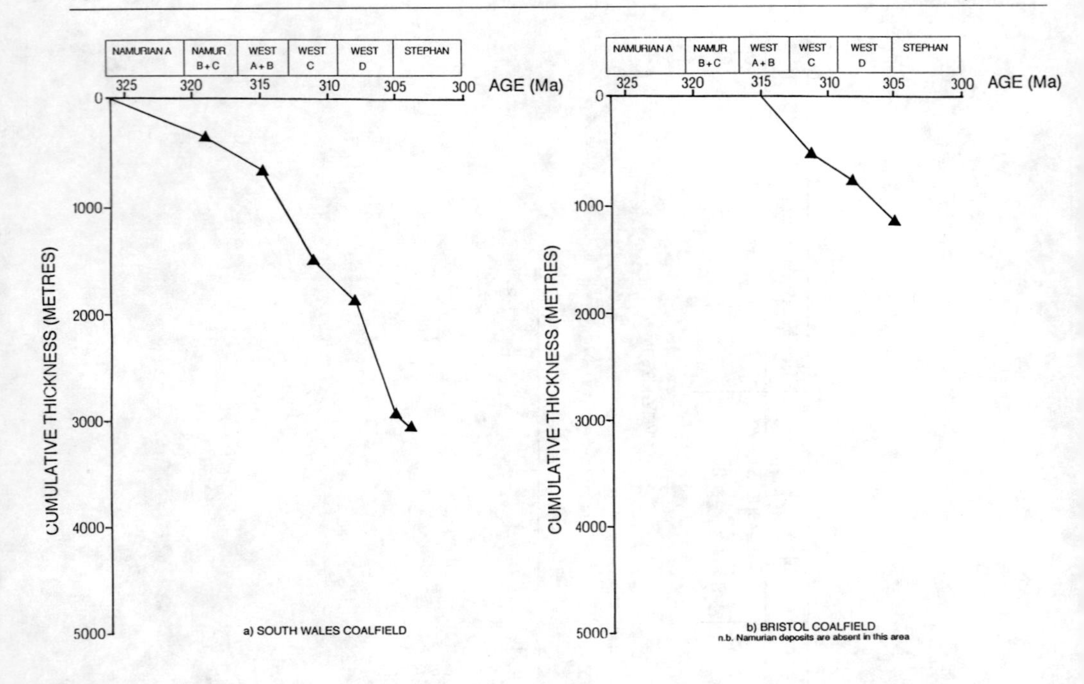
NAMURIAN A
NAMUR B+C
WEST A+B
WEST C
WEST D
STEPHAN
AGE (Ma)
CUMULATIVE THICKNESS (METRES)
a) SOUTH WALES COALFIELD
b) BRISTOL COALFIELD
n.b. Namurian deposits are absent in this area

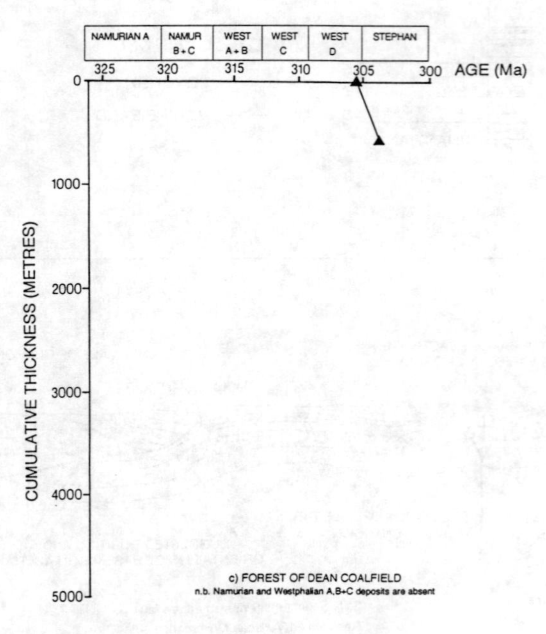
NAMURIAN A
NAMUR B+C
WEST A+B
WEST C
WEST D
STEPHAN
AGE (Ma)
CUMULATIVE THICKNESS (METRES)
c) FOREST OF DEAN COALFIELD
n.b. Namurian and Westphalian A,B+C deposits are absent

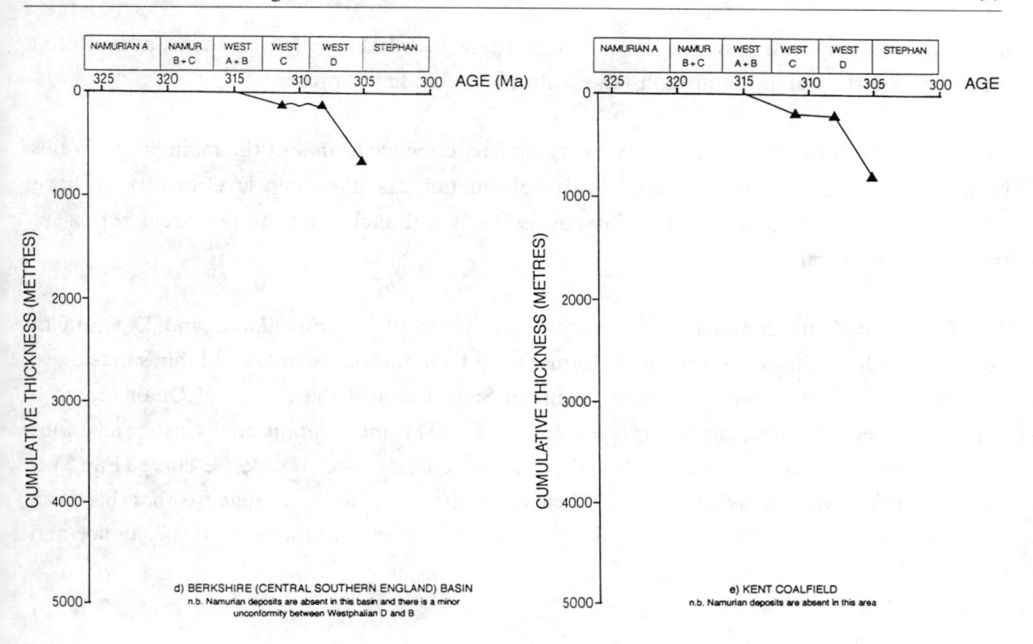

FIGURE 3
Subsidence curves for the Silesian of the coal basins along the northern Variscan margin in Britain. The curves are derived from the maximum stratigraphic thickness developed in each basin, and the time-scale is after Lippolt et al. (1984). No correction for syn-depositional compaction has been made.

Tonsteins are very rare in South Wales having been recorded in only four seams at scattered localities. Apart from these rare tonsteins, volcanic rocks are not present.

The Pembrokeshire coalfield contains a very similar sequence to that of the main South Wales basin, although the intense fold and thrust deformation and the deep level of erosion have resulted in a less complete record (Thomas 1974). It is thought that the two areas represent parts of the same basin.

The Forest of Dean coalfield preserves only c. 600 m of Westphalian C and D strata of Pennant Sandstone facies resting unconformably on Dinantian carbonates and thin sandstones. Like the main South Wales basin, the Pennant Sandstones of the Forest of Dean record a northerly directed palaeocurrent (Gayer & Stead 1971) and contain coal clasts and other material thought to be derived from southern tectonic lands. The subsidence curve (Fig. 3) is almost identical with the Westphalian C/D curve for the main basin and suggests that this basin was developed as part of the South Wales basin during a period of more rapid subsidence and transgression.

ii) Deformation

Deformation of the main South Wales basin developed three distinct structural elements.

Firstly, E-W striking thrusts and thrust-related folds. These have a characteristic style of extensive layer-parallel thrust flats connected by short frontal ramps (e.g. Jones 1989b; Frodsham et al. this vol.). They occur most prominently along the southern margin of the basin where the thrusts verge southwards. Jones (1989b) estimated that locally shortening exceeded 50 % in this zone. North of a prominent south-downthrowing normal strike fault (the Moel Gilau Fault) the thrusts and folds verge northwards. In the west of the coalfield these structures result in up to 62 % shortening (Frodsham et al. this vol.), and similar structural styles are recorded in the Pembrokeshire coalfield to the west (Dunne 1983; Coward & Smallwood 1984). In the north of the main South Wales basin both thrusts and thrust related folds, although present, are more variable in trend. For example in the northwest the structures strike NE-SW and verge northwest, whilst in the north-centre of the coalfield they strike variably between N-S verging east, to E-W verging north. This variability has been interpreted as the effect of Variscan reactivation of basement structures in the overlying Carboniferours cover (Cole et al. 1991).

Secondly, superimposed on the thrust related deformation, is a set of cross faults that strike NNW-SSE in the west of the main coalfield and NW-SE in the east. They have been interpreted as dextral strike-slip faults (Trotter 1947; Anderson 1951; Gayer et al. 1973) but slickenside lineations record only normal dip-slip movement (Cole et al. 1991). They have the effect of extending the basin E-W and, although in every case, where age relationships can be demonstrated, they post-date the thrust deformation, the strain is compatible with Variscan N-S shortening. However, the evidence of syn-sedimentary activity along some of the NW-SE striking faults, suggests that these may result from reactivation of basement structures and may not simply reflect the Variscan strain pattern.

Thirdly, there are a number of NE-SW trending major fault zones that cross the basin and show a range of structures (Owen 1953) most easily related to sinistral transcurrent deformation (Anderson 1951). They show evidence of a long-lived movement history and most authors suggest that they represent reactivated Caledonoid basement structures.

Southern England

To the east, the Bristol and Somerset coalfield shows very many similarities with the South Wales coalfield, including age and sedimentation of Coal Measures sequences, subsidence history, and subsequent deformation. It is likely that the coalfield was developed as part of the South Wales basin and was subsequently isolated by post-deformational erosion.

The coalfields to the east of Bristol and South Wales are concealed beneath a Mesozoic and younger cover. Deep boreholes through this cover have revealed Coal Measures in two main areas: the Oxfordshire and Berkshire basins and the East Kent coalfield.

i) The Oxfordshire and Berkshire Basins

Coal basins are developed in two NW-SE trending synclinal structures, forming a NNW-SSE dextral zone across the Wales-Brabant Massif. The more southerly of the two basins, the Berkshire basin, lies along the southern flank of the massif at the northern margin of the Variscan fold and thrust belt (Fig. 1). The Coal Measures rest unconformably on a thin sequence of Dinantian limestones in the south and east of the Berkshire Basin, but overstep onto Devonian Old Red Sandstones and Lower Palaeozoic rocks to the north and west. The Coal Measures are entirely Westphalian in age and have a maximum thickness of 960 m in the Oxfordshire basin (Dunham & Poole 1974) and 650 m in the Berkshire basin (Foster et al. 1989). The Lower and Middle Coal Measures are a relatively thin mud-dominated sequence, thickening northwards into the Coventry area. The presence of at least one marine band in the Westphalian A suggests a lower delta plain environment. In the south and east of the Berkshire basin basic volcanic rocks and high-level dolerite intrusions are developed within the Lower and Middle Coal Measures. Up to 500 m of Upper Coal Measures are present in the Berkshire basin, dated by non-marine faunas as Westphalian D in age (Foster et al. 1989). The sediments are lithic sandstones of typical Pennant Sandstone facies, with stacked channels forming more than 80 % of the sequence. Palaeocurrent directions are variable but are not inconsistent with having been derived from Variscan mountains to the south (Foster et al. 1989). Channel lags contain rounded clasts of coal and siderite, indicating cannibalisation of earliert lithified coal sequences.

Despite the presence of volcanic rocks in the lower part of the sequence, no tonsteins have been recorded from theses basins.

Throughout most of the area dips are less than 25½, consistent with gentle folding across the Wales-Brabant Massif. However, in the southern boreholes of the Berkshire basin dips increase to 70½ and numerous faults are present in the bore cores. This is interpreted as the northern margin of Variscan thrust deformation.

ii) East Kent Coalfield

The East Kent coalfield is developed in a gentle NW-SE trending synform beneath an unconformable Jurassic and Cretaceous cover. The Coal Measures, up to 885 m thick, rest on Dinantian platform carbonates. The coal-bearing sequence starts with 215 m of dominantly mudstones with thin sandstones and workable coals of Westphalian A, B and early C age. Sediments representing the Namurian and the base of Westphalian A are missing. The upper parts of Westphalian C and Westphalian D are represented by a thick (670 m) unit of sandstones of Pennant Sandstone facies. Marine bands are present in the argillaceous Lower and Middle Coal Measures, including the Vander becki and Aegiranum Marine Bands, but no indications of marine influence have been recorded above the Aegiranum Marine Band in the sand dominated Upper Coal Measures. No tonsteins have been reported from the Kent coalfield. The Subsidence patterns for the Kent coalfield (Fig. 3) and the Berkshire basin are very similar and probably reflect a similar position relative to the Variscan tectonic wedge.

Information about the deformation within the Kent coalfield is relatively sparse. However, John Rippon (pers. comm.) has suggested that NW-SE zones of sandstone thickening, seen in the mine plans may represent thrust repetition associated with essentially layer-parallel thrust deformation. This would be in accord with the Kent coalfield lying at the northernmost extremity of Variscan thrusting in a similar position to that of the Berkshire basin.

NORTHERN FRANCE, BELGIUM AND AACHEN

The coal basins described in this section are: the French basin, comprising the Pas-de-Calais and Nord sub-basins; the Belgian (or Namur) basin, comprising, from west to east, the Mons,

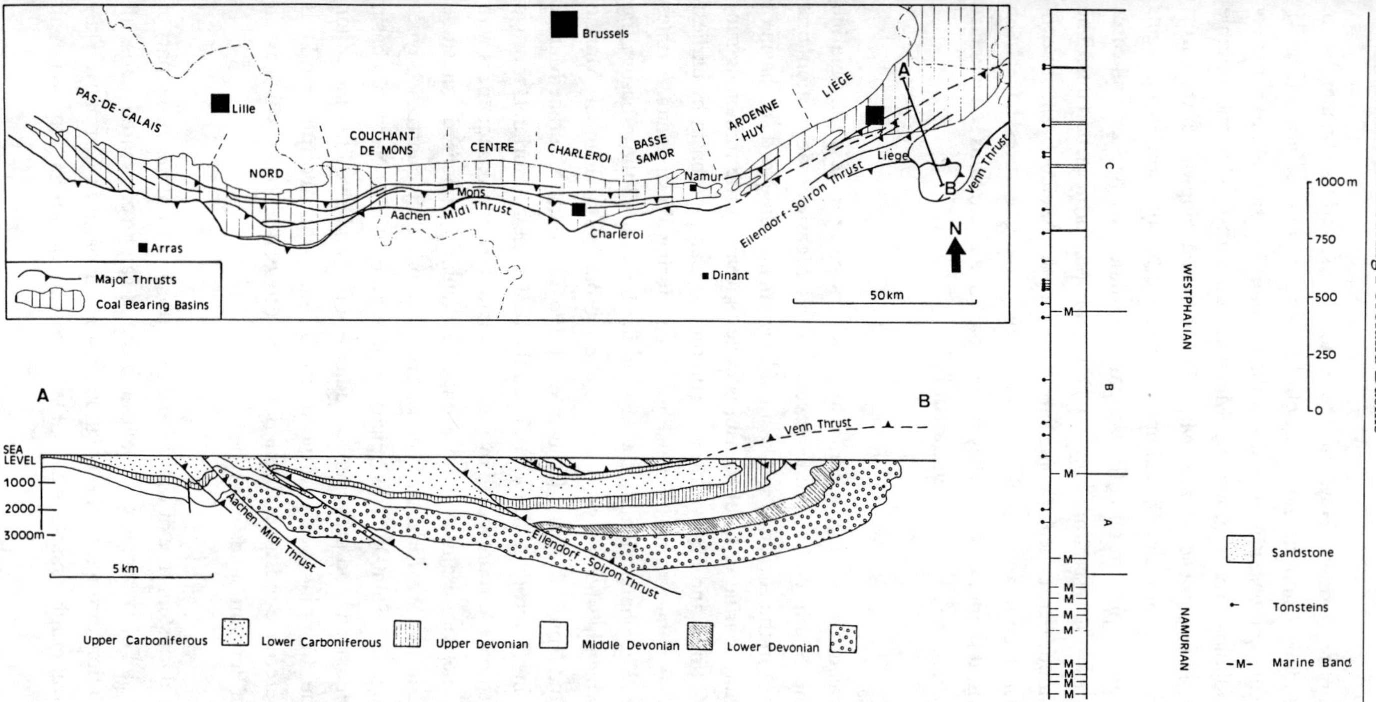

FIGURE 4

A) Map showing the N France, Belgium (Namur & Campine) and Aachen coal basins, and the Midi-Aachen thrust system.

B) Typical stratigraphic column for the Nord sub-basin (N France).

C) N-S cross-section along the line X-Y marked in A) through the Nord sub-basin, showing the principal structures.

Centre, Charleroi, Basse Sambre, Ardenne-Huy, and Liége sub-basins; and the Aachen basin, comprising the Wurm and Inde synclinoria, and the South Limburg area to the north (Fig. 4). These basins are developed along the southern flanks of the Brabant Massif and are generally overthrust by Devonian and older rocks along the Midi-Aachen thrust. The one notable exception is the Inde sub-basin in the Aachen area which lies in the hangingwall of the thrust system. In northern Belgium a further basin, the Campine basin, overlies the northern margins of the Massif, and may possibly represent an entirely separate sedimentary basin. In all cases the clastic Silesian sequences rest disconformably on Dinantian platform carbonates. The French basin, parts of the Namur basin, the Campine sub-basin, and the Aachen basin are concealed beneath transgressive (cenomanian rocks. Although the Coal Measures have been worked in separate coalfields, it seem likely that they formed part of a continuous sedimentary basin now lying in the footwall of the Midi-Aachen thrust.

i) Stratigraphy/Sedimentology

As with the South Wales basin, the Namurian commenced with a more or less fully marine mudstone facies. Unlike the South Wales basin, however, Coal Measures deposition first appeared in Late Namurian A in the Andenne sub-basin of Belgium and the Nord sub-basin of France. The change from fully marine to Coal Measures cycles spread to the remaining sub-basins by Early Namurian C times and finally to the northernmost Campine basin by earliest Westphalian A times. As in Britain, marine influence was great during Westphalian A times but diminished through the Westphalian B until the last marine band in all these basins, the Aegiranum Marine Band, which has been recorded in all the sub-basins (the Rimbert Marine Band of the Pas de Calais and Nord sub-basins (Bouroz et al. 1964), and the Maurage Marine Band of the Belgian sub-basins (Paproth et al. 1983)). In all the basins the Namurian B through to Westphalian C Coal Measures are represented by mud-dominated overbank sequences with relatively thin sheet sandstones representing flood events. Lenticular channelised sandstone bodies have been recorded in the Westphalian C consisting of dominantly quartz-rich types with sub-angular grains (e.g. the Saint-Jacques microconglomerate of the Nord sub-basin (Bouroz et al. 1964)). Palaeocurrent and provenance studies are few and give little indication of palaeogeography. In the Campine basin of northern Belgium the Westphalian D is represented by the Neeroeteren Sandstone Formation (Paproth et al. 1983) which is an immature lithic sandstone of Pennant Sandstone type.

Unlike the British basins, coal tonsteins are of widespread development from the upper part of Westphalian A to the lower half of Westphalian C. These tonsteins can be correlated across all the basins and allow correlation through to the neighbouring Ruhr basin in Germany. They constitute the best chronostratigraphic markers available in the Carboniferous (Paproth et al. 1983).

Subsidence curves for the French and Namur basins are very similar, except that no Westphalian D strata are preserved in the basins along the footwall of the Midi-Aachen thrust (Fig. 5). The curves show a strong similarity with that of the South Wales basin both in total subsidence and in the sudden increase in subsidence rate at the start of Westphalian A. As in South Wales, this is interpreted as the initial flexural subsidence resulting from the Variscan tectonic load. The earlier onset of Coal Measure type sedimentation in these basins probably reflects differences in palaeogeography rather than tectonics, since the change in sedimentation occurs before the increase in subsidence. The Aachen basin shows a distinct subsidence pattern in that the increased subsidence rate, related to flexural subsidence, occurs during Namurian B/C times, earlier than in the other basins. This might suggest an earlier development of the tectonic load in the Aachen basin. The maximum total subsidence in the French, Belgian and Aachen basins is very similar, being about 3 km, comparable to that of the South Wales basin. The thickness varies, however, within each basin. The thickness is lowest in the Pas de Calais sub-basin (c. 1,140 m) thickening towards the east into the Nord sub-basin (c. 2,600 m), and the thickness in each of the sub-basins diminishes towards the Brabant Massif in the north. There are very few records, however, of controls on subsidence by syn-sedimentary structures. The Neeroeteren Sandstone of Westphalian D age in the Campine basin is thought to have been sourced from Variscan tectonic lands to the south, and Paproth (1987) concluded that syn-sedimentary thrusting occurred from Westphalian A times onwards. Within the Dinant Nappes in the hangingwall of the Midi Thrust there is considerable evidence for syn-sedimentary activity along east-west striking extensional faults active during the deposition of Devonian and Early Carboniferous sediments. These faults were inverted as thrusts during the end-Carboniferous stages of the Variscan Orogeny (Bless et al. 1989; Fielitz in press). It appears, however, that these or similar structures were not reactivated during Coal Measures deposition, although Bless et al. 1989) suggest that the northern boundary of the Namur basin with the Brabant Massif is a major fault - the Bordiére Fault - that may have controlled the basin subsidence.

ii) Deformation

There are only a few detailed descriptions of the Variscan structures of the Coal Measures in this area (e.g. Bouroz 1960), although there are many examples of cross-sections through individual coalfields showing the general structure. It is clear that throughout the belt the strata have been intensely deformed by thrusts and thrust-related folds, generally verging northeastwards in the Pas-de-Calais sub-basin, northwards in the Nord sub-basin and northwestwards in the Namur and Aachen basins. These structures are typical of a thin-skinned fold and thrust belt, commonly developing a trailing imbricate fan (Boyer & Elliott 1982) in front of the Midi-Aachen Thrust which may have displacements of 50 - 150 km (Raoult & Meilliez 1987), but far less according to Wrede et al. (this volume). The imbricate thrusts of the fan commonly have displacements of 600 - 700 m and subdivide the coalfields into structural

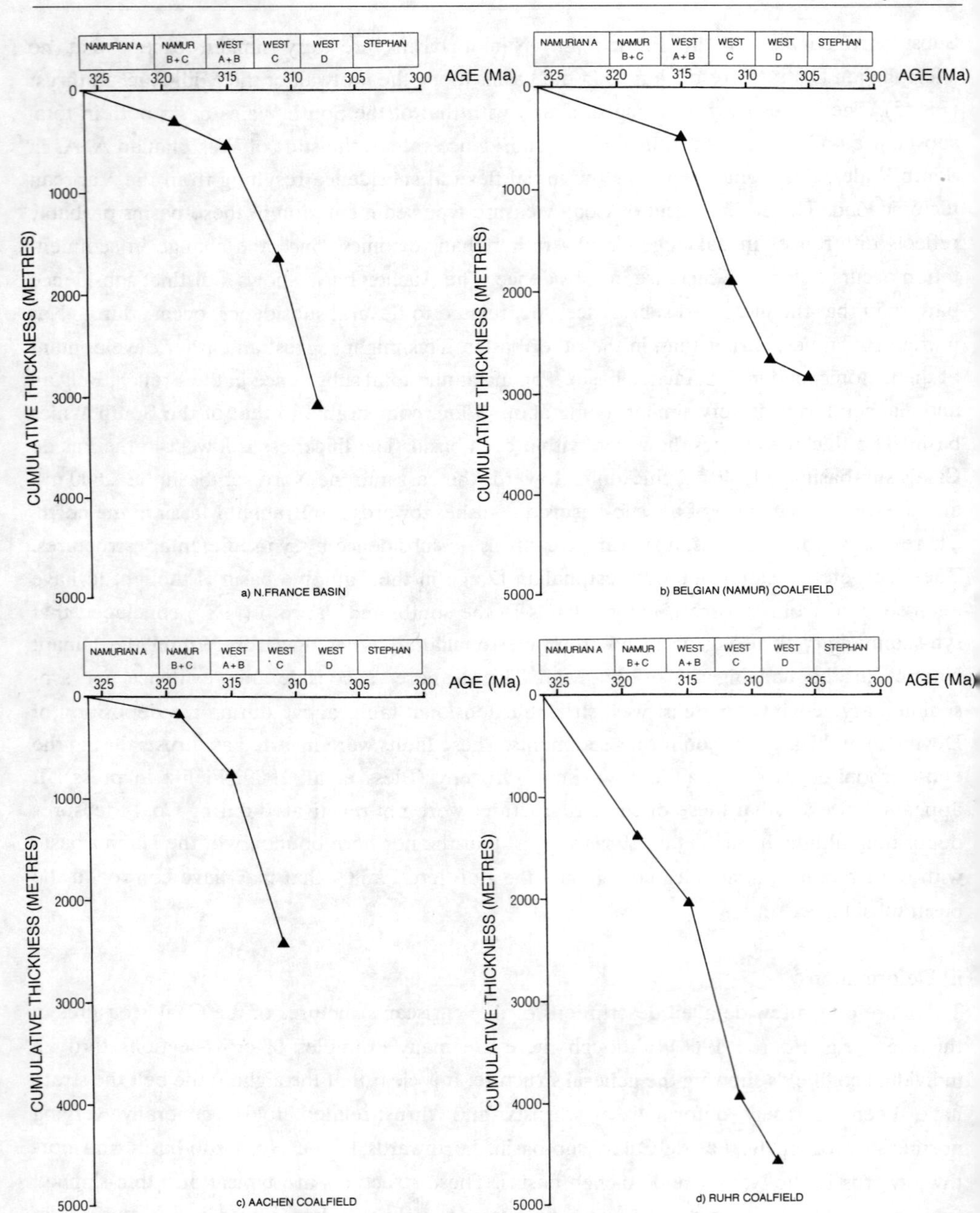

FIGURE 5
Subsicence curves for the Silesian of the coal basins in N France, Belgium, Aachen and the Ruhr.The curves are derived in the same way as those in Fig. 3

and mining entities, in which coals of contrasting rank and structural setting were extracted. These thrusts can be mapped along the length of the entire sub-basins, the larger thrusts sometimes continuing across adjoining sub-basins, e.g. the Pruvost Thrust of the Pas-de-Calais sub-basin extends into the Nord sub-basin as the Chalard Thrust (Bouroz et al. 1964). The structures of the French basin show a number of interesting features that suggest the transport direction during Variscan thrusting was northwards. These include lateral branchlines of imbricate stacks and the presence of oblique dextral ramps striking northwest (see Bouroz et al. 1964; Fig. 6).

In addition to the major imbricate thrusts the coalfields are deformed by a large number of parallel flat-lying thrusts that are intimately associated with folding (Ancion 1942, 1955). In some instances the northward verging folds occur as propagation folds in the hangingwall of the flat-lying thrusts, whereas in other cases the thrusts are folded by later asymmetric folds in such a manner that the thrusts are almost bed-parallel in one fold limb but cut the bedding at a high angle in the opposite limb. The bed-parallel thrusts commonly occur in the roofs of coal seams and are associated with intense shearing and slickenside development (Sax 1946), reminiscent of the style of thrusting in the South Wales basin.

There appears to be no evidence for reactivation of earlier structures during the Variscan thrust-related deformation of these basins. The regional change from a northwards vergence in the west to a nortwestwards vergence in the east may reflect the presence to the north of the Brabant Massif.

NORTHWEST GERMANY

The Ruhr Coalfield

The Ruhr coalfield lies in a NE-SW trending basin (Fig. 6). It contains over 3500 m of productive Coal Measures which range in age from Namurian C to Westphalian C (Bachmann et al. 1971). Namurian A and B sequences consist of upto 1500 m of dominantly marine turbiditic facies, with an incrasing influence from terrestrial flora and fauna towards the overlying Coal Measures. The clastic Silesian sediments rest conformably on marine black shales and cherts of Dinantian age in the south of the basin. A Lower Carboniferous carbonate platform facies may be present beneath the nortwestern parts of the basin. The upper surface of the Carboniferous dips gently to the north and, moving north, is overlain by a progressively thicker cover of post-Carboniferous, mainly Cretaceous rocks. Only in the south of the coalfield do coal-bearing rocks crop out. These are of Namurian C and Westphalian A age.

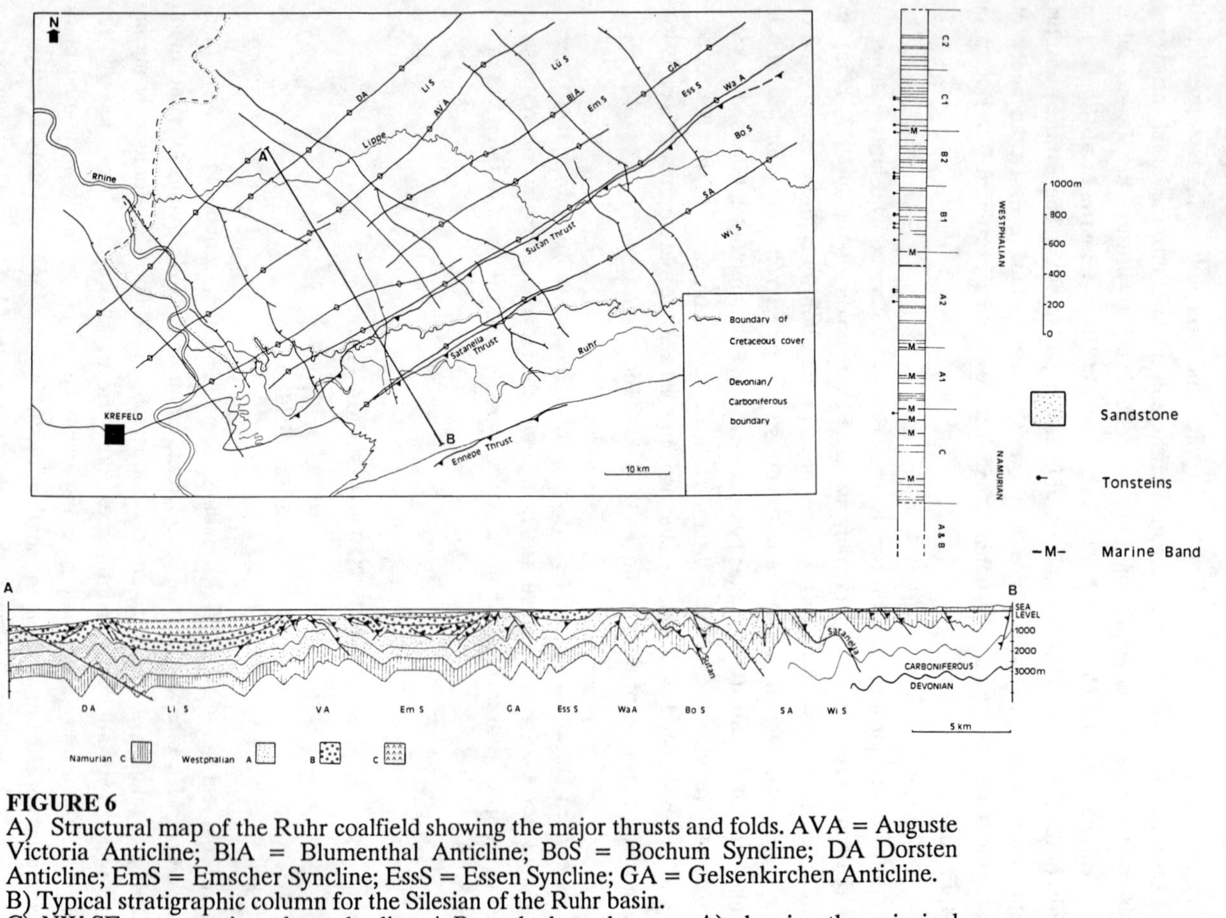

FIGURE 6
A) Structural map of the Ruhr coalfield showing the major thrusts and folds. AVA = Auguste Victoria Anticline; BlA = Blumenthal Anticline; BoS = Bochum Syncline; DA Dorsten Anticline; EmS = Emscher Syncline; EssS = Essen Syncline; GA = Gelsenkirchen Anticline.
B) Typical stratigraphic column for the Silesian of the Ruhr basin.
C) NW-SE cross-section along the line A-B marked on the map A) showing the principal structural style (after Drozdzewski et al. 1982).

i) Stratigraphy/Sedimentology

The stratigrahy of the coalfield has been determined using both biostratigraphic and tonstein correlation. Tonsteins are widespread and numerous in the Ruhr coalfield and produce exellent event stratigraphic marker horizons. The main sedimentological trend through Coal Measures times is from marine to deltaic sedimentation. The basin is interpreted as a non-marine basin, marginal to a cratonic landmass, which was subject to periodic invasion by the sea. Deposition was mainly in the upper part of a highly constructive deltaic system (e.g. Cassyhap 1975). Marginal basin folding and block uplift in front of the basin began in Westphalian C, leading to cannibalisation of earlier strata and fan formation around the basin margins (Jankowski et al. this vol.).

The passage from marine turbidites to thick fluvial channel sandstones occurs in Namurian C where one of the channel sandstones, the Grenzsandstein, marks the beginning of the productive Coal Measures. The lower part of the Coal Measures from Namurian C to Westphalian A consists of thick fluvial channel sands overlain by seat earths and coal seams, with relatively frequent marine bands. Coal contents increases upwards. The strata are thought to represent deposition on a lower delta plain with frequent marine transgressions, at least some of which may have been eustatic (Jankowski et al. this vol.). The Lower Westphalian B contains very few fluvial channels and marks a time when sedimentation in the basin changed considerably. It is the finest grained part of the coal-bearing sequence and represents deposition on a lower delta plain. The Upper Westphalian B contains fluvial channel sandstones overlain by siltstones which grade upwards into seat earths and coals which are overlain by lake or bay sediments. The proportion of sandstone and conglomerate increases upwards, whilst the number and thickness of marine bands decrease until the Aegiranum Marine Band above which, in the Westphalian C, only a few brackish layers are found (Jankowski et al. this vol.). At this stratigraphic level the sandstone channels are nearly all braided.

ii) Deformation

The Ruhr coalfield was deformed during the Variscan Orogeny to produce a series of SW-NE striking thrusts and folds as well as NW-SE striking extensional faults which divide the area into horsts and graben (Hoyer & Pilger 1971). There are also strike-slip faults which form a conjugate set generally striking WNW-ESE and N-S. The thrusts have displacements of upto 2.5 km and are distributed at various stratigraphic levels (Brix et al. 1988). The transport direction is mainly to the northwest, but also occasionally towards the southeast (Hoyer & Pilger 1971). Folds are the dominant tectonic element in the coalfield, with wavelengths varying from decameters to over 10 km. The folds are considered to have formed contemporaneously with the thrusting (Drozdzweski et al. 1980), and to have dictated the geometry of the thrusts. Most of the folds are upright, with no overriding sense of vergence.

Layer-parallel thrusting and shearing is not a characteristic feature of the deformation, and when it occurs is clearly related to flexural slip processes in the fold limbs. The overall structure of the coalfield is characterised by stockwerk tectonics (Drozdzewski et al. 1980, 1985), whereby the style of deformation changes with depth, but is independent of stratigraphy. There are three recognised structural levels which, from the base upwards, are characterised by: 1) many small-scale folds of constant wavelength and amplitude, plus a few, mainly small-scale thrusts which increase in displacement upwards; 2) large-scale tight antiforms and open synforms, plus folded thrusts; 3) large-scale, high amplitude anticlinoria, broad low amplitude open synforms, and few thrusts (Fig. 6).

The extent of deformation is greatest in the south of the coalfield, with over 50 % shortening, and decreases towards the northwest to 5 - 10 % shortening (Wrede 1987a). The decrease is gradual, and is thought to be due to the absence of a cratonic massif to the north against which deformation might have been buttressed. This contrasts with the coal basins to the west, south of the Brabant Massif, where thrusting and folding is more intense and polarised strongly towards the north. In the southeast of the coalfield anticlinoria and synclinoria are both tight, whereas towards the nortwest anticlinoria remain tight but synclinoria are open. Thrusting dies out both upwards and downwards and is not thought to be linked to an underlying detachment (Brix et al. 1988). There is no clear evidence for the control of later Variscan deformation by pre-existing basement structures as in South Wales.

iii) Basin Evolution

The Ruhr coal basin has long been regarded as a foredeep to the northern branch of the European Variscides (e.g. Engel & Franke 1983 and refs. therein). Unfortunately the erosion of the southeastern margin of the basin prevents a full documentation of the migration of the basin depocentre, which would provide evidence for its foreland basin origin. However, isopach maps for the coalfield (e.g. Strack & Freudenberg 1984, Drozdzewski 1990) show a greater thickness in the southeast of the coalfield from Upper Westphalian A to Westphalian C times, which suggests a depocentre in the southeast of the coalfield that may have migrated northwestward during the Westphalian. In addition the maps show a NE-SW trending region of lower thicknesses that gradually migrated northwestwards from the southeast margin of the coalfield in Early Westphalian A times through Late Westphalian A to Westphalian C times. Coals are thicker and cleaner in this region indicating that it formed a palaeogeographic element. Thicknesses increase both to the northwest and the southeast of this migrating zone of reduced subsidence, which might reflect the development of a syn-depositional tectonic feature. This could represent the foreland tip of the northwestward migrating Variscan deformation (Drozdzewski pers. com.).

Palaeocurrent data and provenance studies indicate that sediment was sourced from the northeast during the Early Carboniferous but changed during the Late Carboniferous to sources from the south and west. The southerly source is thought to be the rising Variscan mountain belt. Westphalian C sandstones contain reworked pebbles of coal of high rank which suggests that uplift of Lower Westphalian strata had taken place to the south.

Subsidence curves for the Ruhr basin show rapid subsidence from Namurian A through to Westphalian C with only a slight increase in subsidence rate at the onset of Westphalian A (Fig. 5). The Westphalian curves are similar to those described for coal basins to the west but suggest that flexural subsidence started earlier, during Namurian A, than in the more westerly basins.

There seems to be very little evidence for any syn-depositional reactivation of basement structures in the coal basin. The isopach maps show no local thickness variations related to the position of later structures within the basin. The positions of coal seam splits do not suggest any structural alignment, but rather a more random distribution that might suggest compaction induced subsidence. Finally, the proportion of sandstones in the sequence does not vary systematically across fold structures. These observations suggest that the Variscan structures entirely post-date the deposition.

VARISCAN TECTONIC LOADS INDUCING FLEXURAL SUBSIDENCE

The three main areas considered in this paper have all been investigated by recent deep seismic reflection profiling: SW Britain - the SWAT Profile (BIRPS & ECORS 1986, Le Gall 1990, 1991); N France - the ECORS Profile (Cazes et al. 1986, Raoult & Meilliez 1987); N Germany - the DEKORP 2 Profile. (DEKORP Research Group 1985, Franke et al. 1990, Meissner & Bortfeld 1990). Thus comparisons between the Variscan structures that formed the potential tectonic loads for the coal-bearing foreland basins can be made. Simplified transport-parallel cross-sections for each of the three areas are shown in Fig. 7. All three sections show major north-directed thrusting. In the SW Britain and N France sections the thrusting is linked and very thin-skinned towards the northern, external margin of Variscan deformation. However, in the N German section the thrusting gradually dies out into a zone of upright folding and related north and south directed thrusting within the Ruhr coal basin. These observations confirm the regional descriptions of the structure given above.

Towards the orogenic hinterland, the three sections show a greater similarity, with pronounced northward directed thrusting, involving at least the upper half of the crust and in the case of the SW Britain section, the entire crust in an 'imbricate stacking wedge'. These more internal parts

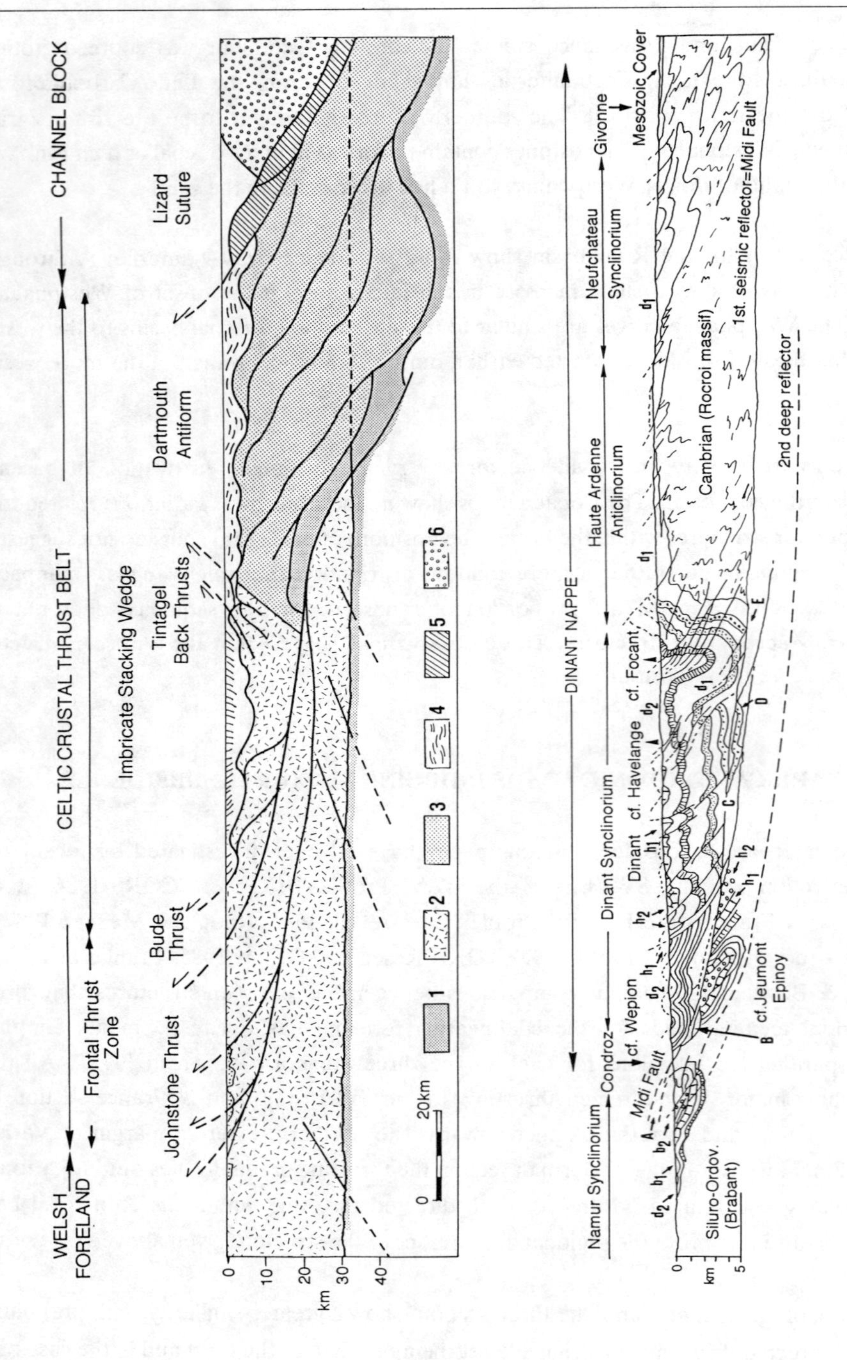
WELSH FORELAND
Frontal Thrust Zone
CELTIC CRUSTAL THRUST BELT
CHANNEL BLOCK
Johnstone Thrust
Bude Thrust
Imbricate Stacking Wedge
Tintagel Back Thrusts
Dartmouth Antiform
Lizard Suture
km
0
10
20
30
40
0
20km
1
2
3
4
5
6
Namur Synclinorium
Condroz
Dinant Synclinorium
DINANT NAPPE
Haute Ardenne Anticlinorium
Neufchateau Synclinorium
Givonne
Mesozoic Cover
Midi Fault
cf. Wepion
Dinant
cf. Havelange
cf. Focant,
Siluro-Ordov. (Brabant)
cf.Jeumont Epinoy
Cambrian (Rocroi massif)
1st. seismic reflector=Midi Fault
2nd deep reflector
A
B
C
D
E
km
0
5

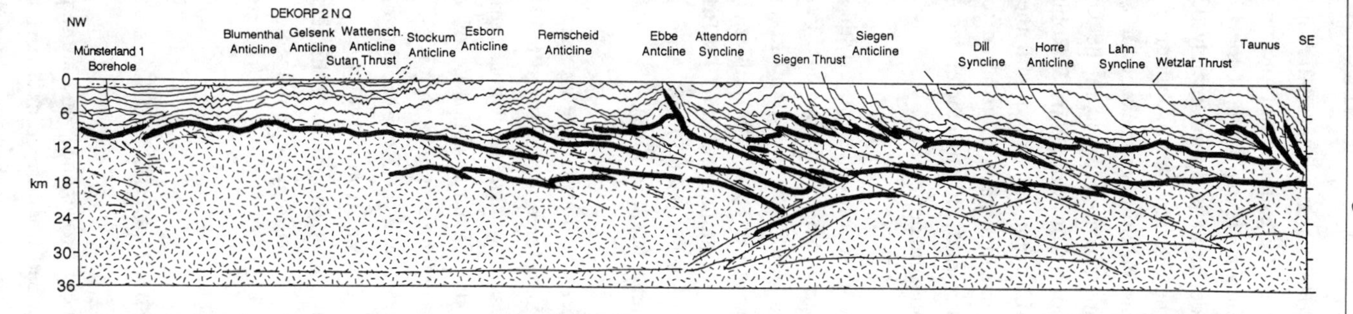

FIGURE 7
Cross-sections oriented approximately N - S (N to the left), based on deep seismic profiling, through the European North Variscan Margin used to estimate crustal-scale shortening of:
A) SW Britain - based on Le Gall (1990), 1 = Upper mantle, 2 = Caledonian basement, 3 = Pre-Devonian continental thinned crust, 4 = Devono - Carboniferous units, 5 = Lizard ophiolitic suture, 6 = Channel - Mancellia Cadomian block;
B) N France - based on Raoult & Meilliez (1987), h_2 = Silesian, h_1 = Dinantian, d_2 = Upper and Middle Devonian, d = Early Devonian, A, B, C, D, E thrust sheets in the footwall of the Midi Thrust; and C) NW Germany - based on Franke et al. (1990), white: Carboniferous and Devonian strata (with Cretaceous cover in the N); random pattern: Silurian and older rocks; black lines: Cambrian level (top) and reflectors at deeper levels.

of the sections are likely to have contributed to the tectonic loading that produced the foreland basin downflexure. A measure of the tectonic load is given by estimates of the tectonic shortening for the sections. For SW Britain the crustal-scale shortening north of the Lizard oceanic suture and south of the Late Carboniferous Bude basin is approximately 35 %. Assuming this shortening was developed entirely before the Silesian basin subsidence (see Pamplin 1990, Warr this volume) and by plain strain, the crust would have thickened to about 50 km. This is in fair agreement with the epizonal metamorphism recorded in the Upper Devonian-Lower Carboniferous Trevone succession from north Cornwall (Warr & Robinson 1990), although this metamorphism is attributed to an M2 event associated with the underthrusting of the Bude basin. An additional structure that may have contributed to the tectonic load is the obducted slice of ophiolite in the Lizard complex, although the effect of this load is probably only slight due to its distance from the downflexure depocentre.

In the case of the N France ECORS section it is difficult to estimate the crustal-scale shortening, as the Dinant nappe, imaged in the profile, is only some 5 - 6 km thick. Further south, beneath the Mesozoic-Cenozoic cover, the seismic profile shows a much thicker Variscan thrust sheet development involving the entire upper crust. Raoult & Meilliez (1987) estimated the shortening in the Dinant nappe to be about 35 %, similar to that of SW Britain, and suggesting a crustal thickness during Variscan deformation of about 50 km. An entirely similar structural situation exists in the Belgian section (Raoult & Meilliez 1985).

The N. German DEKORP 2 profile shows a somewhat different situation. The interpretation of the southern part of the profile through the Rheinisches Schiefergebirge is suggested either to be a series of isolated thrusts cutting down in some cases through the whole crust, or alternatively the thrusts are interpreted to imbricate upwards from a major decollement surface within the middle crust (Franke et al. 1990). The former interpretation, shown in Fig. 7, gives the smaller amount of crustal-scale shortening which, between the Remscheid antiform and the Taunus, is approximately 41 %. This would result in a Variscan crustal thickness of about 60 km, assuming plain strain. It is very difficult to date the development of this tectonic load since the available isotopic dates are cooling (uplift) ages that may be significantly later than the initial tectonic thickening.

DISCUSSION - COMPARATIVE FORELAND BASIN EVOLUTION

Analysis of subsidence rates, depocentre migration, sediment provenance, and subsequent deformation (Tale 1) suggests that each of the basins described was formed as a peripheral foreland basin. The downflexure of the lithosphere was produced by thickened continental crust essentially as a result of Variscan crustal-scale thrust shortening.

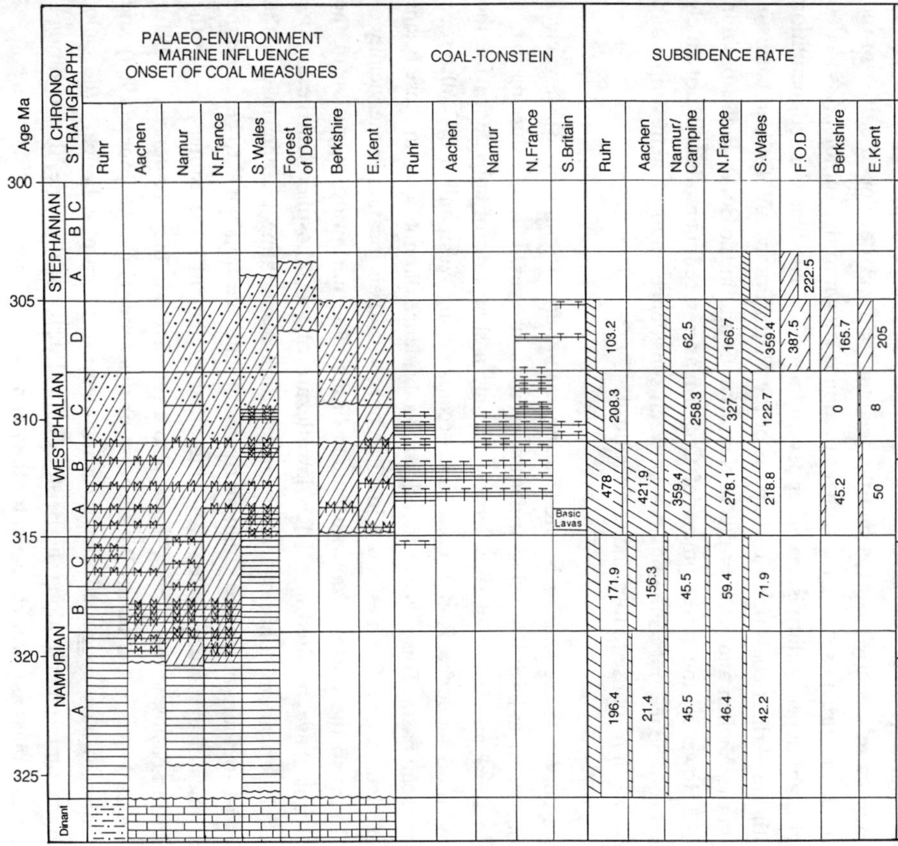

Age Ma	Chronostratigraphy	Subsidence rate: Ruhr	Aachen	Namur/ Campine	N.France	S.Wales	F.O.D	Berkshire	E.Kent
300–305	Stephanian A						222.5		
305	Westphalian D	103.2		62.5	166.7	359.4	387.5	165.7	205
310	Westphalian C	208.3		258.3	327	122.7		0	8
	Westphalian B–A	478	421.9	359.4	278.1	218.8		45.2	50
315	Namurian C–B	171.9	156.3	45.5	59.4	71.9			
320–325	Namurian A	196.4	21.4	45.5	46.4	42.2			

Column groups: PALAEO-ENVIRONMENT, MARINE INFLUENCE, ONSET OF COAL MEASURES (Ruhr, Aachen, Namur, N.France, S.Wales, Forest of Dean, Berkshire, E.Kent); COAL-TONSTEIN (Ruhr, Aachen, Namur, N.France, S.Britain; Basic Lavas); SUBSIDENCE RATE; TECTONICS. Chronostratigraphy: Dinant, Namurian A B C, Westphalian A B C D, Stephanian A B C.

TECTONICS		Ruhr	Aachen	Namur	N.France	S.Wales
VARISCAN DEFORMATION	Stockwerk Tectonics	●				
	North verging thrusts dominant		●	●	●	●
	S. Verging backthrust zones present					●
	N+S verging thrusts equally important	●				
	N. verging folds		●	●	●	●
	Upright folds	●				
	Reactivated basement struct.		●	●	?	●
BASEMENT	Syn-sedimentary basement structures		●	●		●
	Wales-Brabant Massif		●	●	●	●
TECTONIC LOAD	% Shortening	41	40	35	35	35
	Thickness of crustal load (km)	60	58	50	50	50
	Subsidence (km)	>4.6	>2.5	>2.9	>3.1	>3.0

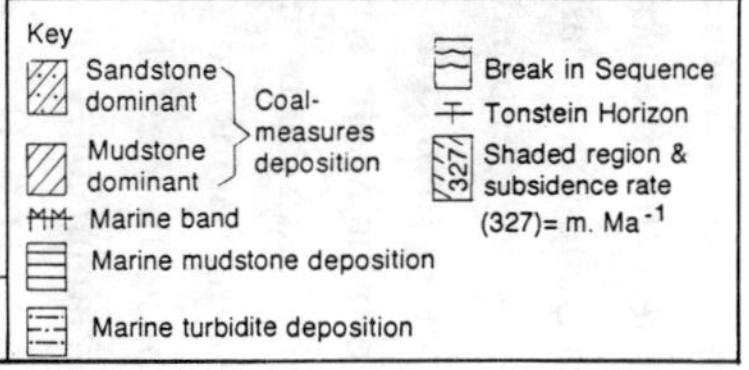

Key: Sandstone dominant / Mudstone dominant — Coal-measures deposition; Marine band; Marine mudstone deposition; Marine turbidite deposition; Break in Sequence; Tonstein Horizon; Shaded region & subsidence rate (327)= m. Ma^{-1}

TABLE 1
Summary of palaeoenvironments, coal-tonsteins, subsidence rates and tectonics for the coal basins considered in the text.

Comparison of the subsidence curves for each of the basins (Fig. 3 & 5) shows a broadly similar subsidence pattern for the South Wales, North France, Belgian, Aachen, and Ruhr basins. These are designated the main coal-bearing foreland basins. Two main trends within these main basins can be identified. Firstly total subsidence within the Ruhr basin (4.7 km) is significantly greater than in the other basins (2.5 - 3.2 km). This may in part be due to a greater tectonic load producing the Ruhr subsidence (Table 1), but it might also reflect differences in the basement beneath the basins - see below. The second trend can be seen in the age at which rapid downflexure begins. In the South Wales, North France and Belgian basins the change occurs at the onset of Westphalian A, in the Aachen basin rapid subsisdence starts at the boundary between Namurian A and B, whilst in the Ruhr basin rapid subsidence affects Namurian A sequences onwards (Fig. 5 & Table 1). The most likely explanation for this earlier subsidence in the more easterly basins is that the tectonic load was developed earlier in the east. Unfortunately no direct evidence for such a situation is known to the authors. Associated with the trend in subsidence pattern is an earlier development of Coal Measures sedimentation in the east, with the earliest coals occuring in parts of the North France and Belgian basins in Namurian A, in the Aachen and Ruhr basins in Namurian C, and in the South Wales basin in Westphalian A. However, the development of Coal Measures sedimentation is not solely related to basin subsidence but to the balance between subsidence and sediment supply and thus is not obviously a response to foreland basin control.

The subsidence curves of the smaller isolated basins in southern Britain (Fig. 3) show clear differences from the main basins, but similarities between themselves. The Berkshire basin and the Kent basin show no Namurian subsidence and only a slow rate of subsidence for the Westphalian A and B sequence. Rapid subsidende occcurs during Westphalian C/D, also seen in the Forest of Dean basin. This suggests that flexural subsidence did not occur in these basins until Late Westphalian C or D times and is in agreement with their location in the vicinity of the Variscan Front to the north of the other main basins. The northward migration of the tectonic load would have resulted in a later downflexure of the lithosphere in these more northern areas. Independent confirmation of this northward migrating tectonic load is the sudden switch to a southern immature sediment source in Late Westphalian C to generate the Pennant Sandstones of the South Wales basin and other similar lithic sandstones, e.g. in the French and Belgian basins, at this time. There is also a suggestion in the subsidence curves for a more rapid subsidence in the South Wales basin at this time. Gayer & Jones (1989) and Hartley (this volume) have suggested that thrusting stepped northwards beneath the Culm basin of North Devon to emerge along the Bristol Channel thrust at this time, thus uplifting a Bristol Channel landmass as a source region for the Pennant Sandstone.

The basins in southern Britain show very little evidence of syn-depositional volcanic activity, almost all of a basic character. In contrast, the French, Belgian, Aachen and Ruhr basins contain, within the Westphalian A-C sequences, numerous coal-tonsteins that can be correlated between the basins (Table 1). These are thought to be derived largely from acid volcanic tuffs, and are most numerous in the Ruhr basin. It is possible that their volcanic source lay to the southeast in the Variscan internal zones, suggesting a tectonically more active region in the eastern area, perhaps linked to greater crustal shortening.

One of the most significant differences in the evolution of the various basins is the effect that pre-existing basement structures had in the development of the basins. This is most clearly demonstrated in the South Wales basin in contrast with the Ruhr basin. Throughout the depositional cycle of the South Wales basin basement structures influenced sedimentation by allowing differential subsidence across the reactivated structures. The effect can be seen in isopach maps, in syn-depositional faults, in coal seam split maps, in stacking of channel sandstone bodies, and in palaeocurrent trends. In the South Wales basin the main reactivated structures strike E-W to SW-NE, with a second set striking NNW-SSE. Many of the same structures were reactivated again during the Variscan deformation, producing the dominant E-W strike of the coalfield folds and thrusts, and the zones of NE-SW disturbance.

By contrast there appear to be no basement structures influencing deposition in the Ruhr coalfield. The only structural control on sedimentation, although not related to basement structures, is the NE-SW zone of reduced subsidence, visible in the isopach maps, that migrated northwestward through the Silesian. This zone may have influenced the position of river channels and of seam splits, but further work is required to establish this. Subsequent Variscan compressional deformation within the Ruhr basin strikes NE-SW perpendicular to the main Variscan shortening direction. The structures, however, show no consistent northwestward vergence and are not connected to a major thrust detachment at depth. There is no evidence for faults re-using older structures.

It thus seems likely that structures within the cratonic Wales-Brabant Massif extending beneath the basins of Southern Britain and of North France and Belgium had a marked effect on the development of the coal-bearing foreland basins. The boundaries of the Wales-Brabant Massif are difficult to define precisely. The massif is a region of crystalline Precambrian rocks together with Caledonian deformed Lower Palaeozoic sequences that formed a stable block during the Late Palaeozoic (Wills 1951). During the Early Carboniferous **ist** southern margin is indicated by the southern edge of the Dinantian carbonate platform (Fig. 1). The absence of this cratonic massif to the north of the Ruhr basin may have allowed the basin to subside both more rapidly and from an earlier date, but without the controlling influence of pre-existing structures. Subsequent deformation was also strongly affected by the massif. The direction of tectonic transport varies around the southern margin of the Massif, both as a result of reactivation of earlier structures but also due to the establishment of local stress fields where the Variscan stresses became buttressed against the massif. Wrede (1987b) has documented one such example at the eastern extremity of the massif in the Aachen basin. Further east in the Ruhr basin the structures represent the Variscan frontal deformation unaffected by a pre-existing cratonic massif (Brix et al. 1988).

CONCLUSIONS

1. A series of coal-bearing sedimentary basins, extending from South Wales across Southern Britain through North France and Belgium to NW Germany, record the signatures of foreland basin development. These are: rapid rates of subsidence; forelandward migrating depocentres; and an immature clastic supply from a tectonic source in the orogenic hinterland. The basins all lie within the northern Variscan margin, recording the latest stages of thrust and fold deformation.

2. Several small coal basins, lying close to, or just north of the Variscan margin in Southern Britain record only the younger stages of subsidence and represent the most external effects of flexural subsidence.

3. The coal basins in Southern Britain, North France and Belgium are developed over the southern margin of the cratonic Wales-Brabant Massif and have inherited the structures present in this basement. Reactivation of these structures has influenced deposition and subsequent deformation.

4. The Ruhr basin of NW Germany lies to the east of the Brabant Massif and is not influenced at any stage of its evolution by basement structures. The only tectonic control of sedimentation is flexural subsidence and a northwestward migrating zone of reduced subsidence thought to be related to the external tip of Variscan deformation.

5. Relative tectonic loads (crustal thickening) calculated from crustal-scale shortening estimates within the more internal tectonic units, suggest an approximately equal load for SW Britain, North France and Belgium, in close agreement with similar subsidence curves for the respective coal basins. The load associated with the Rheinisches Schiefergebirge is relatively larger and may explain both the earlier and greater subsidence of the Ruhr basin, although the absence of the stable Wales-Brabant Massif in this area may also have influenced subsidence.

6. Coal-bearing basins are excellent candidates for the study of the controls of foreland basin evolution. Further investigation is likely to produce significant advances in understanding of foreland basin processes.

ACKNOWLEDGEMENTS

We are grateful for support from various agencies who sponsored aspects of this study. Work on the South Wales coal basin was funded by a Special Topics - Basin Dynamics NERC Research Grant to RAG, and a NERC studentship to JJ. Numerous geologists at British Coal have participated in valuable discussions and we are particularly grateful to Eilian James, Steve Rhodes and John Rippon. JC acknowledges a NERC studentship for work on the Ruhr coal basin.

Discussions with geologists at Geologisches Landesamt Nordrhein-Westfalen, Krefeld, in particular V. Wrede and G. Drozdzewski were very helpful. The University of Heidelberg provided funds for CH through "Graduiertenförderung" and supported field work by ROG.

REFERENCES

Ancion, Ch. 1942. L'évolution tectonique du bassin de Seraing. Annales de la Société géologique de Belgique, 65, 86-132.

Ancion, Ch. 1955. Recherche du rejet longitudinal de la Faille de Seraing dans la partie occidentale du Bassin de Liége. L'Association pour l'étude de la Paléontologie et de la Stratigraphie Houilléres, 21, 205-215.

Anderson, E.M. 1951. The dynamics of faulting. Oliver and Boyd, Edingburgh, 2o6p.

Bally, A.W. & Snelson, S. 1980. Realms of subsidence. In: Miall, A.D. (ed.) Facts and principles of World petroleum occurrence. Memoir of the Canadian Society of Petroleum Geologists, 6, 9 - 75.

Bachmann, M., Michelau, P. & Rabitz, A. 1971. Das Rhein-Ruhr Revier. (a) Stratigraphie. Fortschritte in der Geologie von Rheinland und Westfalen, 19, 19-34.

BIRPS & ECORS. 1986. Deep seismic reflection profiling between England, France and Ireland. Journal of the Geological Society, London, 143, 45-52.

Bless, M.J.M. & Bouckaert, J. 1988. Suggestions for a deep seismic investigation north of the Variscan mobile belt in the SE Netherlands. Annales de la Société géologique de Belgique, 111, 229-241.

Bless, M.J.M., Bouckaert, J., Calver, M.A., Graulich, J.M. & Paproth, E. 1977. Palaeogeography of Upper Westphalian deposits in NW Europe with reference to the Westphalian C North of the mobile Variscan belt. Mededelingen Rijks Geologische Dienst Nieuwe Serie, 28, 101-147.

Bless, M.J. M., Bouckaert, J., Conil, R., Groessens, E., Kasig, W., Paproth, E., Potty, E., Van Steenwinkel, M., Streel, M & Walter, R. 1980. Pre-Permian depositional environments around the Brabant Massif in Belgium, The Netherlands and Germany. Sedimentary Geology, 27, 1 - 81.

Bless, M.J.M., Bouckaert, J. & Paproth, E. 1983. Recent exploration in Pre-Permian rocks around the Brabant Massif in Belgium, The Netherlands and the Federal Republic of Germany. Geologie en Mijnbouw, 62, 51-62.

Bless, M.J.M., Bouckaert, J. & Paproth, E. 1989. The Dinant nappes: a model of tensional listric faulting inverted into compressional folding and thrusting. Bulletin de la Société belge de Géologie, 98, 221-230.

Bluck, B.J. 1961. The sedimentary history of the rocks between the horizon of G. subcrenatum and the Garw Coal in the South Wales coalfield. Unpublished PhD. thesis University of Wales, 130p.

Bouroz, A., Buisine, M., Chalard, J., Dalinval, A. & Dolle, P. 1964. Bassin Houiller du Nord et du Pas-de-Calais. Cinquième Congrès International de Stratigraphie et de Géologie du Carbonifère, Compte Rendu, 1, 3-33.

Boyer, S.E. & Elliott, D. 1982. Thrust Systems. American Association of Petroleum Geologists Bulletin, 66, 1196-1230.

Brix, M.R., Drozdzewski, G., Greiling, R.O., Wolf, R. & Wrede, V. 1988. The N Variscan margin of the Ruhr coal district (Western Germany): structural style of a buried thrust front? Geologische Rundschau, 77, 115-126,

Brooks, M., Trayner, P.M. & Trimble, T.J. 1988. Mesozoic reactivation of Variscan thrusting in the Bristol Channel area, UK. Journal of the Geological Society, London, 145, 439-444.

Calver, M.A. 1969. Westphalian of Britain. Sixième Congrès International de Stratigraphie et de Géologie du Carbonifére, Compte rendu, 1, 233-254.

Casshyap, S.M. 1975. Lithofacies analysis and palaeogeography of the Bochumer formation (Westfal A2) Ruhrgebiet. Geologische Rundschau, 64, 610-640.

Cazes, M., Mascle, A., Torreilles, X., Bois, Ch., Damotte, X., Matte, Ph., Raoult, X, Pham, V.N., Hirn, A. & Galdeano, X. 1986. Large Variscan overthrusts beneath the Paris Basin. Nature, 323, 144-147.

Chadwick, R.A., Kenolty, N. & Whittaker, A. 1983. Crustal structure beneath southern England from deep seismic reflection profiles. Journal of the Geological Society, London 140, 893-912.

Cleal, C.J. 1987. Macrofloral biostratigraphy of the Newent Coalfield, Gloucestershire. Geological Journal, 22, 207-217.

Cole, J., Miliorizos, M., Frodsham, K., Gayer, R.A., Gillespie, P.A., Hartley, A.J., & White, S. 1991. Variscan structures in the Opencast Coal Sites of the South wales Coalfield. Proceedings of the Usher Society, 7.

Collet, L.W., 1927. The structure of the Alps. Edward Arnold & C., London. 289pp.

Coward, M.P. & Smallwood, S. 1984. An interpretation of the Variscan tectonics of SW Britain. In: Hutton, D.H.W. & Sanderson, D.J. (eds.) Variscan tectonics of the North Atlantic Region. Geological Society, London, Special Publication 14, 89-102.

DEKORP Research group. 1985. First results and preliminary interpretation of deep reflection-seismic recordings along profile DEKORP 2-South. Journal of Geophysics, 57, 137-163.

Dickinson, W.R. 1974. Plate tectonics and sedimentation. In Dickinson, W.R. (ed.) Tectonics and sedimentation. Society of Economic Palaeontologists and Mineralogists, Special Publication 22, 1 - 27.

Donato, J.A. 1988. Possible Variscan thrusting beneath the Somerton Anticline, Sommerset. Journal of the geological Society, London, 145, 431-438.

Drozdzewski, G. 1990. The Ruhr coal basin (Germany): Structural evolution of an autochthonous foreland basin. In: Cross, A.T. (ed.) World Class Coal Deposits. International Journal of Coal Geology. In Press.

Drozdzweski, G., Bornemann, O., Kunz, E. & Wrede, V. 1980. Beiträge zur Tiefentektonik des Ruhrkarbons. Geologisches Landesamt Nordrhein-Westfalen, Krefeld, 192p.

Drozdzewski, G., Engel, H., Wolf, R. & Wrede, V. 1985. Beiträge zur Tiefentektonik westdeutscher Steinkohlenlagerstätten. Geologisches Landesamt Nordrhein-Westfalen, Krefeld, 236p.

Drozdzewski, G., Kunz, E., Pieper, B., Rabitz, A., Stehn, O. & Wrede, V. 1982. Geologische Karte des Ruhrkarbons 1:100.000 dargestellt an der Karbonoberfläche. Geologisches Landesamt Nordrhein-Westfalen, Krefeld.

Dunham, K.C. & Poole, E.G. 1974. The Oxfordshire Coalfield. Journal of the Geological Society, London, 130, 387-391.

Dunne, W.M. 1983. Tectonic evolution of SW Wales during the Upper Palaeozoic. Journal of the Geological Society, London, 140, 257-266.

Engel, W. & Franke, W. 1983. Flysch sedimentation: Its Relations to tectonism in the European Variscides. In: Martin, H. & Eder, F.W. (eds-) Intracontinental Fold Belts. Springer-Verlag, Berlin & Heidelberg, 289-321.

Fielitz, W. 1992. Variscan transpressive inversion in the northwestern Rhenohercynian belt of western Germany. Journal of Structural Geology. In Press.

Foster, D., Holliday, D.W., Jones, C.M. Owens, B. & Welsh, A. 1989. The concealed Upper Palaeozoic rocks of Berkshire and South Oxfordshire. Proceedings of the Geologists' Association, 100, 395-407.

Franke, W., Bortfeld, R.K., Brix, M., Drozdzewski, G., Dürbaum, J.J., Giese, P., Janoth, W., Jödicke, H., Reichert, Chr., Scherp, A., Schmoll, J., Thomas, R. Thünker, M., Weber, K., Wiesner, M.G. & Wong, H.K. 1990. Crustal structure of the Rhenish Massif: results of deep seismic reflection lines DEKORP 2-North and 2-North-Q. Geologische Rundschau, 79, 523-566.

Frodsham, K., Gayer, R.A., James, J.E. & Pryce, R. 1991. Variscan thrust deformation in the South Wales Coalfield - a case study from Ffos-Las Opencast Coal Site. This Volume.

Gayer, R.A., Allen, K.C., Bassett, M.G. & Edwards, D. 1973. The structure of the Taff Gorge area, Glamorgan, and the stratigraphy of the Old Red Sandstone-Carboniferous Limestone transition. Geological Journal, 8, 345-374.

Gayer, R.A. & Jones, J. 1989. The Variscan foreland in South Wales. Proceedings of the Ussher Society, 7, 177-179.

Gayer, R.A. & Stead, J.T.G. 1971. The Forest of Dean Coal and Iron-Ore Fields. In: Bassett, D.A. & Bassett, M.G. (eds.) Geological Excursions in South Wales and the Forest of Dean. Geologists' Association, Cardiff. 20-36.

Howard, C.B., Kuznir, N.J., Bamford, M.L.F. and Williams, G.D. 1989. A mathematical model of thrust sheet emplacement and foreland basin formation: rheological, thermal and isostatic constraints. Abstract In: Mountains, Basins and Lithosphere Rheology, Geological Society, London.

Hoyer, P. & Pilger, A. 1971. Das Rhein-Ruhr Revier. (c) Tektonik. Fortschritte in der Geologie von Rheinland und Westfalen, 19, 41-46.

Jankowski, B., David, F. & Selter, V. 1991. Facies complexes of the Upper Carboniferous in Northwest Germany and their structural implications. This Volume.

Jones, J. 1989a. The influence of contemporaneous tectonic activity on Westphalian sedimentation in the South Wales coalfield. In: Arthurton, R.S., Guthridge, P. & Nolan, S.C. (eds.) The Role of Tectonics in Devonian and Carboniferous Sedimentation in the British Isles. Occasional Publication of the Yorkshire Geological Society, 6, 243-253.

Jones, J. 1989b. Sedimentation and tectonics in the eastern Part of the South Wales coalfield. Unpublished PhD. thesis, University of Wales.

Kelling, G. 1974. Upper Carboniferous Sedimentation in South Wales. In: Owen, T.R. (ed.) The Upper Palaeozoic and Post-Palaeozoic rocks of Wales. University of Wales Press, Cardiff. 185-224.

Kelling, G. 1988. Silesian sedimentation and tectonics in the South Wales Basin: a brief review. In: Besly, B. & Kelling, G. (eds.) Sedimentation in a syn-orogenic basin complex: the Upper Carboniferous of Northwest Europe. Blackie, Glasgow and London, 38-42.

Le Gall, B. 1990. Evidence of an imbricate crustal thrust belt in the Southern British Variscides: contributions of South-Western Approaches Traverse (SWAT) deep seismic reflection profiling recorded through the English Channel and Celtic Seas. Tectonics 9, 282-302.

Lippolt, H.J., Hess, J.C. & Burger, K. 1984. Isotopische Alter von pyroklastischen Sanidinen aus Kaolin-Kohlentonsteinen als Korrelationsmarken für das mitteleuropäische Oberkarbon. Fortschritte in der Geologie von Rheinland und Westfalen, 32, 119-150.

Meissner, R. and Bortfeld, R.K. (Eds.) 1990. Dekorp-Atlas. (Springer Verlag) Berlin etc., vii, 19 pp., 80 sections.

Meissner, R., Springer, M. & Flüh, E. 1984. Tectonics of the Variscides in North-Western Germany based on seismic reflection measurements. In: Hutton, D.H.W. & Sanderson, D.J. (eds.) Variscan Tectonics of the North Atlantic Region. Geological Society, London, Special Publication 14, 23-32.

Miliorizos, M. 1991. Variscan structures under the Bristol Channel. Proceedings of the Ussher Society, 7.

Owen, T.R. 1953. The structure of the Neath disturbance between Bryniau Gleision and Glynneath, South Wales. Quarterly Journal of the Geological Society, London, 109, 333-365.

Pamplin, C.F. 1990. A model for the tectono-thermal evolution of north Cornwall. Proceedings of the Ussher Society, 7, 206-211.

Paproth, E. 1987. The Variscan Front north of the Ardenne-Rhenish massif. Annales de la Société géologique de Belgique, 110, 279-296.

Paproth, E., Dusar, M., Bless, M.J.M., Bouckaert, J., Delmar, A., Fairon-Demaret, M. Houllebergh, E. Laloux, M., Pierart, P., Somers, Y., Streel, M., Thorez, J. & Tricot, J. 1983. Bio- and Lithostratigraphic subdivisions of the Silesian in Belgium, a review. Annales de la Société Géologique de Belgique, 106, 241-283.

Raoult, J.F. & Meilliez, F. 1985. Commentaires sur une coupe structurale de l'Ardenne selon le méridien de Dinant. Annales de la Société Géologique, 105, 97-109.

Raoult, J.F. & Meilliez, F. 1987. The Variscan Front and the Midi Fault between the Channel and the Meuse River. Journal of Structural Geology, 9, 473-479.

Sax, H.J.G. 1946. De Tectoniek van het Carboon in het Zuid-Limburgsche Mijngebied. Mededeelingen van de Geologische Stichting, C3, 77p.

Shackleton, R.M. 1984. Thin-skinned tectonics, basement control and the Variscan front. In: Hutton, D.H.W. & Sanderson, D.J. (eds.) Variscan Tectonics of the North Atlantic Region. Geological Society, London, Special publication, 14, 125-130.

Sinclair, H.D., Allen, P.A., Coakley, B. & Watts, A.B. 1989. Simulation of foreland basin stratigraphy using a diffusion model of mountain belt uplift and erosion: an Example from the Central Alps. Abstract in: Mountains, Basins and lithosphere Rheology. Geological Society, London.

Stockmal, G., Beaumont, C. & Boutilier, R. 1986. Geodynamic models of convergent margin tectonics: transition from rifted margin to overthrust belt and consequences for foreland basin development. American Association of Petroleum Geologists, 70, 181-190.

Strack, O. & Freudenberg, U. 1984. Schichtenmächtigkeiten und Kohleninhalte im Westfal des Niederrheinisch-Westfälischen Steinkohlenreviers. Fortschritte in der Geologie von Rheinland und Westfalen, 32, 243-256.

Thomas, L.P. 1974. The Westphalian (Coal Measures) in South Wales. In: Owen, T.R. (ed.) The Upper Palaeozoic and Post-Palaeozoic rocks of Wales. University of Wales Press, 133-160.

Trotter, F.M. 1947. The structure of the Coal Measures in the Pontardawe-Ammanford area, South Wales. Quarterly Journal of the Geological Society, London, 103, 89-133.

Warr, L.N. 1991. Basin inversion and Foreland Basin development in the Rhenohercynian Zone of SW England. This Volume.

Warr, L.N. & Robinson, D. 1990. The application of the illite "crystallinity" technique to geological interpretation: a case study from north Cornwall. Proceedings of the Ussher Society, 7, 223-227.

White, S.C. 1991. Palaeo-geothermal profiling across the South Wales Coalfield. Proceedings of the Ussher Society, 7.

Williams, G.D. & Chapman, T.J. 1986. The Bristol-Mendip foreland thrust belt. Journal of the Geological Society, London, 143, 63-74.

Wrede, V. 1987a. Einengung und Bruchtektonik im Ruhrkarbon. Glückauf-Forschungshefte, 48, 116-121.

Wrede, V. 1987b. Der Einfluß des Brabanter Massivs auf die Tektonik des Aachen-Erkelenzer Steinkohlengebietes. Neues Jahrbuch Geologie Paläontologie Monatshefte, 1987, 177-192.

Wrede, V., Drozdzewski, G. & Dvorak, J. 1991. On the structure of the Variscan front in the Eifel-Ardennes area. This Volume.

Palaeogeographic and Metamorphic Evolution of the Ligerian Belt in Europe

M. S. Oczlon
Geologisch-Paläontologisches Institut, Ruprecht-Karls-Universität, Im Neuenheimer Feld 234
D-6900 Heidelberg, Germany

ABSTRACT

Tectono-metamorphic processes in the Ligerian Belt during Middle/Late Ordovician, Silurian and Early Devonian time can be traced along the North Gondwana margin from N Morocco to the Bohemian Massif and the Alps. The central and eastern parts from NW Spain to the Eastern Alps are characterized by rift-related intrusions at 470-430 Ma accompanied by a phase of HP/HT syn-rift metamorphism around 450-430 Ma. This is followed by HP/LT-metamorphism around 400 Ma and a MP-HT retrograde event with anatexis and granitoid emplacement until 380-375 Ma, accompanied by extensive S-directed overthrusts. Lower Devonian and Eifelian strata are usually missing in Ligerian sedimentation and deformation areas, and a shallow-water carbonate development of the Givetian or Upper Devonian cover is present. The SW part of the Ligerian belt from NW Spain to N Morocco suffered less deformation, and sedimentation may continue in the Early Devonian.

The Ligerian Belt originated as a system of grabens from Morocco to NW Spain with further evolution into an oceanic basin from NW Spain to SE Bohemia. Rifting/ocean formation had its climax during the Early Silurian and led to split-off of the Ligerian Terrane. Subsequent convergence led to establishment of an Upper Silurian sedimentary basin above a northward dipping Benioff-zone beneath the Ligerian Terrane which collided with Gondwana in the Early Devonian. The Ligerian evolution can be distinguished from Cambrian-Early/Middle Ordovician processes of similar extent and history.

1. INTRODUCTION

A Ligerian belt within the Variscan orogen as distinct from other Variscan regions was first recognized by Pruvost (1949) in the Vendée area (24/25 in Fig. 1) of the S Armorican Massif (NW France). He created the term "Ligerian" (Liger = Loire river) for rocks in the Vendée area but related them to a latest Precambrian event. In 1962 Cogné (in Cogné and Lefort 1985) first recognized the importance of the Silurian/Early Devonian Ligerian evolution immediately preceding the Variscan orogeny. Such a "Ligerian" Belt can be traced across Variscan Europe and obviously represents a distinct pre-Variscan palaeogeographic and tectonic unit. As pointed out by Cogné and Lefort (1985), the Ligerian orogeny is contemporaneous with the Caledonian orogeny but not cogenetic. A similar view is expressed by Weber (1984) and Pin (1990) for so-called "Caledonian" processes in Central and Western Europe.

In its type region in the S Armorican Massif the Ligerian succession contains Middle Cambrian-Tremadoc acidic volcano-sedimentary formations followed by sandstone (Arenig) and Middle/Upper Ordovician shale formations (Ters 1979). Intrusion of alkaline magmas around 444 Ma (Le Metour and Bernard-Griffiths 1979) is succeeded by syn-rift HP/HT metamorphism (Weber 1984) at 440-430 Ma (Pin and Peucat 1986). Silurian sediments with black shales and cherts contain basic volcanics indicating early rifting (Llandovery) and a subsequent subduction-related setting (Ludlow/Pridoli, Carpenter et al. 1982). The Early Devonian/Eifelian coincides with cooling ages of HP/LT blueschist metamorphism around 400 Ma and anatectic mobilization with granitoid intrusions until 380-375 Ma (Pin and Peucat 1986), accompanied by south-directed nappe transport (Burg et al. 1987). Post-orogenic sedimentation starts in the Givetian with shallow-water limestones. This tectono-sedimentary pattern is used here to define the Ligerian Belt. It represents a pre-Variscan orogen located in the palaeogeographic realm of the North Gondwana margin (NGM, Paris and Robardet 1990, Cocks and Fortey 1988).

NGM faunal and facies patterns form a distinct palaeogeographic entity that was separated from the South Laurussia margin (SLM, Rhenohercynian zone in part) by the Rheic Ocean until Early Carboniferous time (Paris and Robardet 1990, Cocks and Fortey 1982). Confining the Rheic Ocean, contrasting shelf successions belong to opposite continental margins. Glacially influenced Upper Ashgill peri-Gondwana sections occur in N Africa, Spain, Brittany, Saxothuringia (Robardet and Doré 1988), and the Barrandian area (Storch 1990). They are distinct from contemporaneous subtropical or tropical carbonates found on the British Isles (Avalon Terrane south of the Caledonian Iapetus suture) and the W Baltic margin (Havlicek 1989), regions to which the Rhenohercynian zone is related. Since the NGM displays undeformed shelf/slope successions until Early Carboniferous times, following the Ashgill glacio-marine sections, a collision with the SLM cannot have occurred prior to this time. Over-

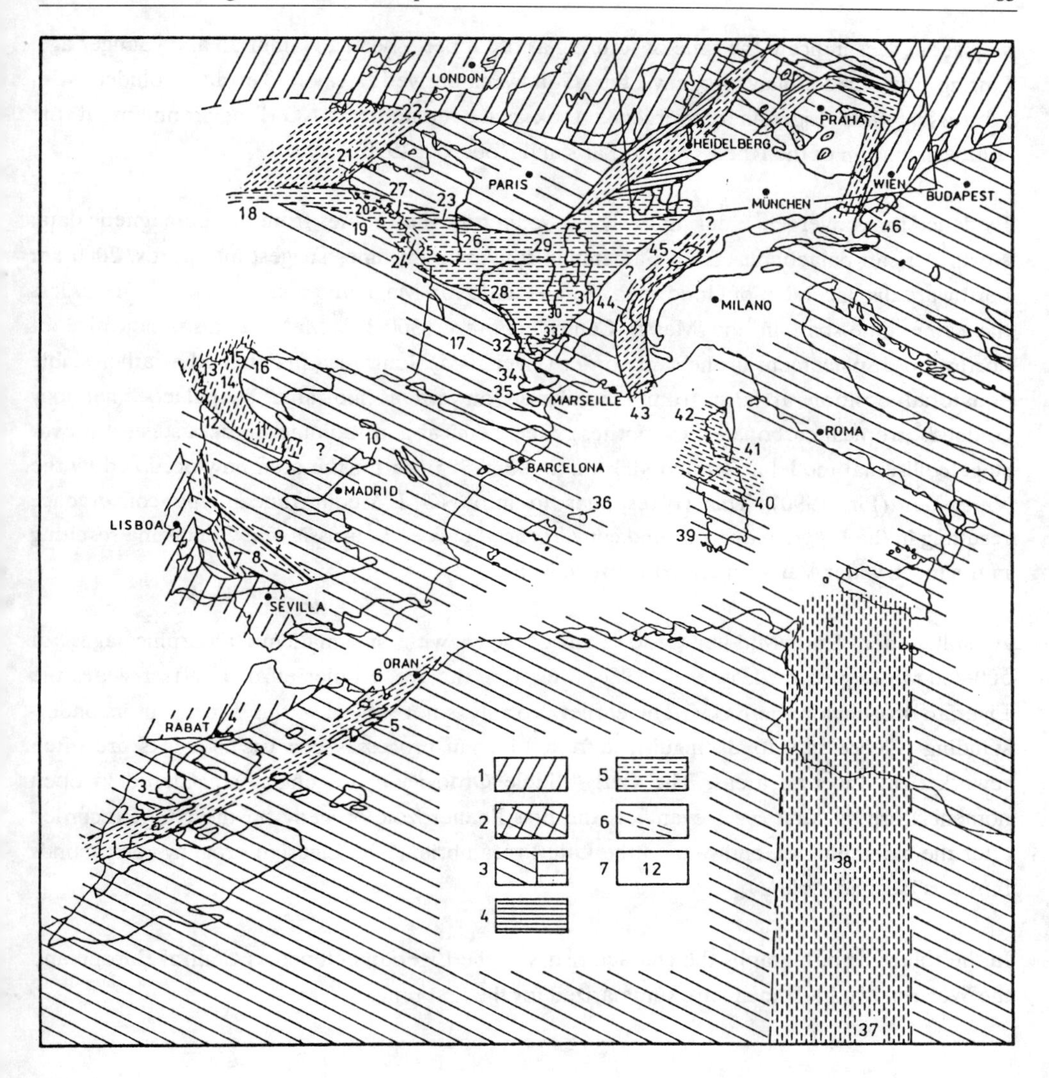

FIGURE 1
Zonation of the Variscan belt in NW Africa and Europe. 1) Units of the South Laurussia margin including the Rhenohercynian zone and its basement; 2) North Gondwana elements thrusted over the Rhenohercynian zone (Oczlon 1991); 3) North Gondwana-type passive shelf/slope deposits of the Variscan orogen (Saxothuringian zone in Fig. 2); 4-6) Lithologies of the Ligerian orogen: 4) Ligerian Terrane with Upper Silurian subduction-related volcanism); 5) Ligerian units of the southern margin of the Ligerian Ocean, in Europe metamorphism at around the Silurian/Devonian boundary, in Africa and SW Spain angular disconformities; 6) Anatectic Zone in the S Armorican Massif NW France. 7) Localities mentioned in the text. Box shows location of Fig. 2.

step sequences indicate the time of suturing as of Upper Visean?, Namurian and younger age. During the Caledonian orogeny, the Gondwana-derived Avalon Terrane collided with Laurussia while Ligerian convergence took place in an intra-NGM environment at the southern margin of the Rheic Ocean (Paris and Robardet 1990).

Evidence for convergence during Early Carboniferous time comes from palaeomagnetic data. North African palaeopoles from Famennian to Namurian times suggest an approx. 2000 km northward displacement of Gondwana (Aifa et al. 1990). Magmatic rocks covering more or less the same time-span in the Massif Central (France, 360-325 Ma) are also indicative of northward displacement in the same order (Edel 1987). Laurussia maintained a rather stable equatorial position at least from the Silurian onward as indicated by palaeoclimatology (redbeds, tropical carbonates, evaporites). Geochemical and radiometric data suggest a two-stage collisional model at around 400 Ma and 360-325 Ma for the area now occupied by the Variscides (Pin 1990). This corresponds to intra-NGM Mediterranean-type convergence resulting in the Ligerian orogeny and convergence between Laurussia and Gondwana resulting in the much wider Variscan orogeny respectively.

A still older Early/Middle Ordovician orogeny with magmatic/metamorphic ages of 500-460 Ma (e.g. Girardeau et al. 1986, Menot et al. 1988, Köhler et al. 1989) precedes the Ligerian orogeny on the NGM. The earlier orogeny is not subject of this paper, but its understanding is necessary to distinguish it from Ligerian processes because its rocks are often reworked in Ligerian areas. The Early/Middle Ordovician orogeny is considered in open nomenclature as "first cycle event" as the first Phanerozoic orogeny on the NGM recorded after the assembly of Gondwana in the Late Precambrian. The Ligerian orogeny corresponds thus to a "second cycle".

In the following paragraphs the characteristics of the Ligerian belt across Central Europe and the Western Mediterranean are summarized for the first time.

2. REGIONAL REVIEW

2.1. NW Africa

Thin (c. 100 m) NGM-type graptolite-bearing Silurian black shales and limestones that continue into the Devonian characterize most of Morocco (Berry and Boucot 1973). In the Jérada-area, Oujda (5 in Fig. 1, NE-Morocco) the Silurian succession is confined to 600 m of conglomerates, quartzites, black shales and cherts (Owodenko 1976) of Middle Llandovery age. The Oued Madakh Massif, Oran (6, NW-Algerian coast) contains pre-Variscan serpentinite masses which are reworked in basal Devonian strata (Centène et al. 1984). North of Marrakech, in the Rehamna region (3, NW-Morocco) and in the southerly adjacent western Jbilet (2), Silurian/Devonian sequences contain Lochkovian conglomerates (Hollard et al. 1982). Lower/Middle(?) Devonian conglomerates in the SW Rehamna (Skhour) are found above Cambro-Ordovician rocks with angular disconformity. In a similar occurrence in the adjacent Atlas Mountains (1, Talmakent), Lochkovian conglomerates ("Vieux grès rouges") overlie Cambrian rocks with an angular disconformity (Michard 1978).

The Jbilet-, Jérada-, and Oued Madakh regions form an elongate zone (900 km) of Ligerian activity, the NE-segment of which is best defined (Jérada-Oued Madakh, 200 km). Ligerian extension and graben-formation is documented by thick coarse clastics of Llandovery age. Pre-Devonian serpentinites of the Oued Madakh Massif belong perhaps to this extensional phase similar to the situation in the NW Iberian Peninsula (2.3.). In Earliest Devonian time, the graben system was inverted leading to uplift and erosion with deposition of conglomerates above angular disconformities.

The Taconian/Acadian magmatic/metamorphic evolution in the Rabat-Tiflet area (4, El Hassani and Zahraoui 1984) is similar to that in the E Appalachians and its offshore areas (comp. Zartman and Naylor 1984, Hermes et al. 1978) suggesting that the Rabat-Tiflet area has no relation to Ligerian processes (Fig. 1).

2.2. Ossa Morena Zone (OMZ, SW Iberian Peninsula)

Ordovician peralkaline granitoids (470 Ma, Rb-Sr whole rock, Garcia Casquero et al. 1985) related to rifting are present along the boundary of the OMZ to the Central Iberian Zone (CIZ), represented by the prominent Variscan strike-slip fault system of Porto-Badajoz-Cordoba (9). Most of the granitoids are now found as orthogneisses and show ages (mainly on Rb-Sr whole rock) which cluster around 510 Ma, 482 Ma, 435 Ma, and 380 Ma (Pinto 1984). These intrusions indicate a long history of rifting (Munha et al. 1990), probably occurring in several pulses.

An angular disconformity between Silurian black shales and Lower Devonian conglomerates is reported from the eastern part of the Aracena region (7, Bard 1966). Only 40 km to the east of the latter area, a complete Silurian-Lower Devonian stratigraphic succession without unconformity is exposed (8, Valle syncline, Racheboeuf and Robardet 1986, Oczlon 1989). These data do not indicate any large-scale Ligerian movements but are conistent with Ordovician rifting and inversion at around the Silurian/Devonian boundary leading to erosion and a minor unconformity.

2.3. NW Iberian Peninsula

2.3.1. Ligerian Crystalline Massifs and their Emplacement

An important element of the NW Iberian Peninsula is represented by the Ollo de Sapo formation which occurs as a (syn-formationally) reworked acidic volcanic formation with giant feldspar crystals of Middle/Upper Cambrian age (Iglesias Ponce de Leon and Ribeiro 1981). In a wider regional extent Ollo de Sapo rocks are found in autochthonous position in a band which follows the Ibero-Armorican arc from Cabo Ortegal (15) to an area NE of Madrid (10, Hiendelaencina, Matte 1968). The calc-alkaline nature of the Ollo de Sapo volcanics suggests a setting at an active continental margin (Iglesias Ponce de Leon and Ribeiro 1981) during the first cycle. Later Ordovician/Early Silurian Ligerian rifting follows the Ollo de Sapo band (Lancelot et al. 1985).

Ligerian rifting can be traced from the northern coast of Galicia (15, Cabo Ortegal) to the Tras-Os-Montes region (11, NE Portugal) and is associated with four major mafic/ultramafic massifs which are contained in the allochthonous zone of Galicia - Tras-Os-Montes as defined by Arenas et al. (1986). From north to south these are the Cabo Ortegal Massif (15), various massifs in the region of Ordenes-Lalin (14, Martinez Catalan et al. 1984), and the massifs of Braganca and Morais (11).

The Morais Massif is part of a sequence of different units first referred to by Ribeiro (1974) as a lower autochthonous unit (A) composed of Ordovician-Silurian sediments followed by an overthrusted unit (B, diameter c. 50 km) of similar lithology but including Upper Ordovician initial rift volcanics (Ribeiro 1987). A catazonal basic/ultrabasic unit with granulites (C, diameter c. 25 km) occupies the central part of the structure. Its allochthonous position has been deduced from the occurrence of mylonites (tectonic breccias) at the contact to unit (B) (Anthonioz 1967). Unit (C) bears on its top a volcano-sedimentary unit (D, diameter c. 10 km) consisting of acidic volcanic rocks with giant feldspar crystals (up to 10 cm) and detrital sedimentary rocks (Ribeiro 1974). Sequence (D) is comparable to the Ollo de Sapo volcano-sedimentary formation which is also present in the sedimentary units (A) and (B). In the Lower Palaeozoic to the east (16, Asturias) of the Ollo de Sapo band and to the west (NW Portugal), the Ollo de Sapo formation is no longer present. These relations suggest that the root-zone for the uppermost unit (D) is found in its own vicinity. This is consistent with an in-situ closed basin and conflicts with a large nappe transport of the basic/ultrabasic massifs (C) and unit (D) over 200 km from the west off the Galician/N Portuguese coast (e.g. Iglesias et al. 1983).

At Braganca (11), the autochthonous unit (A) is overlain by Upper Devonian flysch with pebbles derived from the catazonal unit (C) and from the overthrusted Ordovician and Silurian unit (B) (Ribeiro and Ribeiro 1974). This fact restricts the emplacement to Late Devonian time or earlier. Post-Ludlow/pre-Emsian, probably Lochkovian conglomerates (conodonts, Aldaya et al. 1976) in the autochthonous unit 50-60 km to the east of the Braganca/Morais Massifs contain metamorphic elements like schists and granulitic fragments for which there is no other source than the adjacent massifs. Older conglomerates of Silurian age in that area contain only reworked sediments (Aldaya et al. 1976). The time of emplacement for the catazonal units is thus roughly confined to the Silurian/Devonian boundary. Cooling ages (hornblende) from the Cabo Ortegal complex (15) cluster around 390 Ma (Den Tex 1981) and date progressive retrogradation with mylonitization under amphibolite facies conditions (Martinez Catalan 1990). This coincides with stratigraphically inferred uplift and thrusting of catazonal metamorphics at around the Silurian/Devonian boundary.

In the hypothetical root-zone of the allochthonous units (C-D) off the Galician/N Portuguese coast (Iglesias et al. 1983) there are continuous Silurian - Upper Emsian NGM-sedimentary sequences near the coast in northernmost Portugal (12, Perdigao 1977) exactly to the west of the Morais Massif. These sediments provide good evidence for the absence of widespread tectonic processes in the N Portuguese coastal area for Silurian/Early Devonian times.

2.3.2. Graben Formation

An evolution from rifting towards incipient ocean formation was elaborated by Ribeiro (1987) for volcanic formations of the graben sequence near Morais (unit B). Radiometric dating of this magmatism lies in the range of 460-430 Ma (Rb-Sr, Priem et al. 1970, 1972). During increased subsidence (collapse, accelerated rifting), some 2000 m of diastrophic sediments and volcanics filled the graben during Llandovery time. These sediments are also found to the northwest of Braganca (11) and at Cabo Ortegal (15, Matte 1968). A connecting link is the lithostratigraphically related metamorphic sequence of central Galicia (14, Farias et al. 1987). According to the model of Van Calsteren (in Den Tex 1981) a mantle plume rose into rifted continental crust producing anatexis and sending differentiated magmas into shallow crustal levels. Early granulites and eclogites of the catazonal massifs formed during that time as a result of HP/HT syn-rift metamorphism in the lower crust (Weber 1984). MORB-characteristics of the ultramafic massifs (Martinez Catalan 1990) may have been acquired during that time when the asthenosphere rose to shallow crustal levels.

At Cabo Ortegal rifting had achieved the stage of formation of oceanic crust as indicated by redeposited fossiliferous limestones of Silurian age intercalated in basaltic lavaflows and ultrabasic slices (Van Der Meer Mohr 1975). These little metamorphosed rocks show relics of hydrothermal alteration as found in oceanic crust (Arenas and Peinado 1981). Close association of graben in-fill with the catazonal massifs points to a genetic link most obvious for the Cabo Ortegal and the Branganca/Morais areas. The distance of transport of the catazonal massifs over the graben-sediments is thus in the order of a few km (or a few tens of km at Cabo Ortegal).

Llandovery sediments are missing at a distance of about 50 km to the east of the line Cabo Ortegal - Tras-Os-Montes (Zeitz and Nollau 1984, Matte 1968, Antona and Martinez Catalan 1990). This is due to uplift and erosion during the Llandovery, i.e. during the time of maximum graben subsidence. In places, transgressing Wenlock strata overlie Lower Cambrian limestones or Arenig sandstones (Matte 1968). Farther to the east (Villaodriz syncline, 16, Asturias) Silurian sediments are present from the Lower Llandovery (vesiculosus zone) onwards. This uplift, which has locally reached 2 km or more, is obviously associated with graben-formation during Llandovery time and indicates a setting similar to the present Red Sea coastal uplifts.
A likely scenario for the Early Silurian NW Iberian Peninsula could be:
1) a graben zone with great sediment thickness (2000 m or more), high-level mantle intrusions and limited ocean formation in the north (Cabo Ortegal);
2) the graben shoulders being elevated above sea level and eroded;
3) a zone grading into normal marine NGM shelf sedimentation (several tens of metres of sandstones, black shales and cherts).
This zonation is found only in the eastern part of the graben. The western part, including the Malpica-Tuy unit (3.1.), became tectonized and metamorphosed during the Ligerian and Variscan collisional events. All characteristics of the NW Iberian Peninsula point to a crustal and palaeogeographic environment which is broadly similar to the present Red Sea.

2.3.3. Tectonic Concept

The catazonal massifs were formed as a result of syn-rift HP/HT metamorphism most probably in the Middle/Upper Ordovician. They have subsequently ascended to relatively shallow levels in the central graben area during further extension and maximum asthenospheric rise (Llandovery). A rheologically stiffer behaviour of these dry metamorphics compared to wet crustal rocks led to their ascent during crustal shortening (Weber 1984). The deformation pattern around the catazonal massifs corresponds to that of grabens which are inverted shortly after their formation as outlined by Sengör (1985). Thrusting to all sides, in places with nearly circular arrangement of fold axes in unit (B) is observed in the Tras-Os-Montes area (Ribeiro 1974). Such structures may originate in a thermally unstable lithosphere beneath the graben when deformation starts. Upward squeezing of the catazonal graben-basement and the infill during convergence led to thrusting in various directions. The critical time of lithospheric evolution into equilibrum is estimated to c. 60 Ma (Sengör 1985). However, the time-span between Ligerian graben formation (Late Ordovician/Early Silurian) and inversion (Silurian-Devonian boundary) is approximately 40 Ma.

2.4. Southern Armorican Massif (NW France, Ligerian Type Locality)

2.4.1. Radiometric and Stratigraphic Constraints

Similar to NW Spain, the Ligerian evolution in NW France was preceded by an acidic volcano-sedimentary succession with rhyolites and rhyodacites (Le Metour and Bernard-Griffiths 1979). The succession contains Middle Cambrian trilobites at the base (Cavet et al. 1966). The intrusion of the alkaline hypovolcanic Thouars Massif (26, 444+/-9 Ma, Le Metour and Bernard-Griffiths 1979) is ascribed to Ligerian rifting in this area. Ultramafic rocks from the S Armorican Massif yielded U-Pb zircon ages between 436 and 384 Ma (Peucat et al. 1982). HP-HT metamorphic rocks (440-430 Ma, eclogites, granulites, Pin and Peucat 1986) formed as a result of syn-rift metamorphism in the lower crust in the sense of Weber (1984). The Ile de Groix blueschists (20) represent a Ligerian HP-LT element with radiometric ages clustering around 400 Ma (Pin and Peucat 1986) which coincides with subduction prior to overthrusting at around the Silurian/Devonian boundary. Associated blueschists are found some 150 km to the ESE in the Bois de Cené area (24). Their close relation to the Ile de Groix blueschists within the same zone of convergence is stressed by Guiraud et al. (1987). Collision was followed by a decreasing P/T gradient and anatexis dated at 380-375 Ma (Vidal 1973, Goujou et al. 1990). Anatectic granites occur in the "Anatectic zone" (27) which corresponds to the internal area of the Ligerian orogeny in the S Armorican Massif.

Where the Ligerian zone is thrusted over the central part of the Armorican Massif (Ledru et al. 1986), a Ligerian volcano-sedimentary formation crops out in the Saint-Georges-sur-Loire Syncline (23). Two well-dated volcanic phases include a Lower Silurian one derived from partial melting of plume-type mantle-material and an end-Silurian one with eruption of basalts above a subduction zone (Carpenter et al. 1982). Lochkovian sandstones with abundant plant remains ("Grès à Psilophytes", Moreau-Benoit and Dubreuil 1987) in the Saint-Georges-sur-Loire syncline may correspond to a molasse stage during uplift of the Ligerian orogen in the south. Givetian shallow-water limestones represent the post-Ligerian cover.

2.4.2. Ligerian Deformation and Nappe-Emplacement

Burg et al. (1987) elaborated a southward sense of displacement which is consistent with results from the NW Iberian Peninsula and the Massif Central. Ligerian tectonic transport is always directed towards the inner side of the Ibero-Armorican arc (Burg et al. 1987). The distance of southward overthrusting of HP/LT elements (e.g. Bois de Cené, 24) is in the order of 50 km when their actual distance to the presumed suture zone to the south of the Saint-Georges-sur-Loire syncline (23) with subduction-related Upper Silurian volcanism is considered (Fig. 4). The major deformation phase took place probably during Early Devonian time and was followed by fast cooling (Goujou et al. 1990). The first Devonian sedimentary rocks covering the Ligerian Belt in the Vendée and related areas are of Givetian age. Therefore, uplift and retrograde metamorphism must have ceased prior to Givetian time. The general absence of Lower Devonian and Eifelian rocks in the Vendée reflects the contemporaneous tectonic activity (Ters 1979). The Anatectic Zone is the centre of a thickened Ligerian crust after collision at around the Silurian/Devonian boundary. It is probable that large parts of the Ligerian Terrane and slices of the Ligerian Ocean were thrusted from north to south over the (present) Anatectic zone (27). This is indicated by occurrences of southward overthrusted blueschists and eclogites at Ile de Groix (20) and Bois de Cené (24).

2.4.3. Outline of the Ligerian Terrane

The northern and central parts of the Armorican Massif (Brittany, Normandy) carry a continuous Silurian/Devonian sedimentary cover and were not touched by Ligerian subduction and collision. For example, Upper Silurian/Lower Devonian carbonate facies near Angers (Kriz and Paris 1982) immediately to the north of the Ligerian Saint-Georges-sur-Loire syncline points to an open passive margin environment without clastic influx, not affected by Ligerian processes. These rocks must have occupied a different position on the NGM during Silurian/Devonian time and cannot have formed part of the colliding crustal fragment that previously split off the NGM (Ligerian Terrane, Fig. 3). The Ligerian suture as the southern boundary of the Ligerian Terrane must be located between the Anatectic zone (2.4.2.) and the Saint-Georges-sur-Loire basin implying northward Late Silurian subduction (Fig. 4).

Gravimetrically inferred (ultra)mafic masses offshore to the south and west of Brittany (18+19, Cogné and Lefort 1985) are not related to the Ligerian suture which lies farther north. They belong probably to a Late Devonian subduction zone which led to emplacement of the "Tonalite-belt" of S Brittany/W Massif Central (2.5.4).

2.4.4. Ligerian Crust in NW Brittany and in the English Channel

Still north of the NGM-type shelf rocks of the Central/N Armorican Massif occurs another Ligerian unit in the NW-corner of the Armorican Massif (21, Léon, Fig. 1) with gneisses dated at 385+/-8 Ma (Cabanis in Balé and Brun 1986). It is in fault contact with the NGM-type Palaeozoic of Brittany lying to the south and has been emplaced by dextral strike-slip of > 100 km along the Léon fault zone probably around the Devonian/Carboniferous boundary (Balé and Brun 1986). To the north occurs a probably related Ligerian unit in the basement of the Normannian Sea (22, English Channel). It was reached in a borehole where Givetian limestones were found covering a metamorphic basement in a situation similar to other Ligerian regions (Leveridge, pers. com. 1989). Parts of this basement are formed by gneisses referred to as Eddystone Gneiss (375+/-12 Ma, Holder and Leveridge 1986).

2.4.5. Terrane Concept for the Armorican Massif

The presented data suggest that the area north of the Ligerian S Armorican Massif, considered as relatively autochthonous, formed by an amalgamation of fragments of a disrupted passive NGM and the Ligerian belt (Fig. 1). These fragments are bounded on all sides by faults and must be considered as proximal terranes because of their provenance from the NGM. Juxtaposition involving dextral shear along the NGM (Fig. 3) was initiated at around the Devonian/Carboniferous boundary probably induced by dextral displacement of the Gondwana plate relative to the Rheic plate ("Bretonic Phase"). This setting is not unlike the present W North American margin.

2.5. Massif Central

2.5.1. Radiometric and Stratigraphic Data

In the W Massif Central (28, Limousin) two collisional stages are distinguished by Girardeau et al. (1986). The first cycle collision occurred in the Early Ordovician (ages of 510-460 Ma) while the second one (Ligerian) took place around the Silurian/Devonian boundary (ages of 440-380 Ma) and recycles previously formed basement. Ligerian nappe emplacement took place in a MP-HT environment, then conditions evolved towards lower temperatures. Anatexis followed decompression until 375 Ma in the Limousin. In the east-central area of the Massif Central (30, Haut Allier), a crystallisation age of 432+20/-10 Ma has been found for zircons of the "La Borie" eclogite (Ducrot et al. 1983). Peridotites from the Lyonnais region (31) with HP-metamorphism conditions (Gardien et al. 1988) are of unknown age.

Radiometric ages of 480-500 Ma are present all over the Massif Central (Pin and Lancelot 1982) while ages of 440-430 Ma (intrusive and HP/HT metamorphic) are concentrated in the N Massif Central and in the S Armorican Massif. This may suggest that Ligerian syn-rift HP/HT metamorphism is confined to the N part. A problem is posed by the high pressure of 13-15 kb, sometimes even reaching over 20 kb documented in granulites (Pin and Vielzeuf 1988). The corresponding depth would be 40-60 km which seems too high for average continental crust during rifting. However, such a deep crust could have been generated by the immediately preceding first-cycle collisional event during Early/Middle Ordovician time (e.g. Girardeau et al. 1986).

The oldest sedimentary cover in the NE Massif Central is of Givetian age (Delfourt and Gigot 1985) which sets an upper time limit for Ligerian uplift and retrograde metamorphism. The Massif Central occupies a central position in the Ligerian belt. Its N-S width in the W part is approx. 300 km which is more than in any other Ligerian region presently exposed.

2.5.2. Oceanic Domains

Girardeau et al. (1986) distinguished two oceanic domains, an Early Ordovician "Southern Oceanic Domain", and an Early Silurian "Northern Oceanic Domain" with respect to the deduced location in the Massif Central. The Northern Oceanic Domain is consistent with the "Massif Central Ocean" of Matte (1986). While the Southern Oceanic Domain originated probably as a first cycle back-arc basin (Bodinier et al. 1988), the Northern Oceanic Domain originated from continental rifting without involvement of subduction. Because this ocean is not only confined to the Massif Central, but related to the Ligerian orogeny in general, it is better named the "Ligerian Ocean" (Fig. 3). As outlined above, the Ligerian collision at around the Silurian/Devonian boundary did not involve units north of the Vendée/Massif Central area which would be the Central/N Armorican Massif and its eastern continuation underneath the Paris basin. This is in accord with the deduction of Santallier (1983), who considers Ordovician (Ligerian) rifting "close to an already established oceanic zone" which is thought to be the Rheic Ocean (Fig. 3). The split-off continental zone is only of small width (Ligerian Terrane).

2.5.3. Nappe Transport and Leptyno-Amphibolitic Group

The Massif Central is characterized by large southward transported basement nappes reaching even the SW part (Lévezou area, 32, Matte 1986). Considering the southward direction of nappe-transport, the Ligerian suture must be located to the north of the Massif Central (Matte 1986) in the southern Paris basin. There, basement rocks with Ligerian metamorphism were found in a research borehole (29, Costa and Maluski 1988). The cooling age after HT-metamorphism and anatectic mobilisation is narrowly defined there between 390-380 Ma.

A distinct feature of Ligerian nappes are occurrences of the "leptyno-amphibolitic group" found associated with HP-HT elements (granulites, eclogites). This group consists of a characteristic association of acidic ortho- or paragneisses and amphibolites. It may represent the transition of oceanic and continental crust which was affected by the final stage of subduction with later overthrusting due to collision (Pin and Vielzeuf 1988). However, if HP-HT elements are considered as pre-orogenic syn-rift metamorphics their emplacement together with the leptyno-amphibolitic group is easier to explain by overthrusting of lower crustal parts from the transition to oceanic crust southward over the NGM (Fig. 4). This process would not require syn-collisional subduction of continental crust to explain granulite formation. Bodinier et al. (1988) assume an origin of parts of the leptyno-amphibolitic group in an Early Ordovician back-arc basin with formation of new oceanic crust based on geochemical studies in the Marvejols area (33). The age of 484+/-7 Ma found in gabbros in the Marvejols area suggests that they are at least partly associated with the first cycle.

2.5.4. Superimposed Late Devonian Radiometric ages

Late Devonian radiometric ages (mainly 370-350 Ma) are due to subduction-related magmatism described as "Tonalite Belt" (Bernard-Griffiths et al. 1985) with subduction dipping to the NE under the Massif Central (Cogné and Lefort 1985). The Tonalite Belt extends from the S Armorican Massif to the SW Massif Central and is tentatively assigned to a "third cycle". Its origin may be due to intra-NGM convergence of the present Aquitanian basin (17) and the Massif Central/S Armorican Massif. Tournaisian rocks in the E part of the Aquitanian basin (next to the Massif Central) were found unconformably overlying Lower Palaeozoic rocks (Paris et al. 1988).

An overprint related to this convergence is, for example, present in the Leptyno-Amphibolitic Group of the Lévezou area (32). There, calc-alkaline rocks could be shown to be much younger (367+/-10 Ma) than the apparently associated tholeiitic series (485+/-30 Ma) (Pin and Piboule 1988). This superimposition of younger radiometric ages led to the deduction that

Ligerian deformation and metamorphism would last until Carboniferous time (Matte 1986, Pin and Peucat 1986, Cogné and Lefort 1985). This is questioned here because uplift of the Ligerian belt was terminated in the Givetian, as indicated by basement/cover relationships (e.g. 2.5.1.).

2.6. Montagne Noire

The nappes in the S Montagne Noire (35) also contain typical NGM-type Palaeozoic rocks with continuous Silurian/Devonian successions. However, a local angular disconformity between Lochkovian and Arenig strata (Feist 1978) may indicate Ligerian and/or first-cycle influence marginal to the Massif Central. In contrast to the S Montagne Noire nappe-units, the Axial Zone (34) which is considered autochthonous, contains migmatites and mica-schists with cooling ages between 460 and 420 Ma (Pin and Peucat 1986). Intrusion ages of granitoids in the Axial Zone (now orthogneisses) range between 532 and 500 Ma (Duthou et al. 1984), thus being contemporaneous with first cycle intrusions of the Iberian Penisula, the S Armorican Massif, and the Massif Central. These autochthonous basement rocks of the Montagne Noire are overlain by Devonian? sediments (Ellenberger and Santarelli 1974). The Montagne Noire nappes to the N and S of the Axial Zone may be related to Palaeozoic sediments found in the basement of the Aquitanian basin (17, Paris et al. 1988). Nappe-transport is from N to S as deduced from structural data (Matte 1986).

2.7. Southern Branch of the Ligerian Orogen

2.7.1. Libya

A "Caledonian tectonic activity" in W Libya (37, Mourzouk basin) is reported by Massa and Nicol-Lejal (1971). Siegenian Tadrart sandstones are found to overlie Upper Llandovery beds with angular disconformity. Llandovery beds consist of several 100 m of sandstones and shales (Acacus sandstone and Tanezzuft shale). They contain a cruziana ichnofacies (Berry and Boucot 1973) which occurs also in the Ligerian graben sequence of NE Portugal (Ribeiro 1974). Most European/African NGM sections contain thin (several tens of metres) and anoxic Llandovery black sandstones, shales and cherts without benthic life remains. The occurrences of cruziana ichnofacies may be due to better aereated environments (currents?) along with higher sedimentation rates favourable for sediment eaters in graben- or subsiding coastal NGM areas.

The Ligerian unconformity is defined at the Silurian/Devonian boundary in the Ghadames basin (38, NW Libya) where the sandy-conglomeratic Tadrart formation starts already in the Lochkovian (chitinozoans, Jaglin in Moreau-Benoit and Massa 1988) and overlies a sandy-clayey formation of Pridoli age. The area of increased subsidence during Llandovery time and later uplift with disconformably overlying Lower Devonian sediments covers wide areas of W Libya, SE Algeria and N Tschad (Berry and Boucot 1973). Although the tectonic/epeirogenic pattern is a Ligerian one, there is no evidence for faulting in the Llandovery, but only for a broadly subsiding environment. The nearby Gondwana coast is obviously the source for the shedding of sandstones into the basin.

2.7.2 Corsica and Sardinia

In NW Corsica (42) an Upper Ordovician/Lower Silurian formation of conglomerates and sandstones (100 m) is overlain by black shales (150 m) with Early/Middle Silurian and reworked Early Ordovician microfossils (Baudelot et al. 1976). The following Upper(?) Silurian quartzite-shale formation is separated from Frasnian-Famennian flysch deposits by a tectonic phase (Baudelot and Durand-Delga 1981). A succession of Ligerian or first cycle rift volcanics may be present in central Corsica (3.2.).
A Lower Palaeozoic metamorphic succession with sandstones, phyllites, lenses of marbles, and metavolcanics is known from NE Sardinia (40). This sequence is associated with micaschists, paragneisses, and orthogneisses (the latter dated at 458+/-31 Ma, 441+/-33 Ma, data from Carmigniani et al. 1987). Associated with (Ligerian) rifting are Middle/Upper Ordovician volcano-sedimentary successions in S Sardinia and the famous Sardic Conglomerate in SW Sardinia (39) deposited during Caradoc time (Exel 1986). It overlies unconformably Cambrian-Lower Ordovician sediments.

2.7.3. Western Alps External Massifs

In the Belledonne Massif (44) a Late Silurian/Early Devonian stage of HP-metamorphism followed by (Early?) Devonian Barrovian metamorphism which led to anatexites (Menot 1988) represent Ligerian elements. The Chamrousse Ophiolite (496+/-6 Ma, Menot et al. 1988), exposed nearby, is related to a back-arc basin environment as are ophiolites in the S Massif Central (based on geochemical data, Pin and Carme 1987). It was obducted soon after its formation, probably already during the Early/Middle Ordovician first cycle event (Menot et al. 1988). This indicates two separated cyles, the Ligerian cycle not being a simple continuation of Cambrian/Early Ordovician rifting. The Gotthard Massif (45) yields also evidence for a first cycle event with abundant U-Pb data on zircons (but also Sm-Nd on whole rock) that indicate a

470-460 Ma age for metamorphism (Gebauer et al. 1988). Geochemical data show that the protoliths have island arc or ocean floor affinity. A zircon age of 400+/-3 Ma may reflect Ligerian overprinting. Ligerian rifting in the Gotthard Massif is indicated by U-Pb zircon-dating on orthogneisses (440 Ma, Grauert and Arnold 1968, 436+/-17 Ma, Arnold 1970).

2.7.4. Southern Branch of the Ligerian Orogen

The outlined areas from N Africa to the Western Alps (also 3.2. and 3.3.) are all entrained in a Ligerian tectono-metamorphic pattern on the NGM. The southern regions (Corsica/Sardinia) are, however, separated from contemporaneous Ligerian areas in NW France and NW Spain by wide undisturbed NGM shelf regions (Fig. 3). The Ligerian area from N Africa to the Western Alps is, therefore, tentatively interpreted as a branch of the Ligerian rift system as, for example, seen in the northern Red Sea rift or in the East Africa rift.

2.8. Vosges and Schwarzwald (Black Forest)

2.8.1. Radiometric and Stratigraphic Data in the Vosges

Ligerian areas occur in the central (48) and N Vosges (47). Both parts of the Vosges are divided by the Lalaye-Lubine zone with slices of ultrabasic rocks (Schwebel 1983). Kinzigitic gneisses, HP granulites and garnetiferous peridotites with metamorphic conditions of 9-10 kb and 700-750½C appear in the central Vosges (Fluck 1984) with an age of c. 430 Ma (Geyer and Gwinner 1986). A Silurian depositional age is suggested by Fluck (1984) for a meta-sand-shale succession found today as biotite-sillimanite gneisses in the central Vosges. Subsequent Early Devonian metamorphism led to formation of sillimanite (on the expense of kyanite in HP metamorphic rocks). Its radiometric age is confined to 386+/-15 Ma (Bonhomme and Fluck 1981). Whole rock Rb-Sr measurements on granite-gneisses also yielded an Early Devonian intrusion age (395+/-18 Ma, Hameurt 1976). Some metamorphic formations of the central Vosges can be compared with the Gföhl Gneisses from the SE Bohemian Massif (Fluck 1984) which form another Ligerian element (2.10.11.). Evidence for Ligerian tectonic processes is found in the N Vosges where Cambrian/Ordovician (schists of Villé, Reitz and Wickert 1989) and Ordovician/Silurian sediments (schists of Steige, Doubinger and Eller 1963) show a deformation which is more intense than that of the Givetian and younger rocks of the Ligerian cover (Hameurt 1976). Lower Devonian/Eifelian strata are missing due to uplift and erosion following the Ligerian orogeny around the Silurian/Devonian boundary. Givetian reef limestones (Geyer and Gwinner 1986) cover the Ligerian orogen in this area.

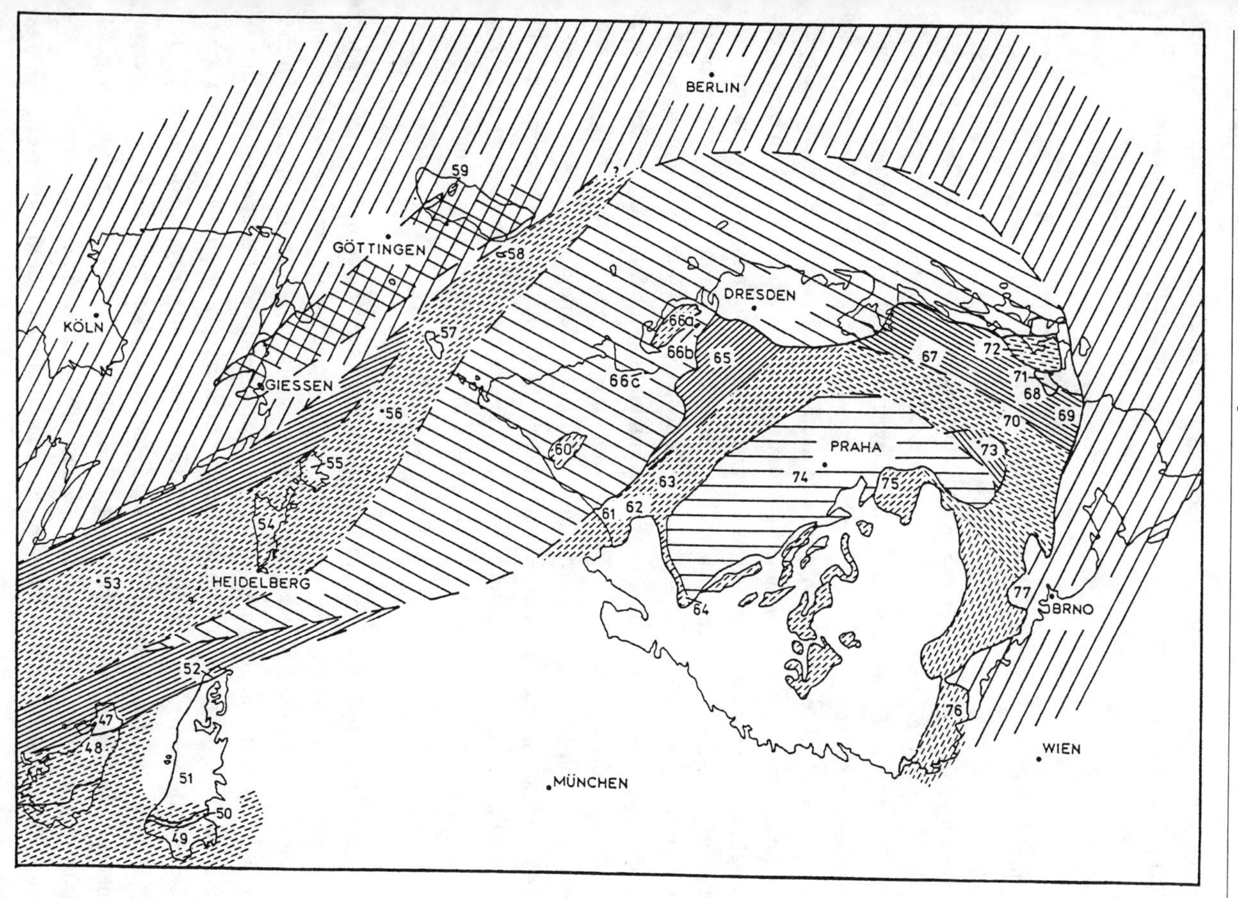

FIGURE 2
Zonation of the Variscan belt in Central Europe. Legend as in Fig. 1; Western extent of the Saxothuringian zone under the S German Mesozoic according to data listed in Trusheim (1964).

The oldest sediments in the S Vosges are Upper Devonian limestones and clastic sediments (Maass 1988) partly associated with basement slices of Devonian or Lower Palaeozoic age (Pin and Carme 1988).

2.8.2. Radiometric and Stratigraphic Data in the Schwarzwald

The Schwarzwald is composed from N to S of four units (Krohe and Eisbacher 1988): a meta-volcano-sedimentary succession (Baden-Baden zone, 52); the Central Schwarzwald Gneiss complex (CSGC, 51); a narrow zone of Upper Devonian/Lower Carboniferous sediments (Badenweiler-Lenzkirch zone, 50); and the Hotzenwald complex (49). The crystalline basement of the Badenweiler-Lenzkirch zone was deeply eroded before the onset of Upper Devonian sedimentation. A syenite of diatectic para-origin with admixed granitic magma from the Hotzenwald Complex yielded an age of 388+/-15 Ma (Geyer and Gwinner 1986). The major part of the CSGC was apparently not subjected to Ligerian metamorphic overprinting because it mainly yields Cambrian-Early/Middle Ordovician ages for basement formation (Geyer and Gwinner 1986). However, zircons from granulites in the S part of the CSGC were dated at c. 410 Ma (Kober et al. 1986). Zircons from an eclogite of the same region show a lower intercept age on the U-Pb discordia also of 410 Ma (Schleicher and Kramm 1986). Further U-Pb dating on zircons from eclogite-amphibolites and Rb-Sr whole rock (small range) dating on an eclogite/gneiss contact yielded an age of 390 Ma (Werchau et al. 1989). The coaxial granulite strain path in the CSGC points to formation during crustal extension (Flöttmann 1990).

At present, Cambrian-Early/Middle Ordovician metamorphics are much wider exposed in the CSGC. They were formed during the first cycle and may have been reworked in a Ligerian nappe complex similar to the lower nappes in the Limousin (28, 2.5.1.). Compared with the wide outcrops of Ligerian metamorphics in the Vosges it seems as if deeper erosion of the Schwarzwald essentially removed a Ligerian metamorphic nappe, now only present in some remnants at the southern margin of the CSGC where it disappears under Upper Devonian/Lower Carboniferous sediments of the Badenweiler-Lenzkirch zone (50). Ligerian metasediments of the N Vosges continue into the Baden-Baden zone as shown by geochemical characteristics of metasediments in both areas (Müller 1989). Tentaculites were found in a volcano-sedimentary succession of the Baden-Baden zone and indicate an Upper Silurian age (Mehl 1989).

2.8.3. Ligerian Evolution and Boundary to the Saxothuringian Zone

The Vosges and the Schwarzwald show a tectono-metamorphic pattern comparable to the Vendée and the Massif Central. The 410 Ma and 390 Ma ages recorded from zircons in HP/HT metamorphic rocks from the CSGC may be Ligerian reset ages from rocks formed during syn-rift metamorphism at 440-430 Ma as found in the central Vosges. An extensional regime is suggested for granulite formation in the CSGS (Flöttmann 1990) which is consistent with syn-rift metamorphism. The Lalaye-Lubine zone between central and N Vosges is the most likely zone of Ligerian convergence and collision. In this sense the Cambro-Silurian sequence of the N Vosges/Baden-Baden zone may belong to the Ligerian Terrane with a subduction-related setting in the Late Silurian (Fig. 4). The Silurian(?) sedimentary protoliths of gneisses in the central Vosges belong to the southern margin of the Ligerian ocean, which was finally overthrusted and loaded from the north and downwarped by units of the Ligerian Terrane. This led to Early Devonian metamorphism, granite intrusion and uplift, followed by transgression of the Givetian and late Devonian sea. The N-S zonation resembles that of the Vendée with an inner, southern cristalline zone in contact with a northern zone of deformed but much less metamorphosed sediments of Silurian or older age finally covered by Givetian rocks.

Since Kossmat (1927) the N Vosges/Baden-Baden Cambro-Silurian sequence is considered as part of the Saxothuringian zone. However, the tectono-sedimentary pattern is clearly a Ligerian one. Today it is known that major parts of the metamorphic domains to the south of the N Vosges/Baden Baden zone are not of Precambrian age, but coincide in age with the closure of the Cambro-Silurian basin which happened in Earliest Devonian time. The stop of sedimentation in the N Vosges/baden/baden zone is therefore related to Ligerian collision, metamorphism, and uplift. Facies of Saxothuringian sediments, like those of the central/N Armorican Massif (2.4.3.) indicate a completely different geotectonic position during Silurian/Devonian times, which is an open passive shelf environment without much clastic input as found, for example, in Morocco or SW-Spain. For this reason, N Vosges/Baden-Baden Cambro-Silurian sediments are not considered as Saxothuringian, but as Ligerian (Fig. 2) both not being related in pre-Early Carboniferous time. A similar view is expressed by Maass (1988) who stresses the strong relations between the N and S Vosges during Late Devonian/Early Carboniferous time, both belonging to the "Moldanubian" and not to the Saxothuringian zone.

2.9. Saarland, Odenwald (Odin's Forest), Spessart, Ruhla and Kyffhäuser Crystalline Complexes

These areas, generally considered as "Mid-German Crystalline Rise" (MGCR), were originally inferred by Brinkmann (1948) as a palaeogeographic high to the south of the Rheinische Schiefergebirge (Rhenish Slate Mts) and the Harz Mts (59) during Devonian/Early Carboniferous times. The MGCR forms most probably a fragment of a Late Devonian/Early Carboniferous magmatic arc along the NGM which was finally thrusted over the Rhenohercynian zone of the SLM (Oczlon 1991). This magmatic arc was established on Ligerian crust as outlined in the following paragraphs.

2.9.1. Stratigraphic Data

The Spessart basement outcrop (55) contains a SE-NW zonal arrangement of ortho-, paragneisses and marbles and the Quartzite-Mica-schist formation (QMF) in the NW (see Hirschmann and Okrusch 1988 for further details). The QMF contains well-preserved spores of Late Silurian age (Reitz 1987). Formations related to the Spessart are known in the Ruhla Crystalline complex (57, Hirschmann and Okrusch 1988) and in the Kyffhäuser basement outcrop (58). Several boreholes between the Spessart and the Ruhla Crystalline complex reached quartzites and mica-schists which are probably equivalents of the QMF of the Spessart (e.g. Hettenhausen, 56, Trusheim 1964). Meta-limestones known as "Auerbacher Marmor" in the Odenwald (54) yielded fossils of Givetian or Early Late Devonian age (Mehl pers. com. 1990) and were metamorphosed during the Variscan orogeny. They probably correspond to unmetamorphic Givetian/Upper Devonian limestones in the Saar 1 borehole (53, Kneuper 1976), which represent a Ligerian cover (2.9.2.). A meta-volcano-sedimentary succession (Alzenauer Gneiss formation) of bimodal character with arc-tholeiites occurs along the NW periphery of the QMF in the Spessart (Okrusch et al. 1985). These authors deduce deposition in a near-coast environment. Primitive (Ligerian) arc-tholeiites are also known from the Ligerian metamorphic basement of the eastern (Böllstein-) Odenwald (Altenberger et al. 1990).

2.9.2. Ligerian Granitoids and Metamorphics

The Saar 1 borehole (53) reached an albite-granite basement with an age of 381+/-24 Ma, which is overlain by Givetian limestones (Lenz and Müller 1976). Orthogneisses in the Böllstein-Odenwald (54) and Spessart (55) ("Rotgneis") have primary Ligerian (intrusion) ages around 400 Ma (Lippolt 1986). The Ligerian granitoids from the Spessart (Okrusch and

Richter 1986) and from the Böllstein-Odenwald (Altenberger et al. 1990) are classified as post-collisional S-type based on geochemical and petrographic data. They immediately follow deposition of Upper Silurian sediments (QMF). Metamorphic basement intruded by the Upper Devonian Frankenstein Gabbro (362+/-7 Ma, Kirsch et al. 1988) in the northern Odenwald can be ascribed to Ligerian or older metamorphism. Dating of zircons from paragneisses in the Böllstein-Odenwald yielded an age of 380 Ma (Todt 1979) indicating Ligerian metamorphism at least for this part of the Odenwald. Later, the Spessart and Odenwald were severely overprinted during the Variscan orogeny (345-320 Ma).

2.9.3. Possible Relationships to other Ligerian Areas

Similar formations are encountered in a wider region. The QMF may correspond to the (meta-)sand-shale succession of the central Vosges for which a Silurian age is claimed by Fluck (1984). It is suggested that these rocks were deposited in the Ligerian Palaeozoic environment of the Bohemian "Varied Group" (2.10.11.) and related successions of the NGM on the southern margin of the Ligerian ocean (Fig. 4).

The magmatic arc-related Alzenauer Gneiss formation/Böllstein Odenwald may correspond to N Vosges/Baden-Baden zone Silurian and older sediments with volcanics of the Ligerian Terrane. An interpretation of the Böllstein Odenwald arc tholeiites and sediments as part of a Silurian/Early Devonian (Ligerian) arc-related accretionary prism is also possible (Altenberger et al. 1990). These data suggest only a short-lived period of subduction without evolution into a mature arc which is consistent with Ligerian subduction restricted to the Late Silurian (2.4.1.). The Ligerian suture in the Spessart may be the S-directed overthrust of the Alzenauer Gneiss formation over the QMF. This direction of thrusting (Okrusch et al. 1985) is opposed to the N-directed Variscan thrusts and is consistent with the general S-directed trend in the Ligerian orogen (Burg et al. 1987). However, an origin as a Variscan backthrust is also possible. Traces of a collisional event at around the Silurian/Devonian boundary (post-collisional intrusions and metamorphism) are found in all areas. The Lower Devonian granite in the Saar 1 borehole demonstrates an Early Devonian/Eifelian period of erosion until subsidence allowed the incursion of a shallow Givetian sea. Boreholes between the N Vosges (47) and the Saar 1 borehole (53) always reached a crystalline basement (Kneuper 1976) which is also encountered in small outcrops along the W margin of the Rhine-Graben, occasionally with Lower Carboniferous magmatic arc-related sediments (Flöttmann and Oncken 1990). There are no Saxothuringian sediments present (Palaeozoic succession in connection with continuous and anorogenic Silurian/Devonian sequences) and the whole area may either be part of the Ligerian orogen or of older basement. Borehole data (Trusheim 1964) confine the westernmost Saxothuringian sediments to the area between the N Schwarzwald (52) and the Spessart (55).

2.10. Bohemian Massif

2.10.1. Zones of Erbendorf-Vohenstrauß (ZEV) and Tirschenreuth-Mähring (ZTM)

The ZEV (61) forms a nappe overlying the Bohemian Massif and occurs in a NW-SE elongated zone (see Kreuzer et al. 1989 for further details). It is the site of the 10-km-deep KTB research-borehole (61) which is located at a distance of c. 8 km SSE of the boundary to the Saxothuringian zone. Metamorphic ages on hornblende of the ZEV cluster around 380 Ma, but there are two other groups of 391-401 Ma and 421-424 Ma probably related to inherited argon (Kreuzer et al. 1989). Amphibolites found in the ZEV display N-MORB (S part) and E-MORB (N part) geochemical patterns (Schüssler et al. 1989b). N-MORB patterns similar to the ZEV are found in amphibolites of the ZTM (62) to the NE of the ZEV (Schüssler et al. 1989b). Amphibolites and paragneisses from the first 480 m of the KTB pilot-borehole yielded a Rb-Sr whole-rock age of 380 Ma (Drach and Köhler in Köhler et al. 1989). A survey of rocks from the first 2000 m (Von Drach 1990) underlines a strong metamorphic overprint at 400-380 Ma. One gabbro yielded a U-Pb zircon age of 494+/-3 Ma. A Rb/Sr whole-rock age of 390 Ma is reported from gneisses in the continental basement which occur outside the ZEV nappe immediately to the SE (Hofmann in Köhler et al. 1989). This Ligerian age is supported by a U-Pb zircon age of c. 400 Ma from an orthogneiss of the same locality. However, Rb-Sr whole-rock data from the same orthogneiss yielded an age of 469+/-20 Ma which is likely to reflect the (Ligerian) intrusion age (Gebauer in Köhler et al. 1989). If the dating is correct, the U-Pb system of zircons was reset during Ligerian anatexis while there was no Sr isotopic exchange.

Ligerian radiometric ages in the ZEV nappe *and* its base suggest that the ZEV is an entirely Ligerian nappe which was sandwiched between the NGM and the Ligerian Terrane in the Early Devonian. Ligerian metamorphic ages are not found farther south in the Bohemian Massif where, similar to the Massif Central, only first cycle Early/Middle Ordovician metamorphic ages occur (Köhler et al. 1989). This suggests an original proximity of the ZEV nappe and its present (Ligerian) base after the Ligerian orogeny but also after Variscan displacements.
Late Silurian and/or Early Devonian microfossils/plant fragments in KTB-paragneisses (Pflug and Prössl 1989), and subsequent metamorphism suggest rapidly increasing P/T conditions

during emplacement of the ZEV nappe. Ligerian MORB in the ZEV may indicate that the nappe formed a Ligerian accretionary prism (Fig. 4) first thrusted over the NGM and then overthrusted by the Ligerian Terrane which led to deep crustal burial.

The gabbro age of 494+/-3 Ma may indicate reworking of first cycle material into the accretionary prism during collision. In view of nearby MORB-type material of the same age (2.10.4.) this gabbro may also correspond to Lower Ordovician oceanic crust not obducted during the first cycle and contained in the Ligerian Ocean until collision in the Early Devonian.

2.10.2. Münchberg Massif

An outlier of the ZEV is represented by the Münchberg Massif (60) which lies in a Variscan overthrusted position some 60 km to the NW in the Saxothuringian zone (Behr et al. 1984). It consists (from top to bottom) of the "Central Gneiss Body" (divided into "Upper series" and "Lower series") overlying the "Marginal Amphibolite series", which rests upon the "Prasinite-Phyllite series". The latter has a history different from the other units and is devoid of Ligerian metamorphic overprinting (3.5.).

Recalculated zircon ages of Gebauer and Grünenfelder (1979) yielded an age of 438+13/-16 Ma for eclogites which coincides with results from the Massif Central (Ducrot et al. 1983). Ducrot et al. (1983) interpret this age as the time of HP/HT metamorphism which would be, in the sense of Weber (1984), a syn-rift metamorphism in the lower crust. Sm-Nd and Rb-Sr whole rock dating on eclogites (P > 13-17 kb) yielded a HP-event at 395 Ma ago (Stosch and Lugmair 1987) which may be related to Ligerian collision and obduction or to resetting. Considering that the eclogites are found as dislodged slices between Upper series and Lower series (Schüssler et al. 1986), they were emplaced during Ligerian obduction and nappe movements in the Early Devonian prior to anatectic overprinting. This is supported by P-T paths of both series which are different until 385 Ma but are identical from that time onward (Müller-Sohnius et al. 1987).

Ages of 500-490 Ma are obtained only from the Lower series (meta-gabbro, 490-500 Ma, U-Pb zircon, Ducrot et al. 1983; meta-granite, 495 Ma, U-Pb monazite, Gebauer and Grünenfelder 1979; ortho- and paragneisses 498+/-10 Ma and 489+/-11 Ma, Rb-Sr- whole-rock, Müller-Sohnius et al. 1987; 499+/-20 Ma, Rb-Sr, whole rock, Söllner et al. 1981). K-Ar data on muscovite and hornblende from the Central Gneiss Body substantiate an amphibolite-facies metamorphism around 380 Ma (Kreuzer et al. 1989) indicating Ligerian reworking. The metamorphic history of the Marginal Amphibolite series appears to be different from that of the Central Gneiss Body but it is similar to amphibolites of the ZEV. It shows two groupings for K-Ar data on hornblende, one around 401+/-4 Ma and another around 384+/-3 Ma (Kreuzer et al. 1989).

From the data outlined above, the three upper nappes (Marginal Amphibolite-, Lower- and Upper series) appear as entirely Ligerian nappes. They form together one Variscan nappe which rests upon two other Variscan nappes, namely the Prasinite-Phyllite series (3.5.) and the NGM-type "Bavarian facies", all emplaced above the Saxothuringian zone.

2.10.3. Relations to the Saxothuringian Zone?

No signs of uplift or erosion are recorded in the Saxothuringian Lower Devonian sediments surrounding the Münchberg Massif. There are pelagic carbonates and cherts of an open, passive NGM shelf/slope environment in the 400-380 Ma interval which approximates the Early Devonian (e.g Greiling 1966). The Münchberg Massif must have been part of a continuous Ligerian belt without influence on the present Saxothuringian zone and the Bavarian facies nappe (2.10.2.) during Early/Middle Devonian metamorphism and ascent.

By Early Devonian time, Saxothuringian sediments found to the north of the ZEV were not yet emplaced at that site (Fig. 3). Saxothuringian Silurian/Lower Devonian sections comprise sediments of a passive open shelf environment similar to N Morocco or SW Spain. A Silurian/Devonian position adjacent to the Ligerian orogen that it holds now or even an incorporation into the Ligerian belt is not indicated by sedimentological and stratigraphic data. It is suggested that the present boundary between the Saxothuringian zone and the ZEV is near-vertical and represents a large-scale Variscan (Early Carboniferous) strike-slipe terrane boundary over which nappes were emplaced during subsequent compression.

2.10.4. Zone of Tepla-Taus (ZTT)

The ZEV and the ZTM continue northeastwards into the area between the Marianské Lazne Massif (63, MLM) in Czechoslovakia and the area of Hoher Bogen (64, Germany), which forms the ZTT (e.g. Kreuzer et al. 1989). Geochemical charateristics of most metabasites and gabbros from the MLM reveal an ocean floor tholeiitic nature (Kastl and Tonika 1984). A U-Pb zircon age of 494+/-2 Ma of a gabbro pegmatite in the MLM (Bowes and Aftalion 1990) may indicate reworked first cycle material or Lower Ordovician oceanic crust not obducted until the Early Devonian. A metamorphic overprint of the MLM during the Early Devonian is indicated by K-Ar mineral data (Schüssler et al. 1989a). K-Ar dating of metamorphism and anatectic intrusions in the Tepla area immediately to the south of the MLM yielded ages around 390 Ma for the Tepla synkinematic granitoid and surrounding meta-sediments (Gottstein 1970).

Adjacent to the southern region of the ZTT appears the Kdyne Massif (CSFR) which apparently continues into basic massifs across the German border (Hoher Bogen, Neukirchen Massif, 64). Geochemical analysis revealed two groups of metabasites with unclear relations: the first is represented by intra-plate tholeiites with alkaline tendency (Neukirchen/Kdyne) and the second comprises sub-alkaline, tholeiitic ocean-floor basalts (Hoher Bogen/Blätterberg/Warzenrieth, Schüssler et al. 1989a). A K-Ar hornblende age of 385+/-12 Ma was found in the Hoher Bogen area (Fischer et al. 1968).

It is here assumed that two unrelated basic massifs are tectonically juxtaposed, the Kdyne/Neukirchen Massif representing an Upper Precambrian or Cambrian intrusion with a well-developed contact aureole in Precambrian rocks (Vejnar 1990). The Hoher Bogen area represents presumably Ligerian or older ocean floor overthrusted and metamorphosed in the Early Devonian. This is consistent with the occurrence of Upper Silurian micaschists to the SE of the Hoher Bogen area (spores, Pflug and Reitz 1987) which do not show any contact aureole as found around the Kdyne/Neukirchen Massif (Vejnar 1990). On the other hand, Ligerian ocean floor basalts are typically associated with Upper Silurian or Lowermost Devonian metasediments (2.10.1., 2.10.10.). In this sense only the Hoher Bogen/Blätterberg/Warzenrieth Massif belongs to the ZTT while the Kdyne/Neukirchen Massif belongs to the Precambrian underlying the Barrandian Palaeozoic (2.10.12.)

2.10.5. Erzgebirge (Ore Mts)

The Erzgebirge (65) forms a crystalline area to the south of the Central Saxonian Lineament, by which it is separated from the Saxothuringian zone (Fig. 2). K-Ar studies on gneisses from the central Erzgebirge (Wolf and Kaiser 1986) revealed ages of 445 Ma, 410 Ma, 395 Ma, and a variety of data that fall into the 380-376 Ma interval. A younger 360-350 Ma group (also recorded from pegmatites intruding the gneisses) is clearly separated (Wolf and Kaiser 1986). This points to a later tectono-metamorphic/magmatic overprint. K-Ar dating on metabasic rocks from the W Erzgebirge yielded ages of 428 Ma and 408 Ma (Wiedemann and Kaiser 1986) which lie in the range of the lower age groups of ZEV and Münchberg metabasites (2.10.1., 2.10.2.). Structural studies in combination with radiometric data revealed a (second) deformation phase in the area of Sayda (E Erzgebirge) which is estimated to fall at about the Silurian/Devonian boundary (Kemnitz 1987). Earliest post-metamorphic granitic intrusions are dated at 380 Ma (Frischbutter 1988). A complex first cycle and earlier (pan-Gondwanian) history for the basement is also evident, and Krentz (1985) estimates an age around 477 Ma

(Rb-Sr, whole rock) for the metamorphic peak in gneisses from the W Erzgebirge. Eclogites of the Erzgebirge give P/T values in the range of 15 kb and 550-630½C and geochemical data indicate MORB characteristics (Werner and Kononkova in Kramer and Werner 1989).

The structural position of the Erzgebirge to the north of the ZEV and ZTT ophiolites suggests that it forms perhaps a crustal part of the Ligerian Terrane. However, the original relations of this crustal block within the Ligerian orogen seem to be intensively reworked during Variscan displacements.

2.10.6. Metamorphic Massifs in SE Germany

2.10.6.1. The Sächsische Granulitgebirge (Saxonian Granulite Massif)

The high-grade metamorphic Sächsische Granulitgebirge (66a) to the north of the Central Saxonian Lineament is - like the Münchberg Massif - a nappe within the Saxothuringian zone. It reaches down to a depth of at least 2000 m as found in a borehole (Wetzel and Gerstenberger 1989). Pan-Gondwanian and first cycle radiometric ages around 600-550 Ma and 450 Ma were recorded as well as ages around 375 Ma for mafic and acidic HP-rocks (Weber 1984, Wetzel and Gerstenberger 1989, Quadt and Gebauer 1990). Eclogites (formation at approx. 15 kb and 600+/-30½C) and other metabasic rocks are geochemically classified as ancient oceanic crust (Werner and Kramer in Kramer and Werner 1989). Basic and intermediate rocks from the Sächsische Granulitgebirge form a series of island arc volcanics (Wetzel and Gerstenberger 1989). A wealth of ages around 350-340 Ma are associated with granite intrusions. Geochemical comparisons suggest that the granites are derived from melting of leuco-granulites (Wetzel 1990). The last metamorphic overprint is recorded at 320 Ma (Wetzel and Gerstenberger 1989).

It is assumed that the Sächsische Granulitgebirge contains Ligerian syn-rift HP/HT metamorphic elements (around 450 Ma, Weber 1984) which were overprinted during the Ligerian orogeny (380 Ma). Later severe overprinting followed during the first steps of the Variscan orogeny at 350-340 Ma. The Early Palaeozoic metamorphic history is comparable to the Münchberg Massif and suggests a common evolution (Quadt and Gebauer 1990). Compared to the Münchberg Massif it represents a deeper level of the Ligerian orogen.

2.10.6.2. The Massifs of Frankenberg and Wildenfels

The amphibolite facies Frankenberg Massif (66b) lies between the SE periphery of the Sächsische Granulitgebirge and the Erzgebirge Ligerian area (Fig. 2). It is devided from both basement units by Saxothuringian rocks. The "Caledonian" Frankenberg orthogneiss is located in the northern part of the massif, where it is in fault contact with the underlying prasinite series (3.5.). According to Werner (in Kramer and Werner 1989) three mafic rock types can be distinguished in the Frankenberg Massif: massive lavas of MORB-type, thin and fine-grained basalt layers varying between low-K tholeiites and calc-alkaline basalts, and thick amygdaloidal andesitic lava flows. The latter two are petrochemically classified as volcanic-arc series. K-Ar dating of amphibolite yielded an age of 390+/-25 Ma and 419+/-20 Ma (whole rock) and 404+/-20 Ma (biotite) for a relictic biotite granite (Kurze et al. 1982).

The Wildenfels Massif (66c) forms a small isolated outcrop (< 1 km) c. 18 km SSW of the Sächsische Granulitgebirge surrounded by Saxothuringian rocks (in the S) and the Permian cover (in the N). It contains basic rock series as in the Frankenberg Massif known as banded amphibole gneisses which are comparable with the Münchberg Massif (Werner in Kramer and Werner 1989). They show petrochemical patterns of low-Ti MORB and high-Ti low-K tholeiites of a volcanic arc.

In analogy to the ZEV and the Münchberg Massif (related rock series, structural position, metamorphic grade and -age), the MOR basalts in both massifs are tentatively interpreted as obducted Ligerian ocean floor thrusted over the Saxothuringian zone during the Variscan orogeny. They may be considered as part of the Ligerian accretionary prism (2.10.1.).

2.10.7. Krkonosze-Klodzko/Snieznik Ligerian Successions in the NE Bohemian Massif

Two phases of deformation postdating deposition of Upper Silurian sediments but predating the onset of Upper Devonian sedimentation are known in the Krkonosze Mts (67, Chaloupsky 1988). The fossiliferous Silurian sediments contain phyllites and limestones associated with a spilite - keratophyre group. Geochemical studies show the spilites to be island-arc tholeiites (Narebski 1980). To the south of the volcanic rocks occurs a glaucophane schist belt which may belong to a Late Silurian arc-trench system with a northwards dipping Benioff zone (Narebski 1980). In the Klodzko Unit (68) of the Sudetes, metamorphic limestones intercalated in phyllites yield Ludlovian corals (Oberc 1980). These rocks are overlain by a meta-volcano-sedimentary succession with diabases, keratophyres, greenschists and greywacke phyllites (Oberc 1980). Unmetamorphic Upper Devonian rocks overlie this complex with a gabbro conglomerate at the base. Don (1990) assigns the Klodzko and Krkonosze areas to a "Caledonian" domain which is separated from the Saxothuringian regions to the north by the Main Intra-Sudetic Fault. In the adjacent Snieznik unit (69), metabasites with amphibolite facies metamorphism at c. 382 Ma are geochemically classified as calc-alcaline series of island arc volcanites (Wojciechowska 1986). Geochemical investigations of metabasites in the Orlice-Klodzko Dome (70) indicates an origin as ocean floor basalts (Domecka and Opletal 1980).

Stratigraphic, geochemical, metamorphic and radiometric data indicate an evolution of the Krkonosze - Snieznik area equivalent to the S Armorican Massif. Subduction-related volcanism during the Late Silurian was caused by northward subduction and indicates that the Klodzko - Snieznik area forms part of the Ligerian Terrane (Fig. 4). Saxothuringian sediments to the north of the Klodzko unit (Bardo Mts., 71, Oberc 1980) have continuous anorogenic Silurian/Devonian outer shelf/slope successions without any trace of a subduction-related or orogenic development.

2.10.8. Basement of the Bohemian Cretaceous Basin

An undated volcano-sedimentary succession with black-grey phyllites, unconformably beneath unmetamorphic Upper Devonian rocks was found in a borehole near Hradec Kralové (73, Chlupac 1976). Another borehole in the Bohemian Creataceous basin (Mlada Boleslav, 50 km NE of Prague) encountered remnants of Silurian or Lowermost Devonian landplants in black phyllites (Pacltova 1980). Such successions with Ligerian tectono-sedimentary patterns are not known in the Barrandian area (74) but may be related to the Krkonosze-Klodzko area (2.10.7.). This may indicate that parts of the Bohemian Cretaceous basin are underlain by crust involved in the Ligerian Terrane accretionary process in Earliest Devonian time.

2.10.9. Gory Sowie (Owl Mountains)

To the north of Saxothuringian rocks of the Bardo Mts. occurs another Ligerian massif (Gory Sowie, 72) with cooling ages of 381 Ma (monazite) and a separated group of 370 - 360 Ma (Van Breemen et al. 1988). Microfossils found in metasediments of the Gory Sowie indicate an Upper Precambrian/Cambrian(?) age for sedimentary protoliths (Gunia 1983). The mode of appearance to the north of Saxothuringian sediments is similar to the Léon Unit in NW Brittany (2.4.4.).

2.10.10. Assignment of the Bohemian Varied Group

Lithologies of the Varied group (gneisses, graphitic mica-schists, calc-silicate rocks, crystalline limestones, metabasites) are found mainly around the NGM-type Barrandian area (74, incl. its Precambrian basement) and in the S Bohemian Massif (76) (e.g. Kreuzer et al. 1989, Tollmann 1985). They form part of the Bohemian polymetamorphic basement which further contains large amounts of Ordovician or older metamorphics and Variscan granitoids.

Available data on microfossils from the Varied group (Pacltova 1980, 1986, Pflug and Reitz 1987, Reitz and Höll 1988, Andrusov and Corna 1976, Konzalova 1980, Pflug and Prössl 1989) indicate rare Late Precambrian and frequent Late Silurian and Earliest Devonian ages for sedimentary rocks from this unit. They contain, apart from acritarchs and chitinozoans, also remnants of vascular land plants which can be ascribed to psilophytes (Pacltova 1980) first known in the Upper Silurian of Wales (Edwards 1979). Woody tissues from the KTB pilot-borehole indicate an age not older than Devonian (Pflug and Prössl 1989). Large parts of the Varied group may have been deposited at the southern margin of the Ligerian Ocean or in the Ligerian accretionary prism (Fig. 4) and were were metamorphosed during the Ligerian collision. Sedimentation stopped when the Ligerian Terrane advanced in Early Devonian time (Lochkovian? Praguian?) and was thrusted over the southern continental margin of the Ligerian Ocean. This process led to metamorphism of all involved units at around 390-380 Ma.

2.10.11. SE Bohemian Massif

Tectonically emplaced above the Varied group are continental basement rocks of the Gföhl nappe sensu Tollmann (1985). Rb-Sr whole rock ages on granulites from this nappe in the S Bohemian Massif (76, Austria) point to an emplacement age of the protolith around 470 Ma and granulite-facies metamorphism at 430 Ma (Arnold and Schabert 1973). Gneisses in the Gföhl nappe are orthogneisses with Ordovician intrusion ages (Fuchs and Matura 1976). The likely position of the Gföhl rocks before thrusting is indicated in Fig. 4. Another part of the Gföhl nappe is represented by the Kutna Hora crystalline area (75) which is associated in its southern part with units of the Varied group (Tollmann 1985). Retrograde metamorphism in Kutna Hora crystalline rocks near the contact to the Varied Group is dated at 401+/-42 Ma (Krylova and Glebovitsky 1985).

Along their E periphery, the Varied Group and Gföhl Gneisses were thrusted during the Variscan orogeny over the Moravicum (77, Fuchs 1976), which forms the basement of the South Laurussia margin (SLM) in this area (Fig. 2). As demonstrated by Fuchs (1976) the nappes in the SE Bohemian Massif are pre-Variscan and have a discordant relation to the

Variscan thrust over the Moravicum. The (actual) direction of Ligerian movements in the SE Bohemian Massif is WNW (Fuchs 1976), i.e. directed to its internal part. This is consistent with the generally S-directed Ligerian overthrusts (in this case directed to the inner part of orogenic arcs, Burg et al. 1987). However, rearrangement of Ligerian thrusts during SE-directed Variscan overthrusting of the NGM over the SLM is probable.

2.10.12. Allochthonous Position of the Barrandian Area?

If the spatial arrangement of some areas with established or likely Ligerian metamorphism is considered (Hoher Bogen (64), ZEV (61), Ore Mts (65), Krkonosze Mts (67), Kutna Hora Massif (75)), they are found to *surround* the unmetamorphic NGM-type Barrandian area (74, incl. its Precambrian basement). Furthermore, no traces of Ligerian processes are found in contemporaneous sediments from the Barrandian area, and its position within a region of metamorphism and nappe movements in the Early Devonain is unlikely. This supports the idea of Tollmann (1985) who assumes that the Barrandian area is a Variscan nappe emplaced on top of the Bohemian crystalline basement. Seismic reflection profiles in the Precambrian NW-part of the Barrandian area (Tomek et al. 1987) revealed SE- and NW-dipping reflectors at shallow depth (1-4 km). They are in accord with Variscan thrusting over the Ligerian basement of the Bohemian Massif.

2.11. Eastern Alps

The Bohemian arc with the NGM/SLM boundary continues into the Eastern Alps where recognition is difficult because of extensive reworking during the Alpine orogeny. Radiometric age data quoted in Neubauer (1988) for the Eastern Alps, where a paragneiss yielded a U-Pb zircon age of 391+/-2 Ma (Veitsch zone, 46), may confirm the presence of Ligerian units. Rb/Sr data also suggest an intra-Devonian metamorphism (Frimmel in Neubauer 1988) and a 400-380 Ma period of granitoid intrusions, now found as orthogneisses in the Eastern Alps. These data suggest a continuation of the Ligerian orogen from the Bohemian Massif into the Eastern Alps basement, now being preserved only in strongly displaced Variscan and Alpine thrust units.

3. REVIEW OF AREAS POSSIBLY RELATED TO THE LIGERIAN BELT

The following regions bear indications of Ligerian influence but the available data are not unequivocal. They form, however, a promising field for future investigations, especially by radiometry, geochemistry, and search for microfossils in low- to medium- grade metamorphic rocks.

3.1. Malpica-Tuy Unit (W Galicia, NW Spain)

The Malpica-Tuy unit crops out along the W coast of Galicia (13) and contains lithologies similar to those encountered in the Cabo Ortegal - Tras-Os-Montes area (2.3.). Three lithostratigraphic groups are distinguished in its northern part (Gil Ibarguchi and Ortega Gironez 1985): The Lower group consisting of mica-schists and paragneisses with intercalated orthogneisses of calc-alkaline affinity which contain K-feldspar megacrysts; the Middle group with alkaline and peralkaline orthogneisses of subvolcanic origin (467+/-10 Ma) associated with minor amounts of metasediments; the Upper group with amphibolites, mica-schists, and lenses of carbonates. Eclogites are found only in the Lower and Middle groups. Their P-T path suggests a HP/HT phase (garnet, omphacite, kyanite) not exceeding 600½C and 12-13 kb, followed by a glaucophane/phengite forming HP phase, and a post-eclogitic amphibolitic stage which is contemporaneous with intense deformation. A following Variscan? overprint led to greenschist grade mineral assemblages. This metamorphic path is equivalent to the Ligerian evolution of Middle/Late Ordovician - Early Devonian age in many European regions. The glaucophane-bearing rocks may form a counterpart of similar occurrences in the S Armorican Massif (2.4.1.) indicating HP-metamorphism preceding obduction around the Silurian/Devonian boundary.

3.2. Central Corsica

Tholeiites related to continental rifting are present in a metamorphic succession near Zivaco (41, central Corsica, Vezat 1988). They are contained in a more than 2000-m-thick volcano-sedimentary succession overlain by metasediments (several thousand metres) with sporadic volcano-sedimentary intercalations ascribed to Early Palaeozoic times (Vezat 1988). A relation to Ligerian rifting and Lower Silurian diastrophic sediments in NW Corsica (2.7.2.) is possible.

3.3. Maures Massif

The Maures Massif (43) has a development comparable to that of the NW Iberian Peninsula with intrusion of mafic/ultramafic material and deposition of Lower Palaezoic volcano-detritic sequences (Bard and Caruba 1981). Tectono-metamorphic data from the E Maures Massif suggest an evolution closely resembling that of the Massif Central (Vauchez and Buffalo 1988): a HP/HT-metamorphism which produced eclogites and granulites is followed by a HT event which produced anatexis leading to the ascent of anatectic granites. Analogy of the P-T path to

Ligerian regions and occurrence of Llandovery age metasediments with graptolites (Gueirard et al. 1970) in the Maures Massif are consistent with a Ligerian evolution. A model proposed by Bard and Caruba (1981) involves rifting and subsequent oceanic crust formation during Ordovician-Silurian times. The direction of closure is E - W (in contrast to a N - S closure for the Massif Central). Contrary to a Ligerian collision around the Silurian/Devonian boundary proposed here, Bard and Caruba (1981) assume Late Devonian - Visean convergence.

3.4. Harz Mountains

An isolated occurrence of probable Ligerian crust is found in the Harz Mts., commonly referred to as Ecker Gneiss (59). It contains metabasalts of an ancient back-arc environment, gneisses, and (quartz-) micaschists (Vinx and Schlüter 1990). Dating of amphibolite facies metamorphic rocks yielded an age of 379+/-10 Ma on Rb-Sr (whole rock, Schoell et al. 1973). U-Pb dating on detrital zircons from paragneisses points to ages of 500-600 Ma interpreted as the metamorphic age (Vinx and Schlüter 1990). However, since zircons may easily survive metamorphic overprinting these ages may well be consistent with the age of the source rocks, reflecting reworking of Pan-Gondwanian and first cycle orogenic material. The Ecker Gneiss occurs in the vicinity of NGM-lithologies which were thrusted over the Rhenohercynian zone (Oczlon 1991). It may have formed part the MGCR (2.9.).

3.5. Nature of the "Prasinite Series"

Prasinite series (containing metabasites with albite, epidote, chlorite, hornblende) and related formations are partly retrograde greenschist facies rock associations found in the footwall of the basement nappes of Münchberg (60) and Frankenberg (66b). The related Erbendorf Greenschist zone (EGZ, 61) occurs at the N margin of the ZEV (2.10.1.) at the boundary to the Saxothuringian zone. Part of the EGZ metabasites were primarily metamorphosed in the lower amphibolite facies and do not strictly correspond to prasinites but are traditionally considered as such (Schüssler 1990). The allochthonous Brévenne unit of the E Massif Central (31, Feybesse et al. 1988) has characteristics similar to the outlined units and is tentatively assigned to the Prasinite series.

Metabasic rocks of the Prasinite series are geochemically classified as part of a volcanic-arc suite (Werner in Kramer and Werner 1989, 66b; Schüssler et al. 1989b, 61; Reitz and Wickert 1988, 31). One gabbro of the EGZ is dated at 365+/-7 Ma (K-Ar on hornblende) which lies in the range of datings on phyllites from the Münchberg Prasinite-Phyllite series (around 366 Ma, Kreuzer et al. 1989). Acritarchs found in phyllites from Münchberg and the Brévenne unit point to an Upper Precambrian age (Reitz and Höll 1988, Reitz and Wickert 1988). Fragments of vascular plants in the EGZ point to an age around the Silurian/Devonian boundary similar to fossils from the KTB pilot-borehole (Pflug and Prössl 1989). These few data may indicate that the Prasinite series contain low grade metamorphic Upper Precambrian-Lower Palaeozoic sediments which were metamorphosed during the Late Devonian. They contain intrusive volcanic arc suites of unknown age.

It is speculated that the Prasinite series formed part of the Ligerian Terrane thrusted over the NGM in a position similar to the N Vosges/Schwarzwald (2.8.3.) or the Saint-Georges-sur-Loire syncline (2.4.3.). These units escaped Ligerian metamorphism due to their position in the overthrusting Ligerian Terrane and a resulting high structural position in the Ligerian collisional belt. Metamorphism around 365 Ma led to greenschist grade overprinting of sedimentary or metamorphic units of the Ligerian Terrane but also of Ligerian high-grade units. For example, the K-Ar system in HP-mica of eclogites from the Münchberg Massif records an overprint at c. 360 Ma (Hammerschmidt and Franz 1988). Overprinting is here assigned to a Late Devonian magmatic arc setting at the NGM raising the geothermal gradient and leading to intrusions like the EGZ gabbro.

During Variscan north-directed nappe emplacement, high grade units from the inner (southern) part of the Ligerian orogen (still located in the central/western part of the Bohemian Massif) were thrusted over sediments of its external (northern) part (e.g. Prasinite-Phyllite series underlying older but higher metamorphic complexes in the Münchberg Massif, 2.10.2.). The Münchberg and Frankenberg Massifs were further thrusted over Saxothuringian sediments, whereas the EGZ remained in a more or less original post-Ligerian position at the N edge of the ZEV and its Ligerian basement (2.10.1.).

4. EVOLUTION OF THE LIGERIAN BELT

From N Morocco to the Eastern Alps, the Ligerian belt extends over some 3000 km. But only from the NW Iberian Peninsula towards the east, the belt takes on the characteristics of a mountain chain during the Early/Middle Devonian (ca. 2000 km). Palaeogeographically it coincides with the zone of ocean-crust formation during Early Silurian time and generation of subduction-related volcanism on the previously split-off Ligerian Terrane during Late Silurian convergence with northward subduction. The cover sequence usually starts with shallow-water carbonates or reef-limestones fringing the cordillera in the setting of a Givetian passive NGM without fault-bounded basins or volcanism (Oczlon 1990). To the south of the NW Iberian Peninsula, rifting never exceeded the stage of syn-rift intrusions, updoming and graben formation. Convergence around the Silurian/Devonian boundary led to local inversion on the NGM

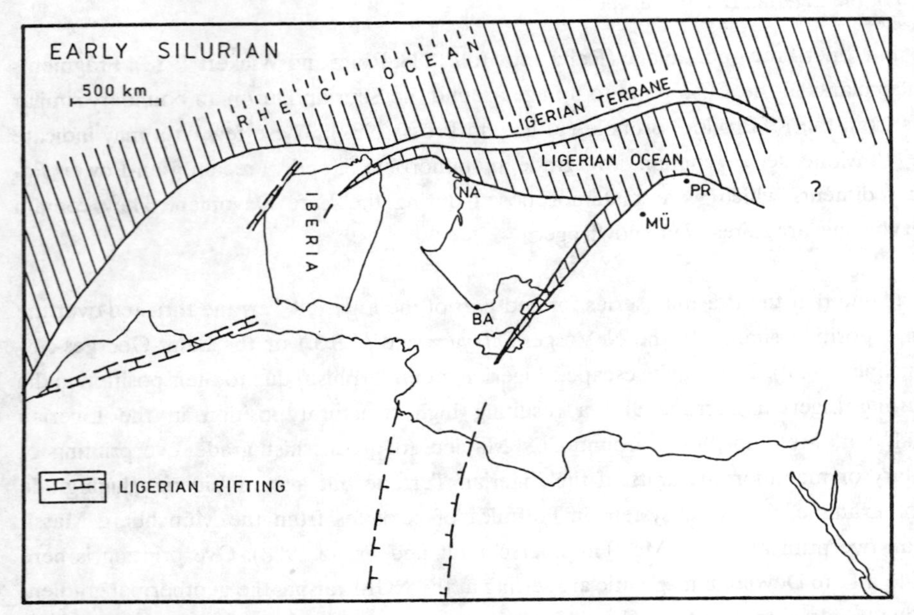

FIGURE 3
Reconstruction of the European/African NGM for the Early Silurian. Original locations of Saxothuringia and Brittany with facies unrelated to Ligerian processes are tentatively located between Spain and Morocco with later Variscan dextral displacement along the North Gondwana margin (2.10.3.). BA-Barcelona, MÜ-München, NA-Nantes, PR-Praha.

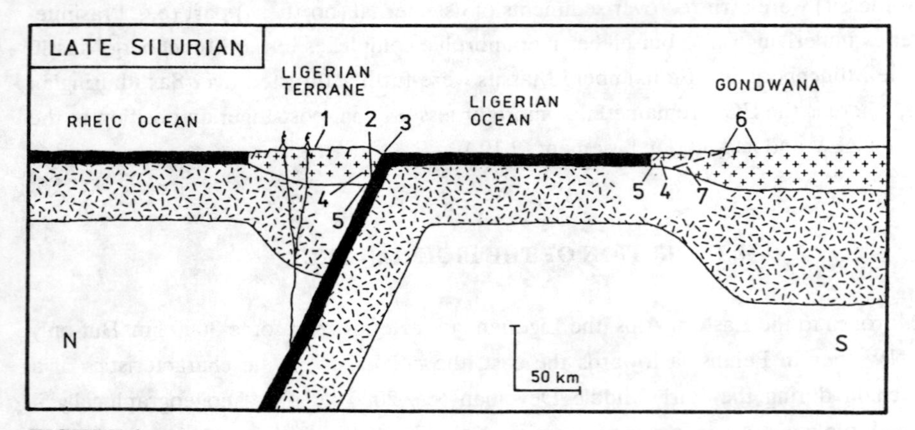

FIGURE 4
Cartoon showing possible Late Silurian geodynamic situation of the Ligerian Ocean. The section is based mainly on data from N Bohemia but may be basically valid also for other regions of the Ligerian Ocean. 1) Upper Silurian subduction-related successions; 2) Blueschists obducted during subsequent collision; 3) Accretionary prism with MOR basalts of the zone of Erbendorf-Vohenstrauß (2.10.1.) and other areas; 4) Middle/Upper Ordovician syn-rift intrusions, now orthogneisses (e.g. Gföhl Gneiss); 5) HP/HT dry syn-rift metamorphic rocks (Late Ordovician/Early Silurian); 4+5 may be present in the North Gondwana basement but also in the Ligerian Terrane; 6) Depositional environment of the Silurian Varied group of Bohemia and associated units on the southern margin of the Ligerian Ocean; 7) Later thrust plane (Earliest Devonian) of basement units 4+5 over the Varied group.

shelf and not to a collisional orogen. This area was already covered by the Early Devonian sea. A second branch of the Ligerian Ocean could have led from the area of the S Vosges/Schwarzwald - NE Massif Central - Western Alps to the south (Fig. 3). As to the extent of the ocean and the branching, an evolution comparable to that of the southern Red Sea is assumed.
Ligerian rifting remained in an initial stage between 470-440 Ma (Middle/Late Ordovician) and reached its climax during accelerated rifting in the Early Silurian with formation of oceanic crust. It is assumed that the Ligerian Terrane was not completely separated from the remaining NGM but that it was still connected in the area of transition from graben-formation to ocean formation in the NW corner of the Iberian Peninsula (Fig. 3). Remnants of the Ligerian Terrane are presently located in the area between the Ligerian suture and the later (Early Carboniferous) juxtaposed Saxothuringian zone (incl. Brittany) to the north. From west to east possible remnants of the Ligerian Terrane are: the Malpica-Tuy unit (3.1.) in the NW corner of the Iberian Peninsula; the area of the Saint-Georges-sur-Loire basin (2.4.3.); an unknown portion of the S Paris Basin basement, all or part of which was displaced during the Variscan orogeny and now appears as the MGCR (2.9.); within the MGCR, the area of the northwesternmost Spessart (Alzenauer Gneiss formation); the N Vosges/Schwarzwald (2.8.); the Central Gneiss Body of the Münchberg Massif which overlies Ligerian ocean floor (2.10.2.) and the Sächsische Granulitgebirge (2.10.6.1.); the Prasinite series (3.6.); the Erzgebirge crystalline basement (2.10.5.) and the Krkonosze-Snieznik area (2.10.7.). The direction of overthrusting is always southwards or towards the inner parts of the Ibero-Armorican and Bohemian arcs.

The original width of the Ligerian Terrane can be estimated only imprecisely because it has been extensively modified during the Ligerian and Variscan collisions. Judging from the preserved remains and the narrow metamorphic belts it can hardly have exceeded a width of 50-100 km of continental crust (Figs. 3, 4). The maximum width of the Ligerian Ocean during Early/Middle Silurian time may have been in the order of 500 km. This value is obtained when complete subduction is considered at an assumed mean rate of 5 cm/a in 10 Ma during the Late Silurian/Earliest Devonian. Such figures are typical for modern-day Mediterranean-type oceanic basins. Closure of the Ligerian Ocean apparently involves a slight diachronism. While there are indications of a molasse stage in the Lochkovian of the Saint-Georges-sur Loire basin (2.4.1.) the pre-collisional Ligerian Varied group (2.10.10.) was still deposited. Thus, closure may have occurred a few Ma earlier in the S Armorican Massif than in the Bohemian Massif.

The continuity of the Ligerian belt still visible implies that it behaved rather like an entity during the Variscan orogeny without large displacements between crustal blocks such as the Massif Central and the "Moldanubian" Bohemian Massif. Large dextral Variscan displacement took place to the north of these areas between the Saxothuringian and related units with continuous NGM passive margin sediments and slices of Ligerian basement between them (e.g. MGCR, 2.9.).

5. LIGERIAN REPERCUSSIONS ON NGM SEDIMENTATION AREAS

A repercussion of Ligerian convergence is an entirely clastic development of the Upper Silurian (greywacke, dark sandstone, and shale) in great parts of Spain and Brittany which stands in sharp contrast to the "normal" shallow water limestone development without clastic input on the NGM. These sequences are interpreted as derived from uplifts created during the time of convergence (inversion) in regions underlain by continental crust. The same areas were covered during the Lower Devonian by bright and red quartztites, sandstones, and, in places, conglomerates interpreted as Ligerian molasse. Basal Devonian layers become more conglomeratic towards the SE Iberian Peninsula (Julivert et al. 1983). Equivalent Silurian-Devonian clastic facies developments are known in NW Libya and in N Algeria except for the Alpine displaced Kabylian Massifs (Berry and Boucot 1973). A relationship of the SE Spanish Lower Devonian conglomerates with Ligerian movements at the Oued Madakh Massif (2.1., NW Algeria) is possible. These palaeogeographic relationships may be indicative for juxtaposition of Spain and N Africa as in Fig. 3. To the west (Morokko) the Upper Silurian/Lower Devonian facies takes on its "normal" limestone development without Ligerian influence comparable to the S OMZ, S Brittany, Saxothuringia, Sardinia, Eastern Alps or S Turkey. These regions were located out of reach of clastic influx from Ligerian sources. A single example of Upper Lochkovian flysch-sedimentation (zone of Monograptus hercynicus) is found on the Isle of Menorca (36, Bourrouilh 1967). The turbidites were probably fed by rising Ligerian domains in the easterly adjacent Maures Massif/Corsica branch.

In contrast, areas which today are located in the vicinity of the Ligerian belt (e.g. Montagne Noire, 35) seem to be devoid of erosional products from the adjacent Ligerian massifs. Yet, considerable uplift during Early Devonian/Eifelian time is testified by Givetian/Upper Devonian limestones covering Ligerian granites or metamorphics. Several reasons may account for this: 1) The time of emergence was relatively short (c. 20 Ma, Early Devonian/Eifelian) which implies that uplift was not as strong as, for example, in the modern Alpine orogen and a large crustal root which causes long isostatic uplift was not developed. This is ascribed to the small width of the colliding Ligerian Terrane. 2) The Ligerian Cordillera was located on the outer edge of the NGM and faced the Rheic Ocean in the north. A much stronger gradient is therefore expected towards this direction than towards the NGM shelf area. Drainage of the Ligerian Cordillera is likely to have had a mainly northern (oceanward) direction. 3) The position on the outer MGM may have caused rainfalls to occur mainly on the oceanward side of the orogen which leads to oceanward flowing river systems. 4) The onset of current-activity on the NGM shelf, comparable in extent to the modern Gulf Stream, can be demonstrated for the time around the Early/Middle Devonian boundary (Oczlon 1990). It is held responsible for non-sedimentation and erosion in many parts of the NGM shelf. 5) Variscan displacement may be responsible for juxtaposition of previously unrelated areas like the Montagne Noire nappes (2.6.) and the Massif Central.

Acknowledgements

I thank I. Chlupac (Prague), A.M. Diez Balda (Salamanca), R.O. Greiling (Heidelberg), K. Weber (Göttingen), and an anonymous referee for the improvement of this work.

REFERENCES

Aifa, T., Feinberg, H., and Pozzi, J.P., (1990): Devonian-Carboniferous paleopoles for Africa: consequences for Hercynian geodynamics. Tectonophysics 179, 287-304.

Aldaya, F., Carls, P., Martinez-Garcia, E., and Quiroga, J.L. (1976): Nouvelles précisions sur la série de San Vitero (Zamora, nord-ouest de l'Espagne). Comptes Rendus Acad. Sci. Paris, D 283, 881-883.

Altenberger, U., Besch, T., Mocek, B., Zaipeng, Y., and Yong, S. (1990): Geochemie und Geodynamik des Böllsteiner Odenwaldes. Mainzer geowiss. Mitt., 19, 183-200.

Andrusov, D. and Corna, O. (1976): Über das Alter des Moldanubikums nach mikrofloristischen Forschungen. Geologické Prace, Bratislava 85, 81-89.

Anthonioz, P.M. (1967): Les brèches tectoniques dans les unités de Braganca et de Morais (provence de Tras-Os-Montes, Nord-Est du Portugal). Comptes Rendus Acad. Sci. Paris, D 264, 233-236.

Antona, J.F. and Martinez Catalan, J.R. (1990): Interpretacion de la Formation San Vitero en relacion con la orogenia Hercinica. Cuaderno Lab. Xeoloxico de Laxe, Coruna 15, 257-269.

Arenas, R., Gil-Ibarguchi, J.I., Gonzales-Lodeiro, F., Klein, E., Martinez-Catalan, J.R., Ortega-Girones, D.E., Pablo-Macia, J.G. and Peinado, M. (1986): Tectonostratigraphic units in the complexes with mafic and related rocks of the NW of the Iberian Massif. Hercynica 2, p. 87-110.

Arenas, R, and Peinado, M. (1981): Presencia de pillow-lavas en las metavulcanitas submarinas de las proximidades de Espasante, Cabo Ortegal, NW de Espana. Cuad. Geol. Ibérica 7, 105-119.

Arnold, A. (1970): On the history of the Gotthard Massif (Central Alps, Switzerland). Eclogae Geol. Helv. 63, 29-30.

Arnold, A. and Schabert, H.G. (1973): Rb-Sr-Altersbestimmungen an Granuliten der südlichen Böhmischen Masse in Österreich. Schweiz. Mineral. Petr. Mitt. 53, 61-78.

Balé, P. and Brun, J.P. (1986): Les complexes métamorphiques du Léon (NW Bretagne): un segment du domaine éo-hercynien sud armoricain translaté au Dévonien. Bull. Soc. Géol. France 2, 471-477.

Bard, J.P. (1966): Quelques précisions sur la lithologie du "Silurien" de la région d'Aracena (Huelva)-Espagne. Notas y Comunicaciones Inst. Geol. Min. de Espana 83, 93-98.

Bard, J.P. & Caruba, C. (1981): Les séries leptyno-amphibolitiques à éclogites relictuelles et sepentinites des Maures, marqueurs d'une paléosuture affectant une croute amincie? Comptes Rendus Acad, Sci. Paris, D 292, 611-614.

Baudelot, S. and Durand Delga, M. (1981): Le Dévonien de Galéria en Corse septentrionale, sa datation et sa place dans le Cadre de la Méditerranée occidentale. Comptes Rendus Acad. Sci. Paris, D 292, 347-354.

Baudelot, S., Doubinger, J., Duran Delga, M., and Vellutini, P. (1976): Caractéristiques et ages des cinq cycles paléozoiques du Nord-Ouest de la Corse. Bull. Soc. Géol. France 18, 1221-1228.

Behr, H.J., Engel, W., Franke, W., Giese, P., and Weber, K. (1984): The Variscan belt in Central Europe: Main structures, geodynamic implications, open questions. Tectonophysics 109, 15-40.

Bernard-Griffiths, J., Gebauer, D., Grünenfelder, M., and Piboule, M. (1985): The tonalite belt of Limousin (French Central Massif): U-Pb zircon ages and geotectonic implications. Bull. Soc. Géol. France 1, 523-529.

Berry, W.B.N. and Boucot, A.J. (1973): Correlation of African Silurian rocks. Geol. Soc. Amer. Spec. Pap. 147, 87 pp.

Bodinier, J.-L.; Giraud, A.; Dupuy, C.; Leyreloup, A., and Dostal, J. (1986): Caractérisation géochimique des metabasites associées à la suture méridionale hercynienne: Massif Central Francais et Chamrousse (Alpes). Bull. Soc. Géol. France 2, 115-123.

Bodinier, J.-L., Burg, J.-P., Leyreloup, A., and Vidal, H. (1988): Reliques d'un bassin d'arrière-arc subducté, puis obducté dans la région de Marvejols (Massif central). Bull. Soc. Géol. France 4, 21-33.

Bonhomme, M.G. and Fluck, P. (1981): Nouvelles données isotopiques Rb-Sr sur les granulites des Vosges. Age protérozoique terminal de la série volcanique calco-alcaline et age acadien du métamorphisme régional. Comptes Rendus Acad. Sci., Paris, sér. II, 293, 771-774.

Bourrouilh, R. (1967): Le Dévonien de Minorque (Baléares, Espagne): Ses limites et sa place en Méditerranée Occidentale. In: Intern. Symposium om the Devonian System, Calgary, 2, 47-60.

Bowes, D.R. & Aftalion, M. (1990): Early Ordovician U-Pb zircon age for an ophiolitic gabbro pegmatite in the Hercynides of western Bohemia. In: 6. Rundgespräch "Geodynamik des europäischen Variszikums" - Zeitmarken im europäischen Variszikum, Clausthal-Zellerfeld, 15.-18.11.1990, Abstracts.

Brinkmann, R. (1948): Die Mitteldeutsche Schwelle. Geol. Rdsch. 36, 56-66.

Burg, J.-P., Balé, P., Brun, J.P., and Girardeau, J. (1987): Stretching lineation and transport direction in the Ibero-Armorican arc during the Silurian-Devonian collision. Geodinamica Acta 1, 71-87.

Carmigniani, L., Cocozza, T., Ghezzo, C., Pertusati, P.C., and Ricci, C.A. (1987): Structural Model of the Hercynian basement of Sardinia. Map at scale 1:500 000. Consiglio Nazionale delle Ricerche, Roma.

Carpenter, M.S.N., Peucat, J.J., and Pivette, B. (1982): Geochemical and geochronological characteristics of Palaeozoic volcanism in the Saint-Georges-sur-Loire synclinorium (S Armorican Massif). Evidence for pre-Hercynian tectonic evolution. Bull. Bur. Rech. Géol. Min. 1982, 63-79.

Cavet, P., Gruet, M., and Pillet, J. (1966): Sur la présence du Cambrien à Paradoxides à Cléré-sur-Layon (M.-et-L.) dans la nord-est du Bocage vendéen (Massif Armoricain). Comptes Rendus Acad. Sci., Paris 263, 1685-1688.

Centène, A., Ciszak, R., and Lacas, J. L. (1984): Un segment de l'orogène Varisque dans les massifs littoraux d'Oran (Algérie). Tectonique polyphasée et chronologie. Comptes Rendus Acad. Sci. Paris, sér. II, 298, 133-136.

Chaloupsky, J. (1988): Caledonian folding in the Bohemian Massif. In: Harris, A. L. and Fettes, D. J. (eds.), The Caledonian-Appalachian Orogen, Geol. Soc. Spec. Publ. 38, 493-498.

Chlupac, I. (1976): The Devonian and Lower Carboniferous in the Nepasice bore in East Bohemia. Vestnik Ustredniho ustavu geologickeho 41, 269-278.

Cocks, L.R.M. and Fortey, R.A. (1982): Faunal evidence for oceanic separation in the Palaeozoic of Britain. J. geol. Soc. 139, 465-478.

Cocks, L.R.M. and Fortey, R.A. (1988): Lower Palaeozoic facies and faunas around Gondwana. In: Audley-Charles, M.G. and Hallam, A. (eds.): Gondwana and Tethys, Geol. Soc. Spec. Publ. 37, 183-200.

Cogné, J. and Lefort, J.P. (1985): The Ligerian orogeny: a Proto-Variscan event related to the Siluro-Devonian evolution of the Tethys I ocean. In: Gee, D.G. and Sturt, B.A. (eds.), The Caledonide Orogen - Scandinavia and Related Areas, John Wiley and Sons Ltd., 1185-1193.

Costa, S. and Maluski, H. (1988): Datations par la méthode 39Ar/40Ar de matériel magmatique et métamorphique provenant de la forage de Couy-Sancerre (Cher, France). Programme G. P. F. Comptes Rendus Acad. Sci. Paris, sér. II, 306, 351-356.

Delfourt, J. and Gigot, C. (1985): Données stratigraphiques apportées par l'étude de conodontes du Dévono-Dinantian de la région de Bourbon-Lancy (Saone-et-Loire). Géologie de la France 1985(4), 31-340.

Den Tex, E. (1981): Basement evolution in the northern Hesperian Massif. A preliminary survey of results obtained by the Leiden Research Group. Leidse geol. Meded. 52, 1-21.

Domecka, K. and Opletal, M. (1980): Metamorphosed Upper Proterozoic Tholeiites of the N.E. Part of the Bohemian Massif. Krystalinikum 15, 55-80.

Don, J. (1990): Interferenz variszischer Strukturen in den Sudeten. In: Bericht, 5. Rundgespräch "Geodynamik des europäischen Variszikums" Rhenoherzynikum/Moravosilesikum, Zentralblatt Geol. Paläont. (in press).

Doubinger, J. and Eller, J.P. (1963): Découverte de Chitinozaires d'age Silurien dans les schistes de Steige (Vallée de l'Andlau - Vosges). Comptes Rendus Acad. Sci. Paris 256, 469-471.

Ducrot, J., Lancelot, J.R., and Marchand, J. (1983): Datation U-Pb sur zircons de l'éclogite de al Borie (Haut-Allier, France) et conséquences sur l'évolution ante-hercynienne de l'Europe occidentale. Earth Planet. Sci. Lett. 62, 385-394.

Duthou, J. L., Cantagrel, J. M, Didier, J., and Vialette, Y. (1984): Palaeozoic granitoids from the French Massif Central: age and origin studied by 87Rb-87Sr system. Phys. Earth. Planet. Interiors 35, 131-144.

Edel, J.B. (1987) Paleomagnetic evolution of the Central Massif (France) during the Carboniferous. Earth Planet. Sci. Lett. 82, 180-192.

Edwards, D. (1979): A Late Silurian Flora from the Lower Old Red Sandstone of South-West Dyfed. Palaeont. 22, 23-52.

El Hassani, A. and Zahraoui, M. (1984): Structure des terrains paléozoiques au Sud-Est de Rabat, Meseta Cotière - Maroc. Royaume du Maroc, Traveaux de l'Institut Scientifique, Univ. Mohammed V, Série Géologique et Géographie Physique 16, 1-20.

Ellenberger, F. and Santarelli, N. (1974): Les "Schistes X" de la Montagne Noire (Caroux - Espinouse). In: 2ème Réunion Annuelle des Sciences de la Terre, Pont-A-Mousson (Nancy), 161.

Exel, R. (1986): Sardinien - Geologie, Mineralogie, Lagerstätten, Bergbau. Samml. geol. Führer 80, Gebr. Borntraeger, Berlin/Stuttgart, 177 pp.

Farias, P., Gallastegui, G., Gonzales-Lodeiro, F., Marquinez, J., Martin-Parra, J., Martinez-Catalan, J.R., De Pablo-Macia, J.G. and Rodriguez-Fernandez, L.R. (1987): Aportaciones al conocimiento de la litoestratigrafia y estructura de Galicia Central. Mem. Fac. Cienc., Univ. do Porto 1, 411-431.

Feist, R. (1978): Das Altpaläozoikum Südfrankreichs. Österr. Akad. Wiss. Schriftenr. Erdwiss., Kommisionsband 3, 191-201.

Feybesse, J.L., Lardeaux, J.M., Johan, V., Tegyey, M., Dufour, E., Lemière, B. and Delfour, J. (1988): La série de la Brévenne (Massif central francais): une unité dévonienne charriée sur le complexe métamorphique des Monts du Lyonnais à la fin de la collision Varisque. Comptes Rendus Acad. Sci. Paris, sér II, 307, 991-996.

Fischer, G., Schreyer, W., Troll, G., Voll, G., and Hart, S.R. (1968): Hornblendealter aus dem ostbayrischen Grundgebirge. N. Jb. Mineral., Mh. 1968, 385-404.

Flöttmann, T. (1990): Aspects of structural an thermal evolution of the Central Schwarzwald Gneiss Complex. In: IGCP 233, Terranes in the circum-Atlantic Paleozoic orogens, Intern. Conf. Paleoz. Orog. Centr. Europe, Göttingen Aug. 1990, Abstracts (Posters).

Flöttmann, T. and Oncken, O. (1990): Constraints on the tectono-thermal evolution of the Mid-German Crystalline Rise west of the River Rhine. In: IGCP 233, Terranes in the circum-Atlantic Paleozoic orogens, Intern. Conf. Paleoz. Orog. Centr. Europe, Göttingen Aug. 1990, Abstracts.

Fluck, P. (1984): Neuere Ergebnisse zur Entwicklung des Varistikums in den Vogesen. Fortschr. Mineral. 62, Beih. 2, 37-47.

Franke, W. (1989): Variscan plate tectonics in Central Europe - current ideas and open questions. Tectonophysics 169, 221-228.

Frischbutter, A. (1988): Erzgebirgische Antiklinalzone. In: Exkursionsführer: Klassische geologische Gebiete in Mitteleuropa - Fundament und Deckgebirge. Akad. Wiss. DDR, Zentr. Inst. Phys. Erde, Potsdam, 105-116.

Fuchs, G. (1976): Zur Entwicklung der Böhmischen Masse. Jb. Geol. B.-A. 119, 45-61.

Fuchs, G. and Matura, A. (1976): Zur Geologie des Kristallins der südlichen Böhmischen Masse. Jb. Geol. B.-A. 119, 1-43.

Garcia Casquero, J.L., Boelrijk, N.A.I.M., Chacon, J. and Priem, H.N.A. (1985): Rb-Sr evidence for the presence of Ordovician granites in the deformed basement of the Badajoz-Cordoba belt, SW Spain. Geol. Rdsch. 74, 379-384.

Gardien, V., Lardeaux, J.-M., and Missier, M. (1988): Les péridotites du Monts de Lyonnais (M.C.F.): témoins privilégiés d'une subduction de lithosphère paléozoique. Comptes Rendus Acad. Sci., Paris, sér. II, 307, 1967-1972.

Gebauer, D. and Grünenfelder, M. (1979): U-Pb zircons and Rb-Sr mineral dating of eclogites and their country rocks. Example: Münchberg gneiss Massif, Northeast Bavaria. Earth Planet. Sci. Lett. 42, 35-44.

Gebauer, D., Quadt, A., Compston, W., Williams, I.S., and Grünenfelder, M. (1988): Archaean zircons in a retrograded, Caledonian eclogite of the Gotthard Massif (Central Alps, Switzerland). Schweiz. Mineral. Petrogr. Mitt. 68, 485-490.

Geyer, O.F. and Gwinner, M.P. (1986): Geologie von Baden-Württemberg. Schweizerbart, Stuttgart, 472 pp.

Gil Ibarguchi, J.L. and Ortega Gironez, E. (1985): Petrology, Structure and Geotectonic Implications of Glaucophane-bearing Eclogites and Related Rocks from the Malpica-Tuy (MT) Unit, Galicia, Northwest Spain. Chem. Geol. 50, 146-162.

Girardeau, J., Dubuisson, G., and Mercier, J.-C. C. (1986): Cinématique de mise en place des ophiolites et nappes cristallophylliennes du Limousin, Ouest du Massif central francais. Bull. Soc. Géol. France 2, 849-860.

Gottstein, O. (1970): K-Ar ages of granitic and ectinitic rocks of the Tepla anticlinorium. Vestnik Ustredniho ustavu geologickeho 45, 201-205.

Goujou, J.C., Guiraud, M., Burg, J.P., and Leyreloup, A. (1990): Two complementary P-T histories in the Variscan External Zone of western Vendée (southern Brittany, France). In: IGCP 233, Terranes in the circum-Atlantic Paleozoic orogens, Intern. Conf. Paleoz. Orog. Centr. Europe, Göttingen Aug. 1990, Abstracts.

Grauert, B. and Arnold, A. (1968): Deutung diskordanter Zirkonalter der Silvrettadecke und des Gotthardmassivs (Schweizer Alpen). Contrib. Mineral. Petrol. 20, 34-56.

Greiling, L. (1966): Sedimentation und Tektonik im Paläozoikum des Frankenwaldes. Erl. geol. Abh. 63, 3-60.

Gueirard, S., Waterlot, G., Gherzi, A., and Samat, M. (1970): Sur l'age Llandovérien supérieur à Tarannonien inférieur des schistes à graptolites du Fenouillet, massif des Maures. Bull. Soc. Géol. France 2, 195-199.

Guiraud, M., Burg, J.-P., and Powell, R. (1987): Evidence for a Variscan suture zone in the Vendée, France: a petrological study of blueschist facies rocks from Bois de Cené. J. metamorphic Geol. 5, 225-237.

Gunia, T. (1983): Microflora from the fine-grained Paragneisses of Jugowice vicinity (Sowie Mts., Sudetes). Geologia Sudetica 18, 7-17.

Hameurt, J. (1976): Le déroulement des événements varisques et de leurs antécédents dans les Vosges. Nova Acta Leopoldina 45, 193-199.

Hammerschmidt, K. and Franz, G. (1988): Eklogitbildung und Deformation der Münchberger Gneismasse: Petrologische und isotopengeochemische Daten. Fortschr. Mineral. 66, Beih. 1, 50.

Havlicek, V. (1989): Climatic changes and development of benthic communities through the Mediterranean Ordovician. Sbornik Geologickych Ved, Geol. 44, 79-116.

Hermes, O.D., Ballard, R.D., and Banks, P.O. (1978): Upper Ordovician peralkalic granites from the Gulf of Maine. Geol. Soc. Amer. Bull. 89, 1761-1774.

Hirschmann, G. and Okrusch, M. (1988): Spessart-Kristallin und Ruhlaer Kristallin als Bestandteile der Mitteldeutschen Kristallinzone - ein Vergleich. N. Jb. Geol. Paläont., Abh. 177, 1-39.

Holder, M.T. and Leveridge, B.E. (1986): A mode for the tectonic evolution of south Cornwall. J. geol. Soc. 143, 125-135.

Hollard, H., Michard, A., Jenny, P., Hoepffner, C., and Willifert, S. (1982): Stratigraphie du Primaire de Mechra-Ben-Abbou, Rehamna. Notes Mem. Serv. géol. Maroc 303, 13-34.

Iglesias, M., Ribeiro, M.L., and Ribeiro, A. (1983): La interpretation aloctonista de la estructura del Noroeste peninsular. In: Geologia de Espana 1, Inst. Geol. Minero Esp. Madrid, 459-467.

Iglesias Ponce de Leon, M. and Ribeiro, A. (1981): Position stratigraphique de la formation "Ollo de Sapo" dans la région de Zamora (Espagne) - Miranda do Douro (Portugal). Comunic. Serv. Geol. Port. 67, 141-146.

Julivert, M., Truyols, J., and Verges, J. (1983): El Devonico en el Macizo Iberico. In: Geologia de Espana, 1, Inst. Geol. Minero Esp. Madrid, 265-311.

Kastl, E. and Tonika, J. (1984): The Marianske Lazne Metaophiolite Complex (West Bohemia). Krystalinikum 17, 59-76.

Kemnitz, H. (1987): Beitrag zur Lithologie, Deformation und Metamorphose der Saydaer Struktur (Osterzgebirge). Veröff. Zentr. Inst. Phys. Erde, Potsdam 94, 89 pp.

Kirsch, H., Kober, B., and Lippolt, H.J. (1988): Age of intrusion and rapid cooling of the Frankenstein gabbro (Odenwald, SW-Germany) evidenced by 40Ar/39Ar and single-zirkon 207Pb/206Pb measurements. Geol. Rdsch. 77, 693-711.

Kneuper, G. (1976): Regionalgeologische Folgerungen aus der Bohrung Saar 1. In: die Tiefbohrung Saar 1. Geol. Jb. A 27, 499-510.

Kober, B., Hradetzky, H., and Lippolt, H.J. (1986): Radiogenblei-Evaporationsstudien an einzelnen Zirkonkristallen zur präherzynischen Entwicklung des Grundgebirges im Zentralschwarzwald, SW Deutschland. Fortschr. Mineral. 64, Beih. 1, 81.

Köhler, H., Propach, G., and Troll, G. (1989): Exkursion zur Geologie, Petrographie und Geochronologie des NE-bayerischen Grundgebirges. European J. Mineral. 1, Beih. 2, 84 pp.

Konzalova, M. (1980): Zu der mikropaläontologischen Erforschung graphitischer Gesteine im Südteil der Böhmischen Masse. Vestnik Ustredniho ustavu geologickeho 51, 233-236.

Kossmat, F. (1927): Gliederung des variscischen Gebirgsbaues. Abh. sächs. geol. L.-A. 1, 1-39.

Kramer, W. and Werner, C.-D. (1989): Earth's Crust - Structure, Evolution, Metallogeny, Projects 2 and 7.3: Prevariscan Mafic Rocks in the Saxothuringian zone. Acad. Sci. GDR, Centr. Inst. Phys. Earth, Potsdam, 108 pp.

Krentz, O. (1985): Rb/Sr-Altersbestimmungen an Parametamorphiten des westlichen Erzgebirgsantiklinoriums, DDR. Z. geol. Wiss. 13, 443-462.

Kreuzer, H., Seidel, E., Schüssler, U., Okrusch, M., Lenz, K.L., and Ratschka, H. (1989): K-Ar geochronology of different tectonic units at the northwestern margin of the Bohemian Massif. Tectonophysics 157, 149-178.

Kriz, J. and Paris, F. (1982): Ludlovian, Pridolian and Lochkovian in la Meignanne (Massif Armoricain): Biostratigraphy and correlations based on Bivalvia and chitinozoa. Géobios 5, 391-421.

Krohe, A. and Eisbacher, G.H. (1988): Oblique crustal detachment in the Variscan Schwarzwald, southwestern Germany. Geol. Rdsch. 77, 25-43.

Krylova, M.D. and Glebovitsky, V.A. (1985): On regional metamorphosm, primary nature and geochemistry of the Moldanubian rocks, the Bohemian Massif. Krystalinikum 18, 29-51.

Kurze, M., Pilot, J., and Kaiser, G. (1982): Erste physikalische Altersdatierungen an Metamorphiten des Zwischengebirges von Frankenberg-Hainichen (Bezirk Karl-Marx-Stadt). Z. geol. Wiss. 10, 531-535.

Lancelot, J. R., Allegre, A., and Iglesias Ponce de Leon, M. (1985): Outline of Upper Precambrian and Lower Paleozoic evolution of the Iberian Peninsula according to U-Pb dating of zircons. Earth Planet. Sci. Lett. 74, 325-337.

Ledru, P., Marot, A., Herrouin, Y. (1986): Le synclinorium de Saint-Georges-sur-Loire: une unité Ligérienne charriée sur le domaine centre Armoricain. Découverte de métabasite a glaucophane sur la bordure sud de cette unité. Comptes Rendus Acad. Sci. Paris, sér II, 303, 963-968.

Le Metour, J. and Bernard-Griffiths, J. (1979): Age (limite Ordovicien/Silurien) de mise en place du massif hypovolcanique de Thouars (Massif Vendéen). Bull. Bur. Rech. Géol. Min. 1979, 365-371.

Lenz, H. and Müller, P. (1976): Radiometrische Altersbestimmungen am Kristallin der Bohrung Saar 1. In: die Tiefbohrung Saar 1, Geol. Jb. A 27, 429-432.

Lippolt, H. J. (1986): Nachweis altpaläozoischer Primäralter (Rb-Sr) und karbonischer Abkühlungsalter (K-Ar) der Muskovit-Biotit Gneise des Spessarts und der Biotitgneise des Böllsteiner Odenwaldes. Geol. Rdsch. 75, 569-589.

Maass, R. (1988): Die Südvogesen in variszischer Zeit. N. Jb. Geol. Paläont. Mh. 1988, 611-638.

Martinez-Catalan, J.R. (1990): A non-cylindrical model for the northwest Iberian allochthonous terranes and their equivalents in the Hercynian belt of Western Europe. Tectonophysics 179, 253-272.

Martinez-Catalan, J.R., Klein, E., De Pablo-Macia, J.G., and Gonzales-Lodeiro, F. (1984): El Complejo de Ordenes: subdivision, descripcion y discusion sobre su origen. Cuad. Lab. Xeol. Laxe 7, 139-210.

Massa, D. and Nicol-Lejal, A. (1971): Le Dévonien à Lycophytes de la Libye sud-occidentale, conséquences paléophytogéographiques. Comptes Rendus Acad. Sci. Paris, D 273, 1182-1185.

Matte, P. (1968): La structure de la virgation hercynienne de Galice (Espagne). Géologie Alpine 44, 155-280.

Matte, P. (1986): Tectonics and plate tectonics model of the Variscan belt of Europe. Tectonophysics 126, 329-374.

Mehl, J. (1989): Biostratigraphische Datierung des nordschwarzwälder metamorphen Paläozoikums mit Hilfe moderner Röntgenuntersuchungen. In: Bericht, 4. Rundgespräch "Geodynamik des europäischen Variszikums", Saxothuringikum, Zenralblatt. Geol. Paläont. 1989, 378-379.

Menot, R.-P. (1988): The geology of the Belledonne Massif: an overview (External crystalline massifs of the Western Alps). Schweiz. Mineral. Petrogr. Mitt. 68, 531-542.

Menot, R.-P., Peucat, J. J., Scarenzi, D., and Piboule, M. (1988): 496 My age of plagiogranites in the Chamrousse ophiolitic complex (external crystalline massifs in the French Alps): evidence of Lower Paleozoic oceanization. Earth Planet. Sci. Lett. 88, 82-92.

Michard, A. (1978): Brève description du ségment Calédono-Hercynien du Maroc. Geol. Surv. Canada, Paper 78-13, 213-230.

Moreau-Benoit, A. and Dubreuil, M. (1987): Confirmation du Silurien et découverte du Dévonien inférieur par la palynoplanctologie dans les schistes et grès de la terminaison orientale du bassin d'Ancenis (sud-est du Massif armoricain). Géologie de la France 1987, 37-54.

Moreau-Benoit, A. and Massa, D. (1988): Palynologie et stratigraphie d'une coupe-type du dévonien inférieur du Sahara oriental (Bassin de Rhadames, Libye). Comptes Rendus Acad. Sci., Paris sér. II, 306, 451-454.

Müller, H.D. (1989): Geochemistry of metasediments in the Hercynian and pre-Hercynian crust of the Schwarzwald, the Vosges and Northern Switzerland. Tectonophysics 157, 97-108.

Müller-Sohnius, D., v. Drach, V., Horn, P., and Köhler, H. (1987): Altersbestimmungen an der Münchberger Gneismasse, Nordost-Bayern. Neues Jahrb. Mineralogie Abh., 156, 175-206.

Munha, J., Mata, J., and Ribeiro, M.L. (1990): Pre-orogenic events in the Iberian Variscides: Magmatism, thermal regime and basement isotopic resetting. In: IGCP 233, Terranes in the circum-Atlantic Paleozoic orogens, Intern. Conf. Paleoz. Orog. Centr. Europe, Göttingen Aug. 1990, Abstracts.

Narebski, W. (1980): Paleotectonic setting of the Circum-Karkonosze Lower Palaeozoic Spilite-Keratophyre Suites based on Geochemistry of Iron Group Elements. Ann. Soc. Geol. Pol. 50, 3-25.

Neubauer, F. (1988): The Variscan orogeny in the Austroalpine and Southalpine domains of the Eastern Alps. Schweiz. Mineral. Petrogr. Mitt. 68, 339-349.

Oberc, J. (1980): Early to Middle Variscan development of the west Sudetes. Acta geol. pol. 30, 27-51.

Oczlon, M.S. (1989): Fazies und Fauna im Silur und Devon des "Valle" (Provinz Sevilla, SW-Spanien). Mit einer überregionalen Betrachtung der Mitteldevon-Schichtlücke. Unpubl. Diploma Thesis, Univ. Heidelberg, 86 pp.

Oczlon, M.S. (1990): Ocean currents and unconformities: The North Gondwana Middle Devonian. Geology 18, 509-512.

Oczlon, M.S. (1991): The Origin of Exotic Variscan rocks in the Rhenohercynian Zone of Germany. Geol. Rundsch. (submitted).

Okrusch, M., Müller, R., and El Shazly, S. (1985): Die Amphibolite, Kalksilikatgesteine und Hornblendegneise der Alzenauer Gneis-Serie am Nordwest-Spessart. Geol. Bavaria 87, 5-37.

Okrusch, M. and Richter, P. (1986): Orthogneisses of the Spessart cristalline complex, Northwest Bavaria: Indicators of the geotectonic environment? Geol. Rdsch. 75, 555-568.

Owodenko, B. (1976): Primaire indeterminé et Silurien. In: Le Bassin houiller de Jereda (Maroc oriental). Notes Mem. Serv. Géol. Maroc, 207, 13-26.

Pacltova, B. (1980): Further micropaleontological data for the Paleozoic age of Moldanubian carbonate rocks. Casopiis mineral. geol. 25, 275-279.

Pacltova, B. (1986): Palynology of metamorphic rocks (methodological study). Rev. Palaeobot. Palynol. 48, 347-356.

Paris, F., Le Pochat, G., and Pelhate, A. (1988): Le socle paléozoique nord-aquitain: caractéristiques principales et implications géodynamiques. Comptes Rendus Acad. Sci. Paris, sér II, 306, 597-603.

Paris, F. and Robardet, M. (1990): Early Palaeozoic palaeobiogeography of the Variscan regions. Tectonophysics 177, 193-213.

Perdigao, J. C. (1977): O Devónico de S. Felix de Laundos. Comunic. Serv. Geol. Port. 61, 13-32.

Peucat, J.-J., Vidal, P., Godard, G., and Postaire, B. (1982): Precambrian U-Pb zircon ages in eclogites and garnet pyroxentites from south Brittany (France): An old oceanic crust in the west European Hercynian belt? Earth Planet. Sci. Lett. 60, 70-78.

Pflug, H.D. and Prössl, K.F. (1989): Palynology in Gneiss - Results from the Continental Deep Drilling Programm. Naturwiss. 76, 565-567.

Pflug, H.D. and Reitz, E. (1987): Palynology in Metamorphic Rocks: Indication of Early Land Plants. Naturwiss. 74, 386-387.

Pin, C. (1990): Variscan oceans: Ages, origins and geodynamic implications inferred from geochemical and radiometric data. Tectonophysics 177, 215-227.

Pin, C and Carme, A. (1987): Sm-Nd isotopic study of 500 MA old oceanic crust in the Variscan belt of western Europe: the Chamrousse ophiolite complex, Western Alps (France). Contrib. Mineral. Petrol. 96, 406-413.

Pin, C. and Carme, F. (1988): Ecailles de matériaux d'origine océanique dans le charriage hercynien de la "Ligne des Klippes", Vosges méridionales, France. Comptes Rendus Acad. Sci. Paris 306, sér. II, 217-222.

Pin, C. and Lancelot, J.R. (1982): U-Pb dating of an early Paleozoic bimodal magmatism in the French Massif central and of its further metamorphic evolution. Contrib. Mineral. Petrol. 79, 1-12.

Pin, C. and Peucat, J.J. (1986): Ages des épisodes de métamorphisme paléozoique dans le Massif Central et le Massif Armoricain. Bull. Soc. Géol. France 2, 461-470.

Pin, C. and Piboule, M. (1988): Age dévonien supérieur de la série calco-alcaline de la ceinture basique du Lévezou (Rouergue). Un exemple de complexe leptyno-amphbolique composite. Bull. Soc. Géol. France 4, 261-265.

Pin, C. and Vielzeuf, D. (1988): Les granulites de haute pression d'Europe moyenne témoins d'une subduction éo-hercynienne. Implications sur l'origine des groupes leptyno-amphiboliques. Bull. Soc. Géol. France 4, 13-20.

Pinto, M.S. (1984): Granitoides Caledonicos e Hercinicos na Zona de Ossa-Morena (Portugal) - Nota sobre Aspectos Geocronologicos. Mem. Not., Publ. Mus. Lab. Mineral. Geol. Univ. Coimbra 97, 81-94.

Priem, H.N.A., Boelrijk, N.A.I.M., Verschure, R.H., Hebeda, E.H. and Verdurmen, E.A.T. (1970): Dating events of acid plutonism through the Paleozoic of the Western Iberian Peninsula. Ecl. Geol. Helv. 63, 255-274.

Priem, H.N.A., Hebeda, E.H., Verdurmen, E.A.T. and Verschure, R.H. (1972): Upper Ordovician-lower Silurian acidic magmatism in the prehercynian basement of Western Galicia, NW Spain. Progr. Rep., Lab. Isot. Geol. Amsterdam, 123-127.

Pruvost, P. (1949): Les mers et les terres de Bretagne aux temps paléozoiques. Annales Hébert et Haug. 7, 345-360.

Quadt, A. and Gebauer, D. (1990): Isotopic ages from the Saxonian Granulite Massif (GDR) and their geodynamic significance. In: IGCP 233, Terranes in the circum-Atlantic Paleozoic orogens, Intern. Conf. Paleoz. Orog. Centr. Europe, Göttingen Aug. 1990, Abstracts.

Racheboeuf, P.R. and Robardet, M. (1986): Le Pridoli et le Dévonien inférieur de la zone d'Ossa-Morena (Sud-Ouest de la Péninsule ibérique). Etude des brachiopodes. Geol. et Palaeont. 20, 11-37.

Reitz, E. (1987): Silurische Sporen aus einem granatführenden Glimmerschiefer des Vorspessarts, NW-Bayern. N. Jb. Geol. Paläontol. Mh. 1987, 699-704.

Reitz, E. and Höll, R. (1988): Jungproterozoische Mikrofossilien aus der Habachformation in den mittleren Hohen Tauern und dem nordostbayerischen Grundgebirge. Jb. Geol. B.-A. 131, 329-340.

Reitz, E. and Wickert, F. (1988): Upper Proterozoic microfossils in low-grade phyllites of the Brévenne-Unit, NE Massif Central (France). Comptes Rendus Acad. Sci. Paris, sér. II, 307, 1717-1721.

Reitz, E. and Wickert, F. (1989): Late Cambrian to Early Ordovician acritarchs from the Villé Unit, Northern Vosges Mountains (France). N. Jb. Geol. Paläontol. Mh. 1989, 375-384.

Ribeiro, A. (1974): Contribution à l'étude tectonique de Tras-Os-Montes Oriental. Mem. Serv. Geol. Portugal 24, 167 pp.

Ribeiro, A. and Ribeiro, L. (1974): Signification paléogeographique et tectonique de la présence de galets de roches métamorphiques dans un flysch d'age dévonien du Tras-Os-Montes Oriental (Nordest du Portugal). Comptes Rendus Acad. Sci. Paris D 278, 3161-3163.

Ribeiro, M.L. (1987): Petrogenesis of early paleozoic peralkaline rhyolites form the Macedo de Cavaleiros region (NE Portugal). Geol. Rdsch. 76, 147-168.

Robardet, M. and Doré, F. (1988): The Late Ordovician Diamictic Formations from southwestern Europe: North-Gondwana Glaciomarine deposits. Palaeogeogr. Palaeoclim. Palaeoecol. 66, 19-31.

Santallier, D. (1983): Main metamorphic features of the Paleozoic orogen in France. In: Schenk, P. E. (ed.), Regional trends in the geology of the Appalachian - Caledonian - Hercynide - Mauretanide orogen, Reidel Publishing Company, 263-274.

Schleicher, H. and Kramm, U. (1986): Altersbestimmungen nach der U/Pb-Methode an mittel- bis hochdruckmetamorphen Gesteinen des Schwarzwaldes. In: 76. Jahrestagung Geol. Ver. Gießen, Abstr., 69.

Schoell, M., Lenz, H., and Harre, W. (1973): Das Alter der Hauptmetamorphose des Eckergneises im Harz aufgrund von Rb-Sr Datierungen. Geol. Jb. A 9, 89-95.

Schüssler, U. (1990): Petrographie, Geochemie und Metamorphose von Metabasiten im KTB-Zielgebiet Oberpfalz. Geologica Bavarica, 95, 5-99.

Schüssler, U., Kreuzer, H., Vejnar, Z., Okrusch, M., Seidel, E., Kopecky, L.jr., and Patzak, M. (1989a): Geochemische Untersuchungen und K-Ar Datierungen in der Zone Tepla-Taus. Europ. J. Mineral. 1, Beih. 1, 175.

Schüssler, U., Oppermann, U., Kreuzer, H., Seidel, E., Okrusch, M., Lenz, K.L., and Raschka, H. (1986): Zur Altersstellung des ostbayrischen Kristallins. Ergebnisse neuer K-Ar Datierungen. Geol. Bavarica 89, 21-47.

Schüssler, U., Richter, P., and Okrusch, M. (1989b): Metabasites from the KTB Oberpfalz target area, Bavaria - geochemical characteristics and examples of mobile behaviour of "immobile" elements. Tectonophysics 157, 135-148.

Schwebel, L. (1983): Analyse pétrostructurale des écailles de la région de Lalaye-Colroy-Climont (Vosges). Les mécanismes de la déformation ductile. Rapport D.E.A. Univ. Strasbourg, 32 pp.

Sengör, A.M.C. (1985): Die Alpiden und die Kimmeriden: Die verdoppelte Geschichte der Tethys. Geol. Rdsch. 74, 181-213.

Söllner, F. Köhler, H., and Müller-Sohnius, D. (1981): Rb/Sr Altersbestimmungen an Gesteinen der Münchberger Gneismasse, NE-Bayern. Teil 1: Gesamtgesteindatierungen. N. Jb. Mineral., Abh. 141, 90-112.

Storch, P. (1990): Upper Ordovician - lower Silurian sequences of the Bohemian Massif, central Europe. Geological Magazine 127, 225-239.

Stosch, H.G. and Lugmair, G.W. (1987): Geochronology and geochemistry of eclogites from the Münchberg Gneiss Massif, FRG. Terra Cognita 7, 163.

Ters, M. (1979): Les synclinoriums paléozoiques et le Précambrien sur la facade occidentale du Massif vendéen. Stratigraphie et structure. Bull. Bur. Rech. Géol. Min. 1979, 293-301.

Todt, W. (1979): U-Pb-Datierungen an Zirkonen des kristallinen Odenwaldes. Fortschr. Mineral. 57, 153-154.

Tollmann, A. (1985): Das Ausmaß des Variszischen Deckenbaus im Moldanubikum. Krystalinikum 18, 117-132.

Tomek, C., Ibrmajer, I., and Cidlinsky, K. (1987): Variscan Thrust Tectonics in the Bohemian Massif - First Results Revealed by Reflection Seismology. In: Vogel, A., Miller, H., and Greiling, R. (eds.), The Rhenish Massif, Structure, Evolution, Mineral Deposits, and present Geodynamics. Vieweg, Wiesbaden, 138-152.

Trusheim, F. (1964): Über den Untergrund Frankens. Ergebnisse von Tiefbohrungen in Franken und Nachbargebieten. Geol. Bavarica, 54, 92 pp.

Van Breemen, O., Aftalion, M., Bowes, D.R., Dudek, Z., Misar, Z., Povondra, P., and Vrana, S. (1982): Geochronological studies of the Bohemian Massif, Czechoslovakia, and their significance in the evolution of Central Europe. Trans. R. Soc. Edinburgh, Earth Sci. 73, 89-108.

Van Breemen, O., Bowes, D.R., Aftalion, M., and Zelazniewicz, A. (1988): Devonian tectonothermal activity in the Sowie Gory gneissic block, Sudetes, southwestern Poland: Evidence from Rb-Sr and U-Pb isotopic studies. Ann. Soc. Geol. Pol. 58, 3-19.

Van der Meer Mohr, C.G. (1975): The Paleozoic strata near Moeche in Galicia, NW-Spain. Leidse geol. Meded. 49, 487-497.

Vauchez, A. and Bufalo, M. (1988): Charriage crustal, anatexie, et decrochements dans les Maures oriental (Var, France) au cours de l'orogenèse varisque. Geol. Rdsch. 77, 45-62.

Vejnar, Z. (1990): The contact aureole of the Kdyne massif, south-west Bohemia. Sbornik Geologickych Ved, Geologie 45, 9-35.

Vezat, R. (1988): Les formations métamorphiques de Zivaco (Corse centrale) et leur signification dans le cadre de l'orogène varisque en Méditerranée. Comptes Rendus Acad. Sci., Paris, sér. II, 306, 725-729.

Vidal, P. (1973): Premières données géochronologiques sur les granites hercyniens du sud du Massif armoricain. Bull. Soc. Géol. France 15, 239-245.

Vinx, R. and Schlüter, J. (1990): Der Eckergneis: Cadomisches Basement im Rhenoherzynicum? In: Bericht, 5. Rundgespräch "Geodynamik des europäischen Variszikums", Zentralbl. Geol. Paläont., Teil 1, 381-382.

Von Drach, V. (1990): Geochronologie und Isotopengeologie an Proben der KTB-Vorbohrung. In: Emmermann, R. and Giese, P. (eds.), KTB-Report 90-4, Niedersächs. Landesamt Bodenf., 96-109.

Weber, K. (1984): Variscan events: early Palaeozoic continental rift metamorphism and late Palaeozoic crustal shortening. In: Hutton, D.H.W. and Sanderson, D.J. (eds.), Variscan Tectonics of the North Atlantic Region, Geol. Soc. Spec. Publ. 14, 3-22.

Werchau, A. and Schleicher, H. (1989): Geochronologische Untersuchungen an Eklogitamphiboliten des Schwarzwaldes. Europ. J. Mineral. 1, Beih. 1, 198.

Wetzel, K. (1990): Über die Herkunft der Granite des Sächsischen Granulitgebirges. Z. geol. Wiss. 18, 743-748.

Wetzel, K. and Gerstenberger, H. (1989): Geochemical and Isotope Constraints on the Evolution of the Saxon Granulite Massif. Z. geol. Wiss. 17, 669-683.

Wiedemann, R. and Kaiser, G. (1986): Neue K/Ar-Datierungen an Metabasiten des Westerzgebirges. Z. geol. Wiss. 14, 607-611.

Wojciechowska, I. (1986): Metabasites in the NW part of the Snieznik metamorphic unit (Klodzko area, Sudetes, Poland). Geol. Rdsch. 75, 585-593.

Wolf, M. and Kaiser, G. (1986): K-Ar-Alter von pegmatoiden Bildungen und Pegmatiten im mittleren Erzgebirge. Z. geol. Wiss. 14, 27-35.

Zartman, R.E. and Naylor, R.S. (1984): Structural implications of some radiometric ages of igneous rocks in southeastern New England. Geol. Soc. Amer. Bull. 95, 522-539.

Zeitz, U. and Nollau, G. (1984): Ordoviz und Silur im Sil-Synklinorium südlich Ponferrada (Prov. León, NW-Spanien). Z. dt. geol. Ges. 135, 211-222.

An Outline of Evolution of the Late Devonian Munster Basin, South-West Ireland (Extended Abstract)

E.A.Williams,/ M.Ford[1]/ H.E.Edwards[2]/ M.J.O´Sullivan
Department of Geology, University College, Cork, Ireland
Present addresses:
1 Geologisches Institut, Sonneggstr.5, ETH-Zentrum, CH-8092 Zurich, Switzerland
2 Department of Geologie, University of Keele, Keele, Staffordshire ST5 5BG, UK

The Variscides of south-west Ireland are the most westerly extension of the Rhenohercynian Zone of the Variscan Orogen in Europe. The central sector of the Irish Variscan fold belt (Zone 1 of Cooper et al. 1986) accommodated ca. 52 % NNW-directed shortening (Ford 1987), mainly by early ductile cleavage processes, buckle folding and late thrusting (Cooper et al. 1984, 1986). Much of the deformed Upper Palaeozoic succession of south-west Ireland, accumulated in the Munster Basin (Capewell 1965), a late Middle to latest Devonian half-graben (Naylor & Jones 1967; Graham 1983; Williams et al. 1989). This basin contains an Upper Old Red Sandstone (ORS) fluviatile (Gardiner & MacCarthy 1981; Graham 1983) and rare aeolianite magnafacies (Carruthers 1987), with exceptionally rare acid and basic volcanics (Penney 1978; Avison 1984). The succession is disposed in a sub-Caledonide trending basin, axially-thickening to the WSW, which culminates in a depocentre accommodating >5,5 km of sediment (Fig. 1). The basin is defined by the 1 km isopac (Fig. 1). The oldest rocks exposed within the basin are late Givetian to early Frasnian (Clayton & Graham 1974; Higgs & Russel 1981). Palinspastic restoration of the fold and thrust belt now representing the Munster Basin (Williams et al. 1989) reveals an original minimum surface area of 38.000 km^2 (Fig. 1).

BASIN FILL CHARACTERISTICS

The alluvial stratigraphy of the Munster Basin lacks basinwide formations and uniform vertical trends, but consists of geometrically complex and areally restricted spreads of coarse (usually sandy) sediment in a matrix of fine-grained clastics (see Williams et al. 1989).

Northern margin

The Munster Basin is predominantly filled by sediment derived from the north, and which prograded southwards across the northern margin i.e. a lateral-dispersal pattern (Figs. 1 & 2). The margin can be considered in two regions. The western region comprises two, co-areal (stratigraphically superimposed) wet-type fan distributary systems of original radius c. 110 km: (i) the older **Chloritic-Sandstone-Slaheny-Gortanimill System**, Fig. 2) and (ii) the Gun Point System (Williams et al. 1989). These coarse-grained fluvial systems show proximal-distal and axial-lateral facies and thickness contrasts which reflect a downcurrent decline in palaeochannel dimensions, changes in fluvial style and antecedent location of the trunk alluvial conduits. The coarse-grained lithosomes have diachronous upper and lower contacts with their fine-grained facies matrix, indicating progradation and decay of the dispersal systems and, in the case of the earlier, **transverse** migration (Fig. 2). The eastern region (Galtee Mountains) saw a more restricted and localised lateral dispersal of clastics southwards into the basin. Basal, fault-related alluvial fanglomerates and interacting erg sandstones (Carruthers 1987) are succeeded by four formations which evolve upwards from a fine-grained "floodplain" to gravelly and sandy braidplain environments, which fine in the palaeoflow direction (Carruthers 1987). These systems **overlap** the northern margin in this region (Carruthers 1987).

The southwestern zone

In this area a condensed and distinct alluvial stratigraphy, sourced from the west (Graham & Reilly 1972), is considered to represent (Williams et al. 1989) part of a large distributary system (the **Sherkin System**, Fig. 2) of low sinuosity braided streams (Graham & Reilly 1972; Graham 1983). This system is demarcated by a WSW-ENE trending line (DCF', Fig. 2) through Sheeps Head. The overlying Castlehaven Formation (siltstones and fine-grade sandstones), which dominates north and east of station 9 in Figure 1, is considered to represent the "decay phase" of the Sherkin System.

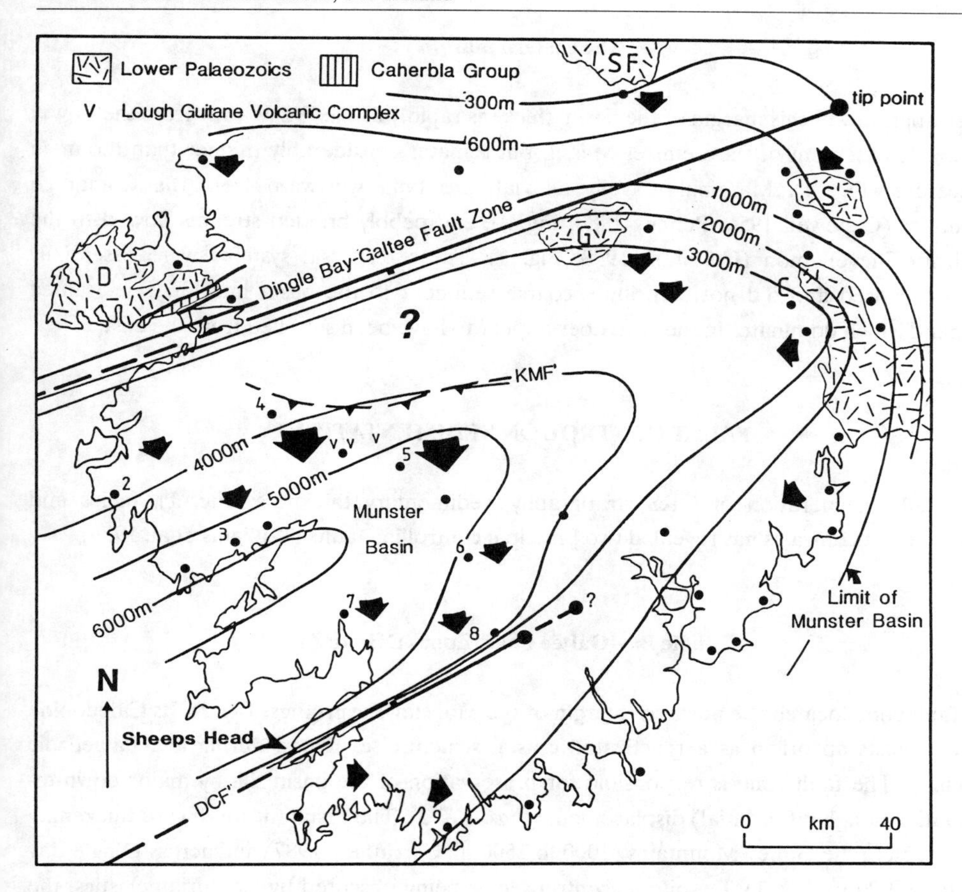

FIGURE 1
Palinspastic restoration of the Munster Basin, showing Upper Old Red Sandstone isopachs, bounding fault zones and generalized palaeoflow. Corrected for c. 50 % NNW-directed shortening. Key: KMF', restored trace of the Killarney-Mallow Fault (part of the northern boundary of the Rhenohercynian Zone; DCF', Dunmanus-Castletown Fault; D, Dingle Peninsula; G, Galtee Mountains; SF, Slive Felim; S, Slievenamon; C, Comeragh Mountains.

The eastern margin

The stratigraphy at this margin of the basin thickens rapidly to the WSW away from the Lower Palaeozoic basement of the Leinster Massif, but remains considerably thinner than the main depocentre (Fig. 1). Marginal gravelly alluvial fans built westward from the Comeragh Mountains (Capewell 1956; Penney 1980; Fig. 2), and pebbly braided streams flowed to the SW from Slievenamon (Colthurst 1978; Fig. 1). These dispersal systems were essentially margin-localised and did not contribute coarse sediment to the basin interior; fine-grained sediment, which dominates in the east, does appear to have been supplied from the east.

FAULT CONTROL ON SEDIMENTATION

Integrated consideration of ORS stratigraphy, sedimentary thickness (Fig. 1), facies and palaeocurrent contrasts has revealed two principal controlling faults (Williams et al. 1989).

Dingle Bay-Galtee Fault Zone (DB-GFZ)

This fault zone located the northern margin of the Munster Basin (Figs. 1 & 2). Its Caledonide trend suggests an origin as a reactivated crustal structure, exploited during late Palaeozoic stretching. The fault zone is responsible for preservation of the basin fill, by major down-to-the-south (though differential) displacement (Figs. 1 & 2). The resultant southward thickening can be seen in the Galtee Mountains (1000 to 2500 m, Carruthers 1987) and across Dingle Bay (850 to >5000 m, Fig. 1). Despite its central sector being obscured by Namurian clastics, the DB-GFZ is extrapolated between these areas, as it is untenable to regard the present-day Killarney-Mallow Fault (Fig. 1) as the faulted margin of the basin (Williams et al. 1990).

The footwall block underwent variable subsidence along its length, to judge from the preserved footwall successions. In the west, the footwall was a site of non-deposition for most of the subsidence history of the Munster Basin. Uplift and back-tilting were negligible, and the fault trace lacked all stratigraphical levels. Punctuated (discrete) episodes of footwall uplift, however, are recorded by the incorporation of thin proximal alluvial pebbly sequences of footwall provenance in the basin fill (Fig. 2). These episodes occurred in the west where maximum fault displacement occurred.

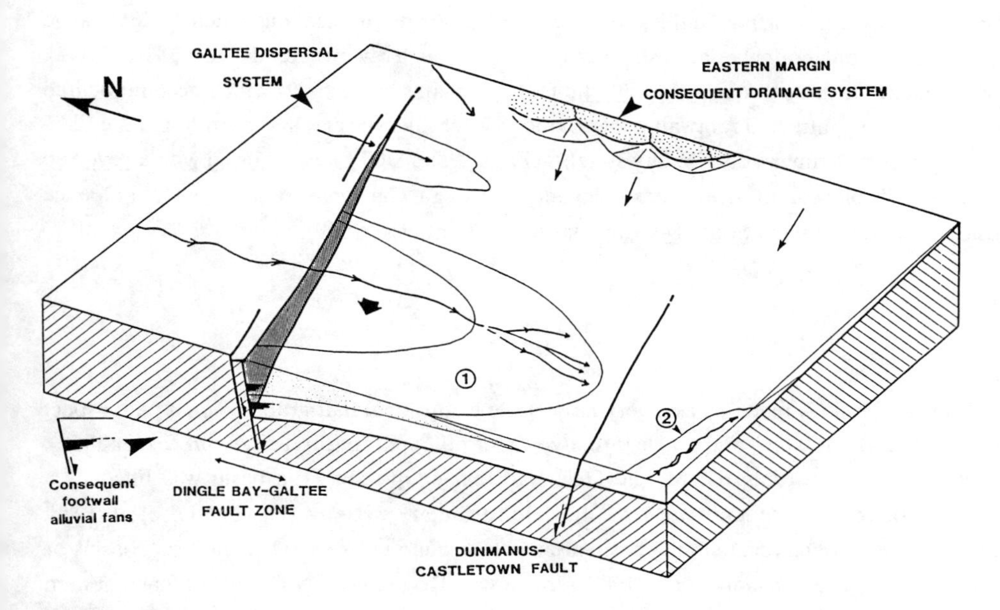

FIGURE 2
Block model of the Munster Basin showing ?Frasnian basin-fill geometry and palaeodispersal patterns. (1) Chloritic Sandstone-Slaheny-Gortanimill System; (2) Sherkin System (oblique-hangingwall dispersal).

Dunmanus-Castletown Fault (DCF')

This is recognised by northward thickening of the ORS, stratigraphical mismatch, facies change and orthogonal palaeoflow relationship across a line through Sheeps Head (DCF', Figs. 1 & 2). Displacement on this sub-parallel antithetic fault (Williams et al. 1989) which accommodated further stretching, affected footwall palaeoslope and subsidence rate, and partitioned the SSE- and E-transporting fluvial distributary systems (Fig. 2). The DCF' was re-tilised with a reversed extensional displacement from latest Devonian through Carboniferous times, to influence subsidence in the younger South Munster Basin (Williams et al. 1989).

SUMMARY

The Munster Basin is revealed as a very large-scale extensional half-graben with a plan aspect ratio of 1.5 (240 x 160 km) and a structure atypical of pull-apart basins (cf. Aydin & Nur 1982). Following initiation during the Middle Devonian (early Givetian), the basin was filled by a series of probable antecedent (wet-type) fluvial distributary systems (e.g. Leeder et al. 1988; Haughton 1989) which resulted in a basin architecture quite unlike smaller tilt-block basins or half-grabens (Leeder & Gawthorpe 1987; Frostick & Reid 1987). No axial drainage system developed, sediment being supplied to the east of the basin (a potential axial entry site) by consequent fans. Facies relationship across the bounding fault zones rule out strike-slip movements, but indicate a N-S to NNW-SSE palaeoextension direction. Sediment accumulation rate in the depocentre was rapid, estimated to be 275 m/Ma
(compacted); in a uniform stretching model, $\beta = 2$.

REFERENCES

Aydin, A. and A. Nur, 1982; Evolution of pull-apart basins and their scale independence. Tectonics, 1, 91-105.

Avison, M. 1984: Contemporaneous faulting, and the eruption and preservation of the Lough Guitane Volcanic Complex, Co. Kerry. Journal of the Geological Society, London, 141, 501-510.

Capewell, J.G. 1956: The stratigraphy, structure and sedimentation of the Old Red Sandstone of the Comeragh Mountains and adjacent areas, County Waterford. Quarterly Journal of the Geological Society, London, 112, 393-412.

Capewell, J.G. 1965: The Old Red Sandstone of Slive Mish, Co. Kerry. Proceedings of the Royal Irish Academy, 64B, 165-174.

Carruthers, R.A. 1987: Aeolian sedimentation from the Galtymore Formation (Devonian), Ireland. In: Frostick, L. and I. Reid (eds) Desert Sediments: Ancient and Modern. Geological Society, London, Special Publication, 35, 251-168.

Clayton, G. and J.R. Graham, 1974: Miospore assemblages from the Devonian Sherkin Formation of south-west County Cork, Republic of Ireland. Pollen et Spores, 16, 565-588.

Colthurst, J.R.J. 1978: Old Red Sandstone rocks surrounding the Slievenamon Inlier, Counties Tipperary and Kilkenny. Journal of Earth Sciences, Royal Dublin Society, 1, 77-103.

Cooper, M.A., D.A. Collins, M. Ford, F.X. Murphy and P.M. Trayner, 1984: Structural style, shortening estimates and the thrust front of the Irish Variscides. In: Hutton, D.H.W. and D.J. Sanderson (eds), Variscan Tectonics of the North Atlantic Region, Geological Society, London, Special Publication, 14, 167-176.

Cooper, M.A., D.A. Collins, M. Ford, F.X. Murphy, P.M. Trayner and M. O'Sullivan, 1986: Structural evolution of the Irish Variscides. Journal of the Geological Society, London, 143, 53-61.

Ford, M. 1987: Practical application of the sequential balancing technique: an example from the Irish Variscides. Journal of the Geological Society, London, 144, 885-891.

Frostick, L.E. and I. Reid, 1987: Tectonic controls of desert sediments in rift basins ancient and modern. In: Frostick, L. and I. Reid (eds) Desert Sediments: Ancient and Modern. Geological Society, London, Special Publication, 35, 53-68.

Gardiner, P.R.R. and I.A.J. MacCarty, 1981: The late Palaeozoic evolution of southern Ireland in the context of tectonic basins and trans-Atlantic significance. In: Kerr, J.W. and A.J. Ferguson (eds.) Geology of the North Atlantic Borderlands. Canadian Society of Petroleum Geologists, Memoir, 7, 342-390.

Graham, J.R. 1983: Analysis of the Upper Devonian Munster Basin, an example of a fluvial distributary system. In: Collinson, J.D. and J. Lewin (eds) Modern and Ancient Fluvial Systems, International Association of Sedimentologists, Special Publication, 6, 473-483.

Graham, J.R. and T.A. Reilly, 1972: The Sherkin Formation (Devonian) of south-west County Cork. Geological Survey of Ireland, Bulletin, 1, 281-300.

Haughton, P.D.W., 1989: Structure of some Lower Old Red Sandstone conglomerates, Kincadineshire, Scotland: deposition from late-orogenic antecedent streams? Journal of the Geological Society, London, 146, 509-525.

Higgs, K. and K.J. Russell, 1981: Upper Devonian microfloras from south-east Iveragh, County Kerry, Ireland. Geological Survey of Ireland, Bulletin, 3, 17-50.

Leeder, M.R. and R.L. Gawthorpe, 1987: Sedimentary models for extensional tilt-block/half-graben basins. In: Coward, M.P., J.F. Dewey and P.L. Hancock, (eds) Continental Extensional Tectonics. Geological Society, London, Special Publication, 28, 139-152.

Leeder, M.R., D.M. Ord and R. Collier, 1988: Development of alluvial fans and fan deltas in neotectonic extensional settings: implications for the interpretation of basin fills. In: Nemec, W. and R.J. Steel (eds) Fan deltas, Sedimentology and Tectonic Settings. Blackie, 173-185.

Naylor, D. and P.C. Jones, 1967: Sedimentation and tectonic setting of the Old Red Sandstone of south-west Ireland. In: Oswald, D.H. (ed.) International Symposium on the Devonian System, Alberta Society of Petroleum Geologists, Calgary, 11, 1089-1099.

Penney, S.R. 1978: Devonian lavas from the Comeragh Mountains, Co. Waterford. Journal of Earth Sciences, Royal Dublin Society, 1, 71-76.

Penney, S.R. 1980: A new look at the Old Red Sandstone succession of the Comeragh Mountains, County Waterford. Journal of Earth Sciences, Royal Dublin Society, 3, 155-178.

Williams, E.A., M.L.F. Bamford, M.A. Cooper, H.E. Edwards, M. Ford, G.G. Grant, I.A.J. MacCarthy, A.M. MacFee and M.J. O'Sullivan, 1989: Tectonic controls and sedimentary response in the Devonian-Carboniferous Munster and South Munster Basins, south-west Ireland. In: Arthurton, R.S., P. Gutteridge and S.C. Nolan (eds.) The role of tectonics in Devonian and Carboniferous sedimentation in the British Isles, Yorkshire Geological Society, Occasional Publication, 6, 123-141.

Williams, E.A., M. Ford, and H.E. Edwards, 1990: Discussion of 'A model for the development of the Irish Variscides'. Journal of the Geological Society, London, 147, in press.

Facies Complexes of the Upper Carboniferous in North-West Germany and their Structural Implications

B. Jankowski/ F. David/ V.Selter
Ruhr-Universität Bochum, Universitätsstr.150, D-4630 Bochum, Germany

ABSTRACT

A section of the Northwest German Upper Carboniferous basin is considered on a large scale under the aspect of facies development. This analysis leads to a subdivision of the basin-fill into several facies complexes, which are defined by their coherent sedimentary patterns. Each of these complexes is interpreted in terms of environmental systems. These are from bottom to top: marine basin, elongate lower delta plain, lobate lower delta plain, fluvial-dominated lower to upper delta plain, fluvial-alluvial plain. Except for the lowermost marine sequence most of the sediments have been accumulated at or near sub-aerial conditions, which indicates that subsidence, rather than filling of a morphological basin, has been the main factor in sediment accumulation. However types and distribution of facies have been periodically determined by eustatically derived marine transgression.

The boundaries of the environmental systems are diachronous on a large scale. A general trend of decreasing marine influence and, in the upper part (Westphalian D - Stephanian), of increasing aridity is obvious.

Three stages of structural evolution can be distinguished:

(1) an early foredeep stage with the depocenter in the Ruhr area,

(2) a late foredeep stage beginning from the middle of Westphalian C, with marginal basin folding and block-uplifting in front resulting in cannibalism and fan formation in the marginal zone of uplifted areas, and

(3) an Atlantic rift stage during Stephanian time with graben and horst systems and a reworking of Namurian and early Westphalian foredeep sediments.

INTRODUCTION

The Upper Carboniferous in NW-Germany is part of the Sub-Hercynian basin and its northern foreland which developed between Poland and Ireland. The width of the considered section is more than 250 km in the N-S direction, extending from the Ruhr area in the south to the North Sea (Fig. 1). The rocks are folded in the southern part and lie almost horizontally in the north (Fig. 2). In the Ibbenbüren-Osnabrück area Carboniferous strata were uplifted during Late Cretaceous time. The coals of this area were heated up to a semi-anthracitic rank by the influence of the Bramsche Intrusion. Farther to the north, the Carboniferous was deposited on a platform which underwent some contemporaneous block faulting and is buried up to 7000 m depth.

The Upper Carboniferous consits of 5000 to 6000 m of sediments. The coal bearing section, approximately 4000 m thick, is subdivided by marine bands and volcanic ash layers preserved mainly in coal seams. The main trend is the development from marine to deltaic conditions (with marine transgressions) and finally to an alluvial plain. The most productive coal measures derive from environments which were dominated neither by marine ingressions nor by large fluvial channels. In Stephanian times the climate changed to semi-arid conditions.

The concept of facies complexes is used to identify the largescale facies architecture in a thick section of sediments and to develop a three-dimensional model of the whole basin fill, including the structural evolution as far as it can be derived from the facies development. We have chosen the names of rivers to identify the facies complexes to distinguish them from stratigraphical sections.

Previous studies on stratigraphy, sedimentology, facies, and basin development of the Upper Carboniferous in Northwest Germany have been carried out by Stille (1930, 1949), Kukuk (1938), Jessen (1961), Fiebig (1969), Hahne (1970), Josten & Teichmüller (1971), Teichmüller (1973), Hedemann et al. (1972), Burger (1980), and Hedemann et al. (1984).
Special sections have been examined by Wachendorf (1965), Malmsheimer (1968), Casshyap (1975), Wendt (1965), Böger (1966), Brauer & Buntfuss (1966), Fiebig & Groscurth (1984), Strack (1989), and many others.

The concept of modern facies analyses (see Reading 1981 and Kleinspehn & Paola 1988) has been recently used by E. David (1989), F. David (1990), Diessel (1988), Selter (1990), and Steingrobe (1990).

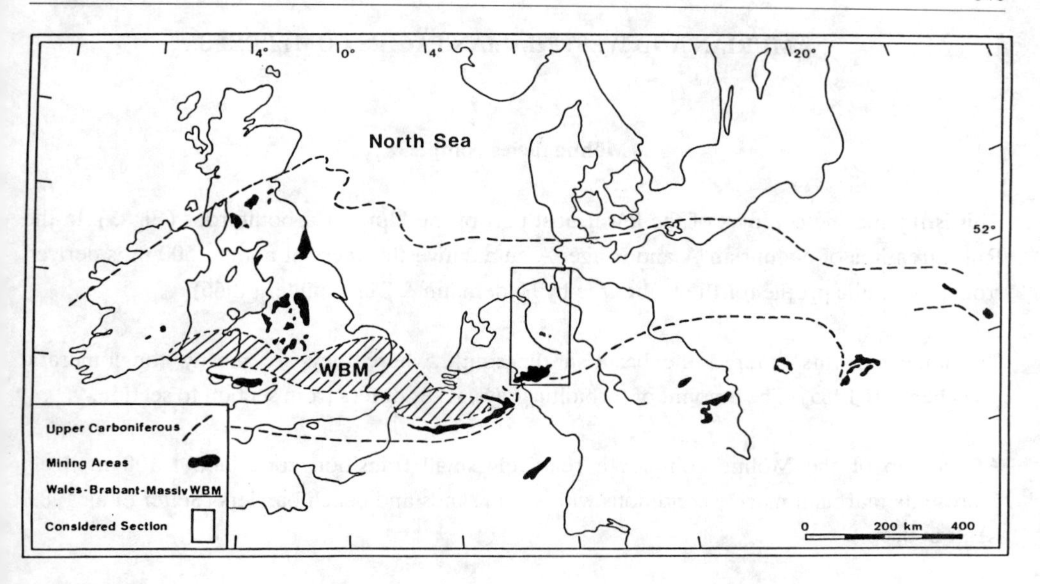

FIGURE 1
Distribution of the Sub-Hercynian Upper Carboniferous from Ireland to Poland, including the considered section in Northwest Germany (simplified after Lippold et al. 1986).

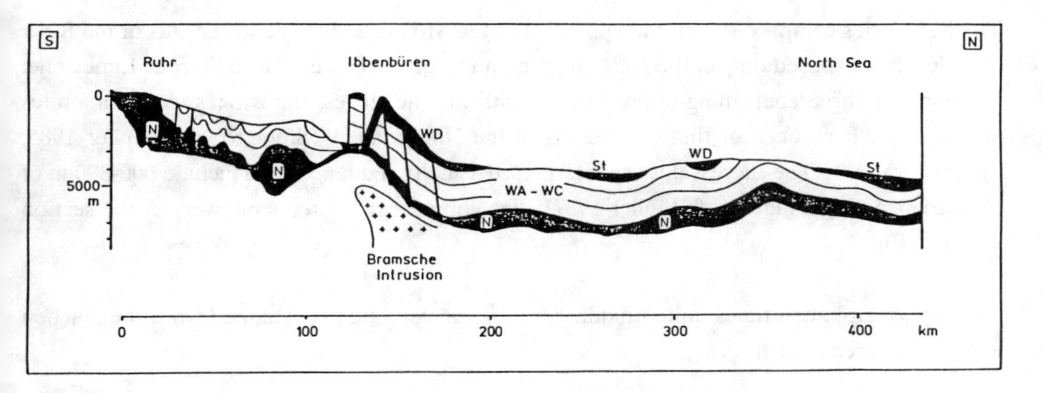

FIGURE 2
Simplified cross-section after Stancu-Kristoff & Stehn (1984) through the Silesian Basin from the Ruhr area to the southern North Sea. The section does not follow a straight south-north line because it is determined by borehole locations. N = Namurian (black), WA-WC = Westphalian A to C, WD = Westphalian D (black), St = Stephanian (black). The strata are underlain by Dinantian and Devonian rocks and overlain by Mesozoic rocks in the south and Permian strata in the north.

NAMURIAN AND WESTPHALIAN FACIES COMPLEXES

Möhne facies complex

This is the marine complex of the lowermost part of the Upper Carboniferous (Fig. 3). In the Ruhr area it is of Namurian A and B age. A cumulative thickness of about 1500 m is derived from a synthetic profile for the Ruhr area by Hedemann & Teichmüller (1966).

Turbiditic currents entered the basin in the south, where they are mostly conglomeratic (Wachendorf 1965). The amount of turbiditic greywackes differs from section to section.

At the top of the Möhne complex a relatively small transition zone, about 100 m thick, represents marginal marine conditions with storm sands and beach barriers (Selter et al. 1989, Kraft in prep.).

The complex is overlain by thick fluvial channel sandstones marking the transition to the fluvio-deltaic Ruhr facies complex.

Inde facies complex

The Inde facies complex is a lateral equivalent to the Möhne and to the lower part of the Ruhr complex. It is exposed only in the Aachen area on top of the Lower Carboniferous Limestone. It comprises three coarsening-upward cycles with marine shales, sandstones, comglomerates and an early occurence of thin coal seams in the Namurian A (Steingrobe & Muller 1985, Steingrobe 1990). The successions are interpreted as delta sequences and include 600-800 m of Namurian A-C (Steingrobe & Müller 1987). The complex is located southwest of the section shown in Fig. 3.

At early Westphalian times quite unique depositional conditions developed from the Aachen to the Ruhr area (Strack 1989).

Ruhr facies complex

In the Ruhr area this complex is about 1650 m thick. It starts in the lowermost Namurian C with the fluvial Kaisberg Sandstone and ends in the middle of Westphalian A2, where the amount of sandstones diminishes (Fig. 3).

The complex consists of marine bands, thick fluvial channel sandstones and palaeosols topped by coal seams. The number and thickness of coal seams increase toward the top of this complex while the marine bands decrease. The complex can be subdivided into several sub-complexes.

The principal sedimentary pattern in this complex is demonstrated by a shift profile (Fig. 4). This profile is based on the theory that coal seams represent a sedimentary situation in which the development leads on the one hand, to a marine bay (left side) and, on the other hand, to a fluvial floodplain or channel (right side). The analysis of the sedimentary record shows a fluvial branch with rooted siltstones and coal seams on top of flood-plain sediments and/or fluvial channel sandstones. After soil and/or peat formation occurred a brackish or marine bay developed. When the bay was filled with sediments, a new swamp was able to establish. This bay-fill sequence may have recurred. Erosional fluvial sandstones occur irregularly in the sequence. Erosion varies from channel to channel and often stops at the roof of a coal seam. The irregularity of these channel sandstones make a formal analysis of sedimentary cycles difficult.
Sandstones within the bay-fills are always fine-grained. They represent mouth bars and splays. The fine-grained marine bands occasionally contain Lingulids or Goniatites and are often characterized by worm traces. Tidal influences probably occurred.

The complex supposedly represents a lower delta plain with frequent (eustatically derived) marine transgressions. Elongated sandstone distribution (Böger 1966, Malmsheimer 1968, Casshyap 1975) points to the existence of a fluvial-dominated elongated delta similar to the birdfoot delta of the modern Mississippi river (e.g. Fisk et al. 1954, Gould 1970 and others).

Emscher facies complex

This complex (about 400-450 m thick in the Ruhr area) is the most fine-grained in the coal-bearing section. It begins with a decrease of fluvial sandstones in the middle to upper part of Westphalian A2 (Fig. 3). Bay fill sequences are most characteristic. They are topped by coal seams and minor fluvial channels as well as crevasse splay sediments (E. David 1989). The included Vanderbecki (=Katharina) Marine Band is overlain by some coarsening-upward cyles indicating delta progradation. The whole complex an be subdivided into three sub-complexes:
(1) coal-rich Emscher-1 sub-complex below Vanderbecki Marine Band,
(2) pro-delta and deltaic Emscher-2 (Vanderbecki) sub-complex, and
(3) coal-rich Emscher-3 sub-complex. It is interpreted as a lower delta plain with minor fluvial sandstone sedimentation. In contrast to the previous birdfoot delta with large open bays, this

seem to be a lobate delta platform with closed bays. The Vanderbecki marine transgression may have been induced by eustatic sea level rise.

Lippe facies complex

Within the Lippe complex the number of marine bands decreases, and the number of fluvial channel sandstones increases. It can be subdivided into three sub-complexes:

(1) a relatively fine-grained Lippe-1 sub-complex with several marine or brackish bands,

(2) a fluvial-dominated Lippe-2 sub-complex with a few thin brackish layers, and

(3) a fluvial-dominated sub-complex with further diminished brackish layers and an increasing content of partly conglomeratic sandstones.

The complex commences with the Domina Brackish Band (Lingula Band). The boundary between sub-complex (1) and (2) is the top of the Aegiranum delta sequence. The top of the Top Marine Band sequence or its equivalent, the pelitic band on top of the Nibelung coal seam group, is the boundary between (1) and (2). The boundary between (3) and the Hase complex is defined by the erosive base of the Dickenberg Sandstone succession (equivalent to the Neeroeteren Sandstone).

a) Lippe-1 sub-complex

Lippe-1 contains coal seams, fluvial channel sandstones, some brackish bands and the widespread Aegiranum Marine Band. In contrast to the Emscher complex it exhibits relatively thick channel sandstones which were deposited mostly by meandering rivers.

Fig. 5 shows a characteristic pattern of this sub-complex. On top of a fluvial channel sandstone siltstones from overbank sedimentation occur. They grade into palaeosols and peats which are overlain by lake or bay sediments. The pattern recurs until another fluvial channel erodes parts of it.

The Aegiranum Marine Band at the top of this sub-complex consists of marine shales which form the base of a coarsening-upward sequence (40-45 m thick). The top is dominated mostly by mouth bar sands. The whole sequence is interpreted as a bay-fill produced by delta progradation. It is overlain by palaeosols and coal seams. The facies of the marine band varies regionally, including the occurrence of mouth bar sands. The preservation of the complete sequence depends on the gradual erosion of the following fluvial channel.

b) Lippe-2 sub-complex

This sub-complex is characterized by thicker and coarser-grained sandstones than in the underlying sub-complex. Marine influence is still present, but no major marine band occurs in the area under examination. It ends at the top of the Nibelung coal seam group including the brackish band which is an equivalent to the Top Marine Band.

c) Lippe-3 sub-complex

This complex differs from the Lippe-2 complex by the higher amount of fluvial channel sandstones, rooted overbank sediments and the lack of thick marine bands. The sandstones were deposited mostly by braided rivers. In the Lippe area this complex is about 400-500 m thick. In the North Sea area it is presumably substituted by the finer-grained Lippe-2 complex.

The overbank deposits with small coal seams are nearly totally rooted. The amount of sandstones increases toward the top. Occasionally there are thin marine incursions with foraminifera and bivalves. The characteristic sedimentary patterns are exhibited by slightly upward-fining fluvial channel sandstones, rooted siltstones, spoiled coal seams which are overlain by rooted siltstones or lacustrine/brackish silty shales. In these sediments fluvial channels have incised, locally eroding the upper part of the deposits to an unknown depth.

In the lower part of the sub-complex (on top of the Nibelung coal seam group) the composition of coal seams change (Strehlau 1988); Desospores disappear, and Lycopodes are replaced by Cordaites as peat-forming flora. The oxidation of organic matter was more intensive, and the coal-seams are consistently contaminated by dispersed clay.

Hase facies complex

The Hase complex contains the coarsest-grained sediments of the coal-bearing Upper Carboniferous. It commences in the uppermost Westphalian C with the Dickenberg Sandstone and is restricted to an area between Ibbenbüren and Oldenburg. The sediments' maximum thickness is about 450 m (F. David 1990). Near Osnabrück the sandstones and conglomerates were silicified by Cretaceous nydrothermal fluids of the Bramsche Intrusion (Stadler 1971, Stadler & Teichmüller 1971).

Regionally the complex is more or less conglomeratic. The less conglomeratic part lies in a more distal position to the source area. Between Osnabrück and Oldenburg the maximum pebble-size is 8 cm. The dominating coarse-grained sediments were deposited by braided rivers. Rooted siltstones (overbank sedimentation), coal seams (swamps), crevasse splays and thin lacustrine and brackish layers occur.

To the south and west the Hase complex becomes less coarse-grained. The maximum pebble-size near Ibbenbüren is 5 cm. It is composed of conglomeratic sandstones, conglomerates, overbank deposits and coal seams. Several thin brackish incursions, less than 1 m thick, with foraminifera occur throughout this complex. In the Ibbenbüren area the whole complex is about 350 m thick.

The restricted occurrence of such conglomeratic sandstones within the basin points to particular tectonic activities within the basin or near the source area at this time. Reworked sandstone particles show the beginning of intra-basional cannibalism (F. Frank, in prep.).

Hunte facies complex

The Hunte Complex includes the red coloured coal-free Westphalian sediments. They are mostly of Westphalian D origin, but red Westphalian C sediments occur in the eastern part of the basin.

The sediments are less conglomeratic than those of the Hase complex. They were deposited in alluvial environments under sub-humid to semi-arid climatic conditions. The complex is composed of braided river channel sandstones, some distal debris flows (up to 2 m thick), sheetflood sandstones (Fig. 6), and fine-grained overbank sediments with palaeosols. Ferralsols and caliches are characteristic. A limestone of 30 cm thickness with calcareous green algae points to the existence of a local lake. Oogonia of Charaphyta have been identified. One hypothesis is that sedimentation was restricted to the northern part of the Upper Carboniferous basin (F. David 1990, Selter 1990). The basin was rimmed by alluvial fans to the south and south-west. A proximal braid plain (south of Oldenburg) graded into an interior distal braid plain (southern Ems area) and an open distal braid plain (northern Ems area), which bordered the marine influenced North Sea area. Different types of palaeosols indicate locally differentiated hydrological conditions (Selter 1990).

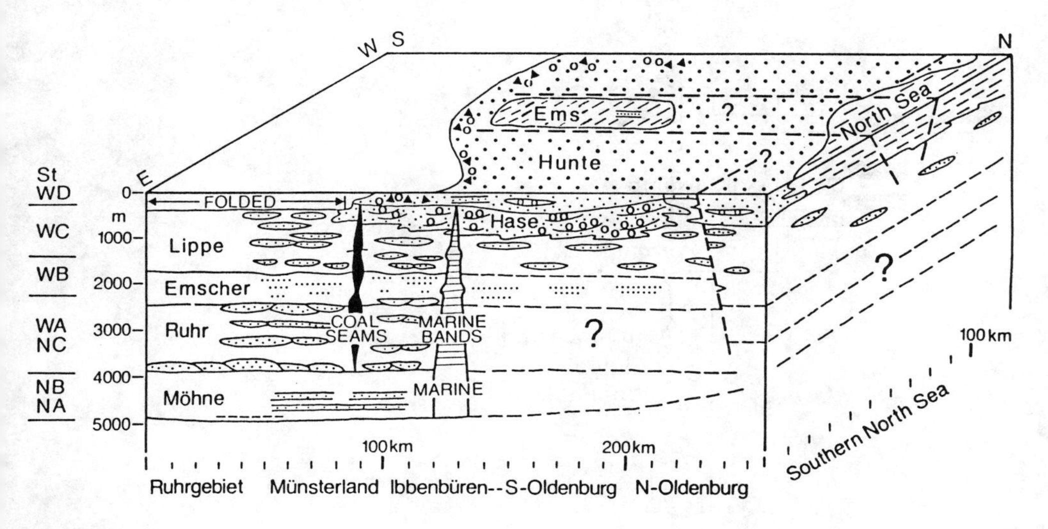

FIGURE 3
Block diagram of the Upper Carboniferous in Northwest Germany. Black triangles in the marginal zone of Hase facies complex indicate alluvial fan sediments.

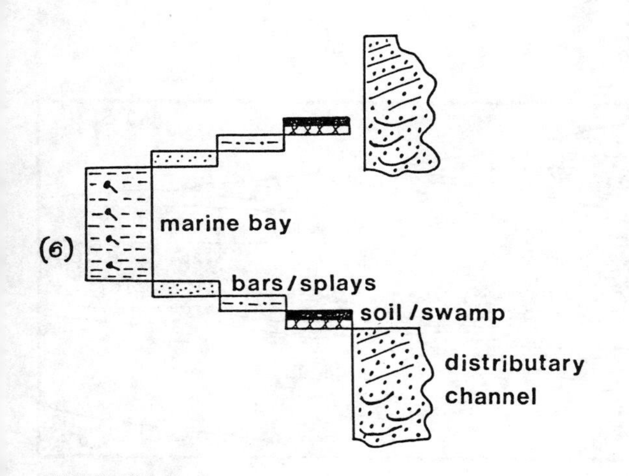

FIGURE 4
Characteristic sedimentary pattern of the Ruhr facies complex. The shift profile shows the environmental development from swamp conditions (black) to brackish or marine bays (with worm traces and/or gastropods on the left side and to fluvial sediments on the right side.

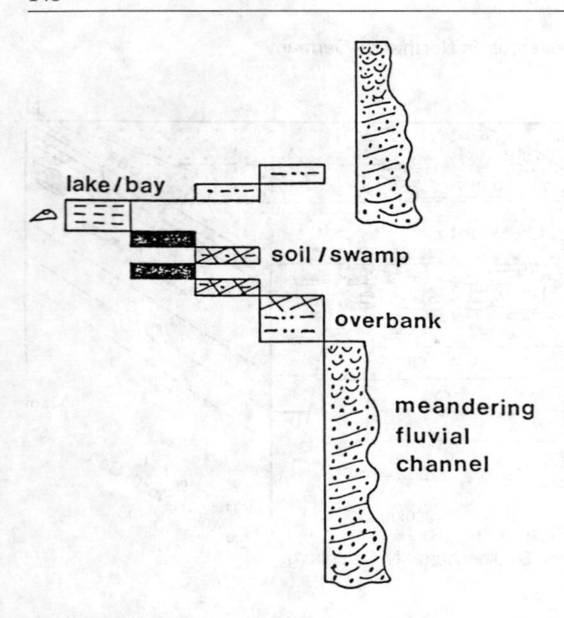

FIGURE 5
Characteristic sedimentary pattern of the Lippe 1 facies sub-complex in the Ruhr area. The shift profile shows the switch of environments from swamp conditions (black) to lacustrine, brackish or marine bays on the left and to fluvial conditions on the right side.

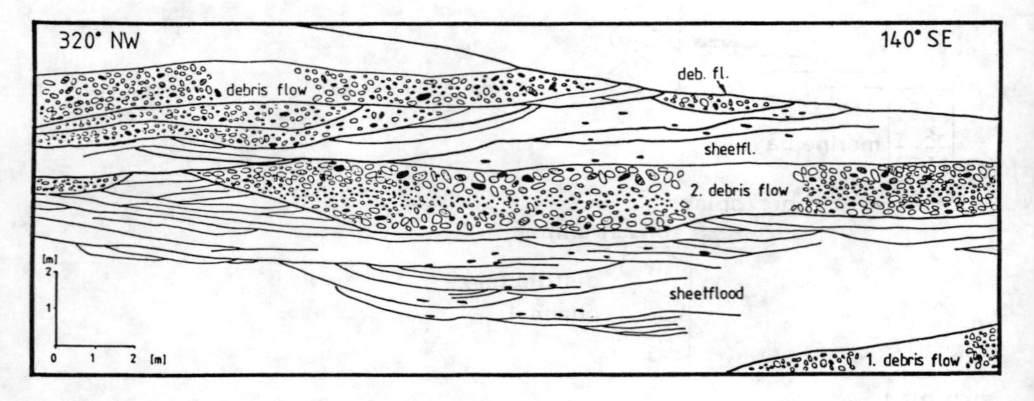

FIGURE 6
Debris flow and sheetflood sediments in the Kälberberg quarry near Ibbenbüren (Hunte facies complex, Westphalian D; from Selter 1990).

STEPHANIAN FACIES COMPLEXES

Ems facies complex

The Ems facies complex is characterized by its red colour and the lack of coal seams. Its alluvial facies repertoir is very similar to the Hunte complex. There is no disconformity either at the base or within this complex; this has been observed in cores and well-logs within the southern Ems graben. Outside the southern Ems graben a widespread discontinuity between Westphalian and Stephanian sediments is inferred by some authors (Drong et al. 1982, Hedemann et al. 1984).

The climate was predominatly semi-arid, which is indicated by the presence of more caliche horizons than in the Hunte complex and by a higher boron content (Selter et al. 1990). The silty Ems complex includes fine-grained overbank sediments, abundant sheetflood and minor fluvial channel sandstones. The sheetflood sediments are laminated, contain clay pebbles, but do not show erosive basinal contacts. The palaeosols occur on top of single flood-plain sequences.

In this area the continuous development from Westphalian D to Stephanian is very clearly shown by a transition between the coal-bearing and the caliche-bearing sequence. The Hunte and Ems complexes are about 500-800 m thick. Hedemann et al. (1984) suppose 260 m thick sediments of Stephanian origin in the southern Ems graben.

North Sea facies complex

In this complex siltstones dominate. They contain fluvial channel sandstones and some marine limestone layers, 20-30 cm thick, with echinoid fragments. The maximum thickness of Stephanian sediments in the North Sea area is about 600 m (Hedemann et al. 1984).

STRUCTURAL IMPLICATIONS

The following preliminary conclusions can be drawn with respect to intra-basinal tectonics:

(1) Most of the coal-bearing and red bed sequences have been accumulated sub-aerially or under shallow water in widespread alluvial fan, fluvial, swamp, and bay environments. Accumulation was controlled mainly by subsidence rather than by pre-existant

morphological basins. Facies development was additionally controlled by eustatic sea level changes. This global mechanism has been shown by Vail and others (see Payton, 1977, Wilgus et al. 1988).

(2) There are transport directions from the south (folt belt) and the west (Mid Netherlands High) in the final stages (F. David 1990, Selter 1990). In Namurian C and Westphalian A times the transport directions, observed in the Ruhr area, was mainly from NE to SW, i.e. parallel to the strike of the Sub-Hercynian folding (Wendt 1965, Malmsheimer 1968, Casshyap 1975, Grube 1978, Strack 1989).

(3) There are no disconformities at all from Namurian A to Westphalian D.

(4) A disconformity between Westphalian and Stephanian sediments exists outside the Ems graben (Drong et al. 1982). Stephanian deposits occur on top of Westphalian sediments of different ages (Hedemann et al. 1984).

(5) The coarse-grained sediments of the Hase and Hunte facies complexes indicate a local source area as evidenced by distal debris flows. Therefore, an uplifted area is postulated for the Westphalian D near the northern margin of the Hercynian orogenic block (F. David 1990, Selter 1990).

(6) Alluvial fans exist along on a line extending from west of the Ems-River to the south near Bentheim then turning east to Ibbenbüren, Osnabrück und Diepholz. This line borders the northern residual basin, in which sedimentation took place up to the end of Stephanian time. Ommissions on top of horst structures are suspected. Source areas were the folded Sub-Hercynian molasse in the south and the uplifted block of the Mid Netherlands High in the west (F. David 1990, Selter 1990). This erosion started in late Westphalian C or early Westphalian D time.

(7) That Namurian and early Westphalian deposits in some parts of the basin were eroded in late Westphalian and Stephanian times is indicated by the occurence of:

(a) coal clasts in sandstones of Westphalian C and D,
(b) reworked sandstone lithoclasts in Westphalian C/D conglomerates with typical early Upper Carboniferous composition (F. Frank, in prep.)
(c) grey siltstone clasts and reworked caliche nodules in red bed facies and
(d) debris flows surrounding a residual basin with continuous sedimentation.

(8) At the same time as tectonic activity started with block uplifting in Northwest Germany, the Brabant Massif subsided (from mid Westphalian D; after Besly 1988). Possibly the Nee-

roeteren Sandstone of the Campine (Dusar 1989) marks the beginning of a basin-wide activity. The Dickenberg Sandstone in the Osnabrück area is an equivalent.

(9) Stephanian horst and graben systems show different orientations to the previous Sub-Hercynian basin structures. Their development is due to Atlantic rift systems (see Ziegler 1982).

(10) The rank of coals in the northern Ruhr area points to a previous 2000 m of Carboniferous on top of Westphalian C (Buntebarth & Koppe 1984, Scheidt & Littke 1989). Consequently, there is no reason to assume a special depocenter in the Osnabrück area for Westphalian D sediments.

(11) By comparing formation thicknesses Drozdzewski (1990) deduced that the depocenter of Westphalian time had always been in the Ruhr area and that the basin enlarged gradationally to the north.

These observations lead to the conclusion that there was cannibalism within the basin in at least two stages.

A Middle-Westphalian-C activity exposed the Sub-Hercynian fold-belt south of the modern Ruhr river. Erosion produced the coarse-grained sediments of the Lippe-3 sub-complex. Today this source area is deeply eroded and cut off to the south by a big fault (Ennepe-Störung). Therefore, the southern marginal facies of the Upper Carboniferous is not preserved.

At the same time there must have been an uplifted area south-east of Ibbenbüren and Osnabrück feeding the coarse-grained Hase complex (F. David 1990). This area is then the northern margin of the Hercynian orogenic block. Presumably, it was bordered by an active tectonic lineament.

The debris flows of the Upper Westphalian C and Lower Westphalian D in the area surrounding the northern residual basin point to another structural event, a Late Westphalian C/Early Westphalian D activity. One provenance area was the Mid Netherland High as documented by alluvial fans and fluvial transport from the west into the southern Ems graben.

Possibly, the main formation of the later North Sea Stephanian basin started at the same time. South of this basin, in the Ibbenbüren-Osnabrück-Oldenburg area (indicated in Fig. 3 by a fault), no Stephanian sediments exist.

The succession of these Middle- and Late-Westphalian-C activities in time and space leads to the conclusion that progradation of tectonic activities from the Hercynian fold-belt into the northern foreland occurred. This postulate by Stille (1930, 1949), Kukuk (1938), and Hedemann & Teichmüller (1971) is now clearly demonstated. In contrast to this progradation the subsidence center during Namurian and most of Westphalian times lay relatively stable in the Ruhr area.

BASIN EVOLUTION

The basin evolution as a whole can be derived from overall trends in the sedimentary record and from facies development:

(1) Marine incursions decrease from bottom to top.

(2) An increasing content and grain-size of sandstones and conglomerates from lower Westphalian B (Emscher complex) to Westphalian D (Hase complex) is obvious.

(3) The maximum coal development occurs in strata with only minor marine influences and minor fluvial channel sedimentation. Many coal seams are developed in a very widespread pattern, which can be shown by enclosed volcanic ash layers ("tonsteins", Burger 1985).

(4) In the Ruhr-Aachen area (Namurian C to Westphalian C) facies changes are perpendicular to the axis of the sedimentary basin (Strack 1989). Fluvial transport, however, frequently occured parallel to this axis.

(5) A climatic change from humid to semi-arid conditions in the upper part is indicated by palaeosol evolution.

(6) A source area southeast of Osnabrück is postulated for the coarse-grained Hase and Hunde complexes.

(7) A very stable depocenter is assumed in the Ruhr area from Namurian to Westphalian C times. The depositional area gradationally enlarged to the north.

(8) A progressive cannibalism of Carboniferous strata from south to north and from the Mid Netherlands High occurred from middle Westphalian C to lower Westphalian D times.

CONCLUSIONS

The trends in f a c i e s d e v e l o p m e n t can be explained by a five-stage model:

I. Marine Basins
Möhne complex: marine basin with siltstone sequences containing turbidites. Regionally black shales were deposited at the base. Marginal marine conditions are characteristic for the top.

II. Elongate lower delta plain
Ruhr complex: braided rivers, large prograding shallow-water deltas, widespread marine floodings. Eustatic changes in sea level was one controlling factor in facies development.

Inde complex: fluvial and marine sedimentation in the Aachen area coexisting with the Möhne complex and the lower part of the Ruhr complex. It developed on a lower delta plain, but its shape is still not known.

III. Lobate lower delta plain Emscher complex: large marine-influenced delta plain with domination of bay-fill and lake sedimentation. The influence of meandering channel sedimentation increases in the upper part.

IV. Fluvial-dominated lower to upper delta plain
Lippe complex: increasing fluvial sedimentation leads from lower delta plain to braided river plain environments. The widespread Aegiranum Marine Band represents the last important marine transgression.

V. Fluvial-alluvial Plain
Hase, Hunte and Ems complexes: The coarsening-upward tendency beginning within the Lippe complex and the occurrence of alluvial fans shows tectonic activities in the source area and cannibalism in southern exposed parts of the former basin. In Stephanian times the rest-basin is divided into graben and horst structures. Marine influence is rare, except in the North Sea area. The fall of sea-level and the transition to semi-arid climatic conditions supported the formation of red beds.

Comparisons to modern delta systems show many similarities in details of individual facies e. g. fluvial channels, overbank sediments, bays and mouth bars. However no modern model for the Upper Carboniferous in Northwest Germany as a whole has so far been identified. Presumably the Ruhr complex can be compared to the modern birdfoot delta of the Mississippi river. The peculiarity is a combination of foredeep subsidence in front of an ending orogeny moving towards the basin and eustatic sea-level fluctuations which are presumably due to glaciations and deglaciations on the Gondwana continent. Modifiers were climatic changes, block-faulting, and cannibalism within the basin.

The structural evolution can be explained by a three-stage model:

I. Early foredeep stage
A very stable depocenter in the Ruhr area existed, especially during Namurian and Westphalian times. At the same time the depositional area enlarged more or less continuously to the north.

II. Late foredeep stage
Prograding tectonic activities from south to north, from the middle of Westphalian C to Westphalian D, resulted in coarser-grained sediments and a narrowing of the basin. Block-uplifting within the basin (e. g. Mid-Netherlands High) produced intra-basinal source areas.

III. Atlantic rift stage
The Atlantic rift systems encroached upon the northern part of the Sub-Hercynian foredeep forming horst and graben systems; a new structural pattern was the consequence. Outside the North Sea area, sedimentation was determined by local sources.

Acknowledgements

This paper is based on first results of a basin anaysis of the Upper Carboniferous in Northwest Germany funded by the German oil and coal industry, represented by the DGMK (German Society for Petroleum Sciences and Coal Chemistry), and the Ministry of Research and Technology. The research was carried out in the Geological Institute of the Ruhr-University in Bochum. We are indebted to all the helpful members of the industrial companies, to Dr. Albertsen from the DGMK and to the Ministry. In particular, we thank Prof. Füchtbauer for discussion and advice. We are obliged to the members of the Geological Institute who supported us.

REFERENCES

Besly, B.M., 1988: Palaeogeographic implications of late Westphalian to early Permian red-beds, Central England. - p. 200-221 in: Besly, B.M. & Kelling, G. eds): Sedimentation in a synorogenic basin complex: the Upper Carboniferous of northwest Europe. Blackie, Glasgow & London.

Böger, H., 1966; Diemarinen Niveaus übr denFlözen Schieferbank und Sarnsbank (Grenze Namur C - Westfal A) im Ruhrgebiet. Fazies, Fauna und Ökologie. - Fortschr. Geol. Rheinland u. Westfalen, 13, 1-38, Krefeld.

Brauer, J. and Buntfuss, J., 1966: Sedimentologische Untersuchungen im Oberen Westfal C und Unteren Westfal D des Ibbenbürener Karbons. - Fortschr. Geol. Rheinland u. Westfalen, 13, 1095-1108, Krefeld.

Buntebarth, G. and Koppe, I., 1984: Abschätzung von Westfal C-Mächtigkeiten im Ruhrrevier aus paläogeothermischen Untersuchungen. - Fortschr. Geol. Rheinland u. Westfalen, 32, 269-273, Krefeld.

Burger, K., 1980: Kaolin-Kohlentonsteine im flözführenden Oberkarbon des Niederrheinisch-Westfälischen Steinkohlenreviers. - Geol. Rundschau., 69, 488-531, Stuttgart.

Burger, K., 1985: Kohlentonsteine im Oberkarbon NW-Europas. Ein Beitrag zur Geochronologie. - Congr. Internat. Strat. Geol. du Carbonifère, Madrid 1983, Compte Rendu, 3, 433-447, Madrid.

Casshyap, S.M., 1975: Lithofacies analysis and paleogeography of Bochumer Formation (Westfal A2), Ruhrgebiet. - Geol. Rundschau, 64, 610-640, Stuttgart.

David, E., 1989: Faziesanalyse der Essener Schichten in der Bohrung Bergbossendorf 1 (Westfal B1, nördliches Ruhrgebiet). - DGMK-Berichte, Forschungsbericht 384-2, 76 p., Hamburg

David, F., 1990: Sedimentologie und Beckenanalyse im Westfal C und D des nordwestdeutschen Oberkarbons. - DGMK-Berichte, Forschungsbericht 384-3, 267 p. Hamburg.

Diessel, C.F.K., 1988: A vertical sedimentological profile through the Upper Carboniferous coal measures of the Ruhr Basin, Germany. - Advances in the study of the Sydney Basin. Proc. of the 22nd symposium 15th-17th April, University of Newcastle, p. 15-25, Australia (NSW).

Drong, H.J., Plein, E., Sannemann, D., Schuepbach A. and Zimdars, J., 1982: Der Schneverdingen Sandstein des Rotliegenden - eine äolische Sedimentfüllung alter Grabenstrukturen. - Z. dt. geol. Ges., 133, 699-725, Hannover.

Drozdzweski, G., 199o: Neue Erkenntnisse zur strukturrellen Entwicklung des Ruhrbeckens als Vortiefe des varistischen Orogens. - Abstract, 1 p., Sediment 90, 6.-7.6.90, Bonn.

Dusar, M., 1989: Non-marine Lamellibranchs in the Westphalian C/D of the Campine Coalfield. - Bulletin de la Société belge de Géologie, 98 (3/4), 483-493, Bruxelles.

Fiebig, H., 1969: Das Namur C und Westfal im Niederrheinisch-Westfälischen Steinkohlengebiet. - 6. Congr. Internat. Strat. Géol. du Carbonifére, Sheffield 1967, Compte Rondu, 1, 79-89, Maestricht.

Fiebig, H. & Groscurth, J., 1984: Das Westfal C im nördlichen Ruhrgebiet. - Fortschr. Geol. Rheinland u. Westfalen 32: 257-267, Krefeld.

Fisk, H.N., McFarlan, E. (Jr.), Kolb, C.R. and Wilbert, L.J. (Jr.), 1954: Sedimentary framework of the modern Mississippi delta. - J. Sedim. Petrol., 24, 76-99.

Frank, F., in prep.: Das primäre Liefermaterial der Sandsteine des Oberkarbons in Nordwest-Deutschland. - Diss. Ruhr-Universität Bochum.

Gould, H.R., 1970: The Mississippi-Delta Complex. - pp. 3-30 in J.P. Morgan, ed.: Deltaic Sedimentation Modern and Ancient. - Society of Econ. Paleont. Mineral., Spec. Publ. 15, Tulsa/Oklahoma.

Grube, H., 1978: Sedimentologie der Bochumer und Essener Schichten des Ruhrkarbons aufgrund von Gesamtmächtigkeiten, Sand- und Kohleanteilen. - Mitt. Westf. Berggewerksschaftskasse Bochum, 40, 118 p., Bochum.

Hahne, C., 1970: Zur Genese des Ruhrkarbons. - Mitt. westf. Berggewerkschaftskasse, 29, 24 p., Bochum.

Hedemann, H.-A. and Teichmüller, R., 1966: Stratigraphie und Diagenese des Oberkarbons in der Bohrung Münsterland 1. - Z. dt. geol. Ges., 115, 787-825, Krefeld.

Hedemann, H.-A. and Teichmüller, R., 1971: Die paläogeographische Entwicklung des Oberkarbons. - Fortschr. Geol. Rheinland u. Westfalen, 19, 129-142, Krefeld.

Hedemann, H.-A., Fabian, H.J., Fiebig, H. and Rabitz, A., 1972: Das Karbon in marin-paralischer Entwicklung. - 7. Congr. Intern. Strat. Géol. du Carbonifére, Krefeld 1971, Compte Rondu, 1, 29-47, Krefeld.

Hedemann, H.-A., Schuster, A., Stancu-Kristoff, G. and Lösch, J., 1984: Die Verbreitung der Kohlenflöze des Oberkarbons in Nordwestdeutschland und ihre stratigraphische Einstufung. - Fortschr. Geol. Rheinland u. Westfalen, 32, 39-88, Krefeld.

Jessen, W., 1961: Zur Sedimentologie des Karbons mit Ausnahme seiner festländischen Gebiete.- 4. Congr. Avanc. Et. Strat. Géol. du Carbonifére, Heerlen 1958, Compte Rondu 2, 307-322, Maestricht.

Josten, K.-H. and Teichmüller, R., 1971: Zusammenfassende Übersicht über das höhere Oberkarbon im Ruhrrevier, Münsterland und Ibbenbürener Raum. - Fortschr. Geol. Rheinland u. Westfalen, 18, 281-292, Krefeld.

Kleinspehn, K. L. and Paola, Ch., eds, 1988: New Perspectives in Basin Anaysis. - Springer-Verlag, 453 p., New York etc.

Kraft, Th., in prep.: Faziesentwicklung vom flözleeren zum flözführenden Oberkarbon (Namur B-C) im südlichen Ruhrgebiet. - Diss. Ruhr-Universität Bochum.

Kukuk, P., 1938: Geologie des niederrheinnisch-westfälischen Steinkohlengebietes. - Verlag Glückauf, 706 p., Essen.

Lippolt, H.J., Hess, J.C. and Burger, K., 1984: Isotopische Alter von pyroklastischen Sanidinen aus Kaolin-Kohlentonsteinen als Korrelationsmarken für das mitteleuropäische Oberkarbon. - Fortschr. Geol. Rheinld. u. Westf., 32:119-150, Krefeld.

Malmsheimer, W.K., 1968: Zur Sedimentation und Epirogenese im Ruhrkarbon. Die Sandsteine im Liegenden von Flöz Mausegatt. - Westdeutscher Verlag, Forschungsberichte des Landes Nordrh.-Westf., 2000, 74 p., Köln, Opladen.

Payton, Ch.E., ed., 1977: Seismic Stratigraphy - applications to hydrocarbon exploration. - Am. Ass. Petrol. Geol., Memoir 26, 516 p., Tulsa/Oklahoma.

Reading, H.G., ed. 1981: Sedimentary Environments and Facies. - Blackwell Scientific Publications, 569 p., Oxford etc.

Scheidt, G. and Littke, R., 1989: Comparative organic petrology of interlayered sandstones, siltstones, mudstones and coals in the Upper Carboniferous Ruhr basin, Northwest Germany, and their thermal history and methane generation. - Geol. Rundschau, 78, 375-390, Stuttgart.

Selter, V., 1990: Sedimentologie und Klimaentwicklung im Westfal C/D und Stefan des nordwestdeutschen Oberkarbon-Beckens. - DGMK-Berichte, Forschungsbericht 384-4, 310 p., Hamburg.

Selter, V., David, F. and Kraft, T., 1989: Zusammenhänge zwischen Fazies und Borgehalten im nordwestdeutschen Oberkarbon-Becken. - Geol. Paläont. Mitt. Insbruck, 16, 1o8-111, Innsbruck.

Stadler, G., 1971: Die Vererzung im Bereich des Bramscher Massivs und seine Umgebung. - Fortschr. Geol. Rheinland u. Westfalen, 18, 439-500, Krefeld.

Stadtler, G. and Teichmüller, R., 1971: Zusammenfassender Überblick über die Entwicklung des Bramscher Massivs und des Niedersächsischen Tektogens.- Fortschr. Geol. Rheinland u. Westfalen, 18, 547-564, Krefeld.

Stancu-Kristoff, G. and Stehn, O., 1984: Ein großregionaler Schnitt durch das nordwestdeutsche Oberkarbon-Becken vom Ruhrgebiet bis in die Nordsee. - Fortschr. Geol. Rheinland und Westfalen, 32, 35-38, Krefeld.

Steingrobe, B., 1990: Fazieseinheiten aus dem Aachen-Erkelenzer Oberkarbonvorkommen unter besonderer Berücksichtigung des Inde-Synklinoriums. - Diss. RWTH Aachen, 325 p., Aachen.

Steingrobe, B. and Muller, A., 1985: Sedimentology of selected Upper Carboniferous profiles from the Inde- und Wurm-Syncline (Aachen Coal District). - N. Jb. Geol. Paläont. Abh., 171 (1-3), 267-279, Stuttgart.

Steingrobe, B. and Muller, A., 1987: Deltaic Development of Namurian Strata in the Aachen Coal District. - Deltas: Sites and Traps for Fossil Fuels, 21.-22.4.1987, Abstract, p. 41, London.

Stille, H., 1930: Die subvariscische Vortiefe. - Z. Deutsch. Geol. Ges., 81, 339-354, Berlin.

Stille, H., 1949: Anbau und Fortbau im mitteleuropäischen Variszikum. - Forschungenund Fortschritte, 25 (23/24), 289-293, Berlin.

Strack, Ä., 1989: Stratigraphie in den Explorationsräumen des Steinkohlenbergbaus. - Mitteilungen der Westfälischen Berggewerkschaftskasse, 62, 210 p., Bochum.

Strehlau, K., 1988: Fazies und Genese von Kohleflözen im nordwestdeutschen Oberkarbon. - DGMK-Berichte, Forschungsbericht 384-1, 367 p., Hamburg.

Teichmüller, R., 1973: Die paläogeographisch-fazielle und tektonische Entwicklung des Kohlenbeckens am Beispiel des Ruhrkarbons. - Z. dt. geol. Ges., 124, 149-165, Hannover.

Vail, P.R.: See in Payton, ed., 1977 and Wilgus et al., eds., 1988.

Wachendorf, H., 1965: Wesen und Herkunft der Sedimente des westfälischen Flözleeren. - Geol. Jahrbuch, 82, 705-754, Hannover.

Wendt, A., 1965: Der Finefrausandstein - Sedimentation und Epirogenese im Ruhrkarbon. - Westdeutscher Verlag, Forschungserichte d. Landes Nordrh.-Westf., 1386, 62 p., Köln, Opladen.

Wilgus, Ch. K., Hastings, B.S., Posamentier, H., vanWagoner, J., Ross, Ch.A. and Kendall, Ch.G.St.C., eds, 1988: Sealevel changes: an integrated approach. - Soc. Econ. Paleont. Mineral., Spec. Publ., 42, 407 p., Tulsa/Oklahoma.

Ziegler, P.A., 1982: Geological Atlas of Estern and Central Europe. - Shell Inter. Petrol. Mij., Elsevier, 130 p., Amsterdam.

Silesian Sedimentation in South-West Britain: Sedimentary Responses to the Developing Variscan Orogeny

A. Hartley
Department of Geology, University of Wales College Cardiff, P.O. Box 914, Cardiff, UK
Present address: Department of Geology and Petroleum Geology, University of Aberdeen, King's College, Aberdeen AB9 2UE, UK

ABSTRACT

Silesian sediments in SW Britain are preserved in the Culm Basin of SW England and the South Wales Basin. These basins developed following a regional compressional deformation event related to the early stages of the Variscan Orogeny. As such, they represent foreland basins formed between a northwardly migrating Variscan thrust 'front' and the cratonic Wales-Brabant Massif.

The Namurian to early Westphalian C Culm Basin includes: the distal turbidites of the Crackington Formation, the shallow marine/brackish clastic Bude Formation and the deltaic Bideford Formation. The South Wales basin includes:
1) high energy, shallow marine sandstones of the early Namurian Basal Grit,
2) low energy, shallow marine mudstones of the late Namurian Shale Group,
3) mudstone dominated coastal plain and tidal sediments of Westphalian A to early C age (Lower and Middle Coal Measures), and 4) sandstone dominated alluvial plain sediments of Westphalian C to early Cantabrian age (Pennant Measures).

Palaeogeographic reconstructions based on thickness, facies and palaeocurrent variations indicate that the South Wales Basin was intermittently sourced from all sides during the Namurian, and that axial (E/W) sedimentation took place in the Namurian Culm Basin. By Westphalian A-B times uplift of a landmass to separate and source the Culm and South Wales Basins had taken place. Northern and eastern sources continued to intermittently supply material to the South Wales Basin. Westphalian C sediments indicate that this landmass (including Devonian sandstones) had been uplifted to comprise a source for the southerly derived Pennant Measures.

The periodic northern source for the South Wales Basin is attributed to uplift related to inversion along pre-existing extensional faults driven by the compressional tectonic regime. The intermittent easterly supply was influenced by movement on the N/S trending Usk Axis.

A model to explain the palaeogeographic reconstructions is proposed in which the southerly dipping (25½) Bristol Channel Fault represents the emergence of a basal thrust (the position of which may have been controlled by a pre-existing extensional fault) from beneath the Culm Basin. Movement on the fault resulted in uplift of the landmass to source the Culm and South Wales Basins during Westphalian A-B times. Further movement detached the Culm to form a thrust-sheet-top basin. By Westphalian C times Variscan deformation had migrated northwards resulting in increased subsidence within the South Wales Basin. The Pennant Measures were therfore sourced directly from the Variscan thrust wedge and deposited in a foredeep basin, prior to the final phase of deformation in post-early Cantabrian times.

INTRODUCTION

The Silesian Culm and South Wales Basins including the Forest of Dean Coalfield (Fig. 1) represent the final stages of sedimentation prior to the late Carboniferous phase of the Variscan Orogeny in SW Britain. They represent foreland basins (Allen et al. 1986) formed to the north of the northwardly migrating Variscan orogen and south of the cratonic Wales-Brabant Massif. A study of these basins using detailed facies, thickness and palaeocurrent analysis will allow the sedimentary response to the northward migration of the compressional orogen to be elucidated.

Recent workers have applied a lithospheric loading model (basins formation through loading due to thrust nappe emplacement-Beaumont, 1981) to the Silesian Culm (Thomas, 1988) and South Wales Basins (Kelling, 1988; Gayer and Jones, 1989). Although Hartley and Warr (1990) favour basin formation related to inversion of pre-existing mid-Devonian to Lower Carboniferous extensional basins. However, attempts to present a coherent model for the Silesian evolution of SW Britain have been hampered by a lack of knowledge regarding the precise relationship of the Culm and South Wales Basins. In particular, problems have arisen due to the possibility that syn- or post-Silesian strike-slip movement may have taken place along the Bristol Channel Fault Zone (BCFZ) (Kellaway and Hancock, 1983; Holder and Leveridge, 1986; Fig. 1). However, recent seismic reflection profiles presented by Brooks et al. (1988), showed that the BCFZ has the form of a shallow (25½) southward dipping reflector which they interpreted as a reactivated Variscan thrust. This observation permits a thin-skinned tectonic link to be proposed between South Wales and SW England (Gayer and Jones, 1989), thus allowing a direct correlation between Silesian sediments in the two areas and negating the necessity to invoke strike-slip movement along the BCFZ. Links between the two areas can therefore be inferred for despositional environments, source area location and composition, thus allowing the Silesian palaeogeography of SW Britain to be reconstructed.

In the following section the sedimentology of Silesian sequences in SW Britain is reviewed (biostratigraphical relationships and thickness variation are shown in Fig. 2). This is followed by a regional synthesis, palaeogeographic reconstructions and a discussion on the implications for the evolution of the Variscan Orogeny in SW Britain.

THE CULM BASIN

The Culm Basin exposed in Devon and Cornwall is bounded to the south by the Rusey Fault Zone (Fig. 3a). It developed conformably on Brigantian marine cherts, impure limestones and black shales (Edmonds, 1974). Following the stratigraphic revision of Edmonds et al. (1979)

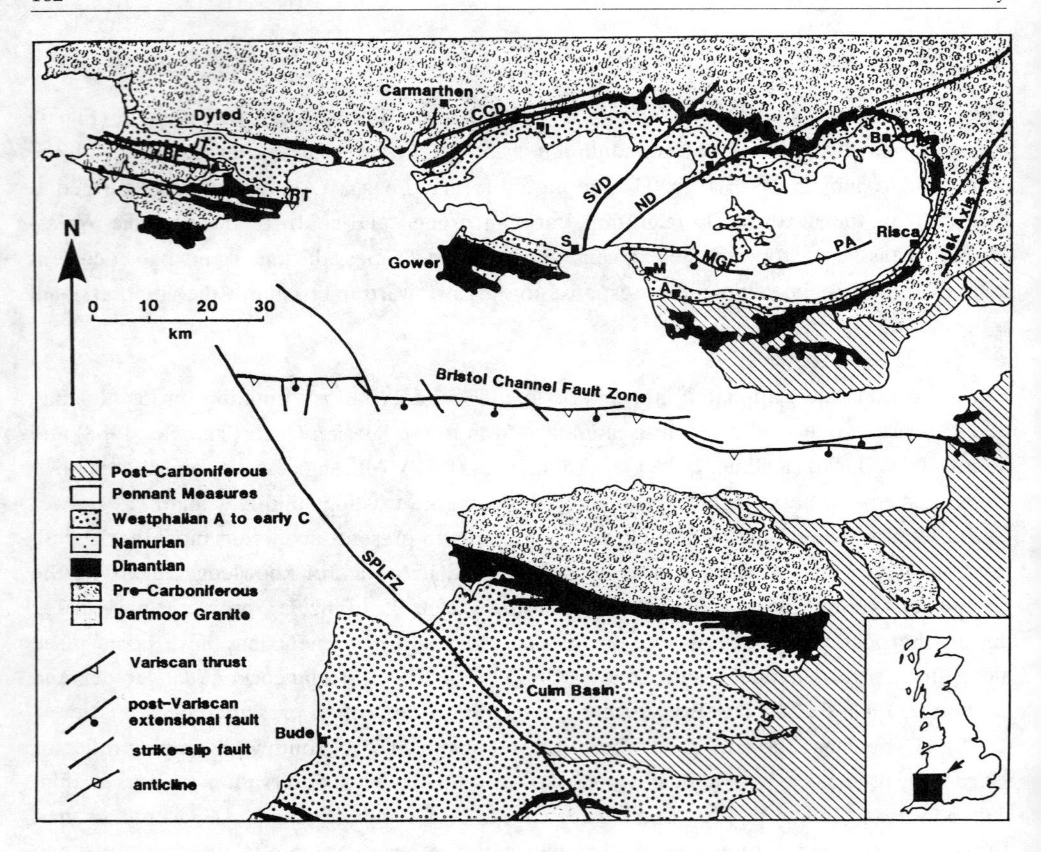

FIGURE 1
General geological map of SW Britain, showing Silesian sediments of South Wales, the Culm Basin of SW England and important structural features. CCD = Carred Cennan Distrubance, SVD = Swansea Valley Distrubance, ND = Neath Disturbance, PA = Pontypridd Anticline, SPLFZ = Sticklepath- Lustleigh Fault Zone, MGF = Moel Gilau Fault, JT = Johnston Thrust, BF = Benton Fault, RT = Ritec Thrust, A = Aberkefnig, M = Margan, S = Swansea, G = Glynneath. Based on B.G.S. 1:625,000 maps. Namurian and Westphalian strata in the Culm Basin have not been distinguished, see Fig. 3a. Arrow on inset indicates position of the Forest of Dean Coalfield.

three Silesian formations are recognised in the basin (Fig. 2): the Crackington (lower Namurian-early Westphalian A), Bude (early Westphalian A-early Westphalian C) and Bideford (Westphalian A) Formations (Edmonds, 1974). However, recently acquired palaeontological data (Xu Li, pers. comm. 1990) suggests that the base of the Bideford Formation may extend downwards into the top of the Namurian. Edmonds et al. (1979) used goniatite marine bands (MB) to demonstrate that despite their large age range the formations were coeval during early Westphalian A times (Fig. 3b). However, due to the structural complexities of the basin and poor exposure inland of the coast, the precise spatial relationships of the formations remains speculative.

Crackington Formation

The Crackington Formation forms the basal sequence throughout the Culm Basin and is conformable on Dinantian sediments. The thickness of the formation is unknown due to structural problems but is probably 500 m+ (Edmonds et al. 1979) and may be significantly greater. Detailed descriptions of the formation can be found in McKeown et al (1973); Edmonds et al (1979); Melvin (1986) and Thomas (1988). The formation comprises crudely parallel laminated black shales. The shales are interbedded with laterally continuous fine to medium grained sandstone beds (Fig. 4). The sandstones are commonly 0.1 - 0.5 m thick, massive with a thin, normally graded (fine sandstone to mudstone), ripple cross-laminated top. Sandstone bases often display flute casts and groove and prod-marks which indicate bimodal E/W flow directions (Melvin, 1986; Fig. 4). Rare synsedimentary slumps have been observed. The percentage of sandstone increases towards the top of the formation (Edmonds et al., 1979) giving an overall coarsening upward profile. The shales occasionally contain goniatite rich faunas. These faunas are scattered throughout the lower parts of the formation (lower-mid Namurian), but become restricted to particular horizons in the upper parts of the formation (late Namurian-early Westphalian A).

Ashwin (1957) first interpreted the sandstones of the Crackington Formation to have been desposited by turbidity currents. Melvin (1986) considered deposition to have taken place in broad shallow channels. However, the lack of extensive channelisation and large lateral extent of individual sandstone beds suggests that the turbidite flows were mostly unconfined.

The goniatite-bearing black shales of the formation have no apparent lithological or compositional difference to the non-fossiliferous shales. On this basis, Thomas (1988) considered the formation to have been deposited in an entirely marine environment. This hypothesis is supported by the development of pyrite within the shales, a feature which commonly reflects early marine pore water compositions (cf. Postma, 1982).

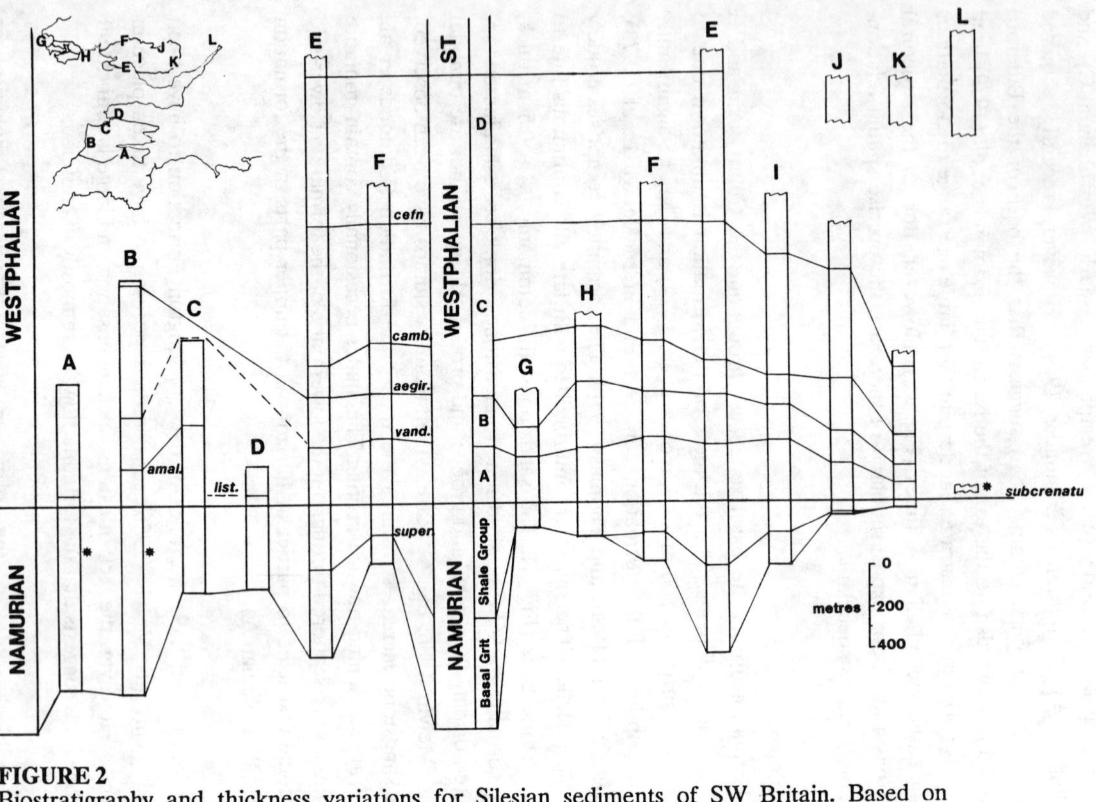

FIGURE 2
Biostratigraphy and thickness variations for Silesian sediments of SW Britain. Based on Jenkins (1962), Woodland and Evans (1964), Williams (1968), Squirrell and Downing (1969), Ramsbottom et al. (1978), Barclay et al. (1988). A = Okehampton, B = Hartland to Boscastle, C = Westward Ho, D = Barnstaple, E = Swansea, F = Ammanford, G = St. Brides Bay, H = Marros, I = Garw, J = Myrthyr Tydfil, K = Rudry, L = Forest of Dean, amal. = G amaliae, List. = G. Listeri, Super. = R. Superbilingue, vand. = A. vanderbeckei, aeigr. = A. aegiranum, camb = A. cambriense, cefn = Cefn Glas coal seam. ST = Stephanian, * thickness uncertain.

The presence of unconfined turbidite flows with E/W palaeocurrents suggests an axial drainage system (see below). An ultimate northerly source is favoured by Thomas (1988) as fine grained orthoquartzitic sandstones are more likely to come from a relatively mature hinterland rather than the coarse grained synorogenic sediment which would be expected from a tectonically active southern source area (Selwood and Thomas, 1986).

The section of the Crackington Formation exposed at Westward Ho! (Fig. 3a) (originally the Westward Ho! Formation) contains slightly different facies to those described above. Walker (1970) described a series of cycles where the thin bedded turbidite facies common to the Crackington Formation (see above) passes upwards into channelised siltstone and fine grained sandstone bodies interbedded with ripple cross-laminated siltstones and streaked mudstones. He interpreted this to represent a series of shallowing upward cycles from turbidites deposited below wave base to 'agitated water' conditions represented by the cross-laminated siltstones and streaked mudstones which are inferred to have been deposited above wave base.

Bude Formation

The formation conformably overlies the Crackington Formation and is coeval with the top of the Bideford Formation (Edmonds et al. 1979; Fig. 3b). Freshney et al. (1979 a, b) claim that Bude Formation type sandstones can be traced laterally into the Bideford Formation.

The Bude Formation is approximately 1300 m thick and consists of massive to crudely laminated dark grey shales, ripple cross-laminated siltstones and thin (0.1 - 0.5 cm) laterally extensive, fine to medium grained sandstones similar to the Crackington Formation. However, it also contains sandstone beds up to 5 m thick which may be amalgamated to form sandstone bodies up to 20 m thick (Fig. 4). Mudstone clasts are common, particularly at the base of thicker beds. The sandstones are more variable in thickness than the Crackington Formation, some are laterally impersistent and show evidence of channelisation. They are commonly structureless although large scale trough cross-stratification (10's of cm; see Freshney et al. 1979a) and ripple cross-stratification (Higgs, 1984) can be locally important. Also, Higgs (1983) tentatively identified hummocky cross-stratification (HCS) in some beds. Sandstone tops are commonly rippled, whilst bases may display flute casts, groove and tool marks which indicate supply from all quadrants except the south (Freshney et al., 1979 a; Higgs, 1986; Melvin, 1986). Synsedimentary slumps are relatively common within the formation, they usually affect beds greater than 3 m in thickness and can affect strata up to 22 m in thickness (Freshney et al. 1979).

The depositional environment of the Bude Formation has been the subject of much debate (see Higgs, 1984, for details) centering on whether the formation was deposited in deep or

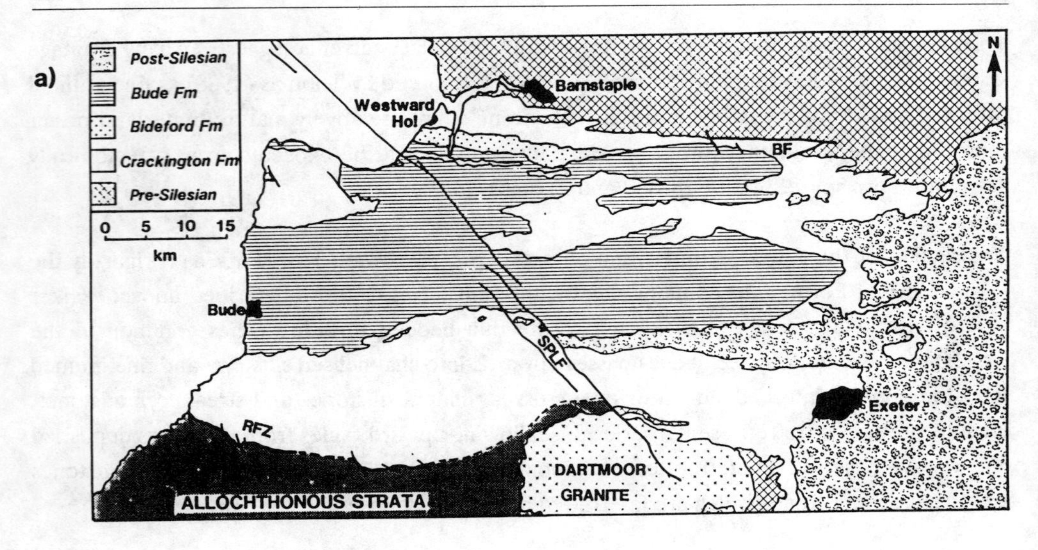

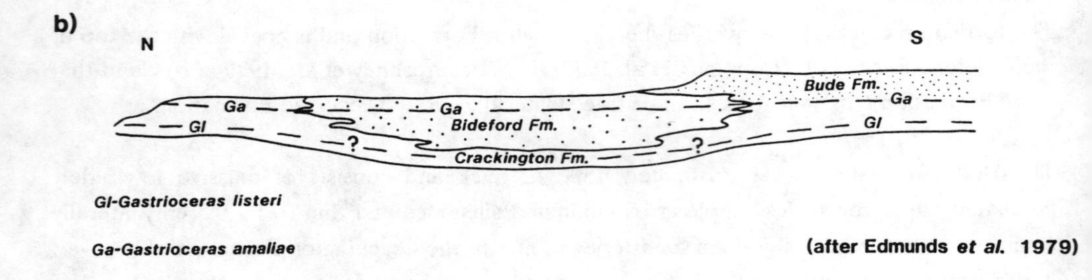

FIGURE 3
(a) Geological map of the Culm Basin. R.F.Z. = Rusey Fault Zone, which froms the southern margin of the Basin, S.P.L.F. = Sticklepath-Lustleigh Fault Zone. Based on Thomas (1988) from B.G.S. 1:50,000 maps.
(b) Schematic age relationships for formations within the Culm Basin (after Edmonds et al., 1979).

shallow water and under marine or freshwater conditions. The presence of thick, crudely channelised, massive sandstones with sole marked bases suggests deposition by turbidity currents in poorly confined channels and led Melvin (1986) to propose a direct fluvia-deltaic supply for the sandstones, deposition occurring in relatively straight broad channels on a gently sloping submarine fan. A direct fluvio-deltaic supply is also supported by the presence of large-scale trough cross-strata (the wedge bedding of Freshney et al. 1979 a), which indicate the development of stable bedforms with a lower but much more constant sediment supply than the massive turbidite beds of the formation. These bedforms may represent mouth bar deposits.

The presence of HCS was taken by Higgs (1984, 1986) to indicate deposition in relatively shallow water between storm and fair weather wave base. Although, it should be noted that a recent study has questioned the bathymmetric significance of HCS (Allen und Underhill, 1989). Higgs took the presence of HCS together with the extensive ripple cross-laminated tops to sandstones to represent primary storm (event) deposits although it is possible that storm-induced wave reworking of previously deposited turbidites was also significant.

Faunas and ichnofaunas are rare within the Bude Formation. Marine faunas are restricted to a few black, sulphurous, nodular shales which have a wide occurence (termed'index' shales by Freshney and Taylor, 1972) and have been correlated with the marine bands of the South Wales Coalfield (see Freshney et al. 1979 a). The shales show a transition from a fish coprolite band (and rare fish remains) to a goniatite rich band, returning to a coprolite band and interpreted by Freshney et al. (1979 a) to record a transition from brackish to marine to brackish conditions representing a marine transgression. Trace fossils within the formation include Planolites, Diplocraterion, fish traces and limulid (Xiphosurid) tracksways (King, 1965; Goldring and Seilacher 1971). With the exception of the latter, the traces are not diagnostic, however, by analogy with the present day occurrence of Limulids the trackways indicate that conditions were largely marine within the basin.

Sedimentological and palaeontological evidence appears to indicate that the Bude Formation is largely shallow marine in nature. Large-scale trough cross-strata and channelisation of the thick turbidite beds suggest a direct fluvio-deltaic influence. Wave reworked tops to sandstones and HCS probably signifies deposition above storm wave base. The absence of extensive faunas and limited bioturbation suggests deposition in oxygen deficient waters with restricted circulation. Where diagnostic, the fauna and ichnofauna indicate marine conditions predominated, although their limited distribution cannot preclude the presence of brackish or freshwater conditions.

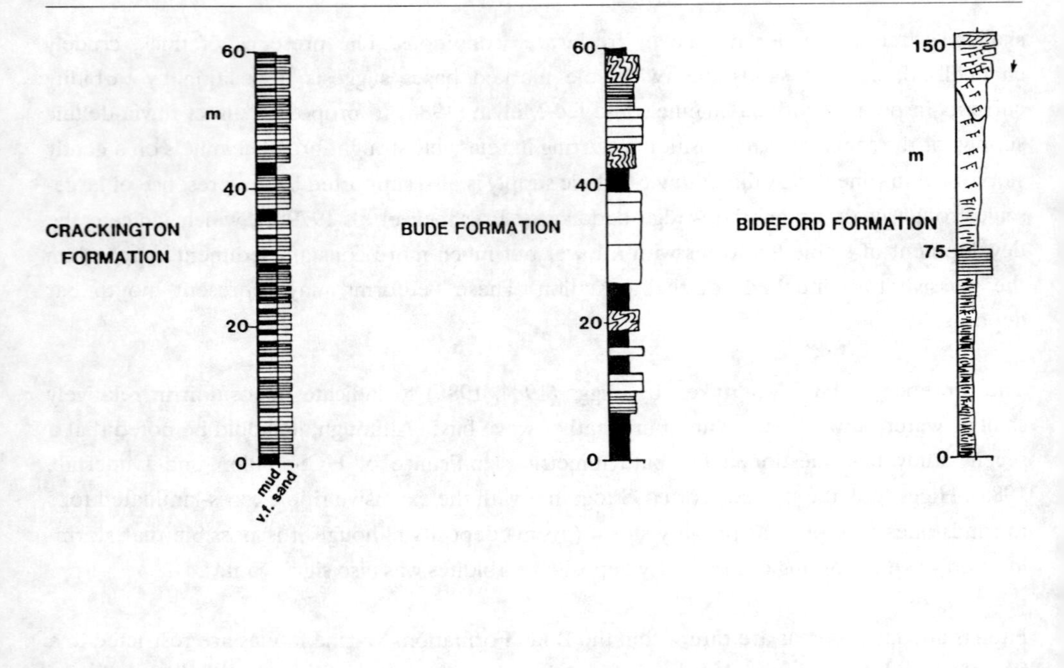

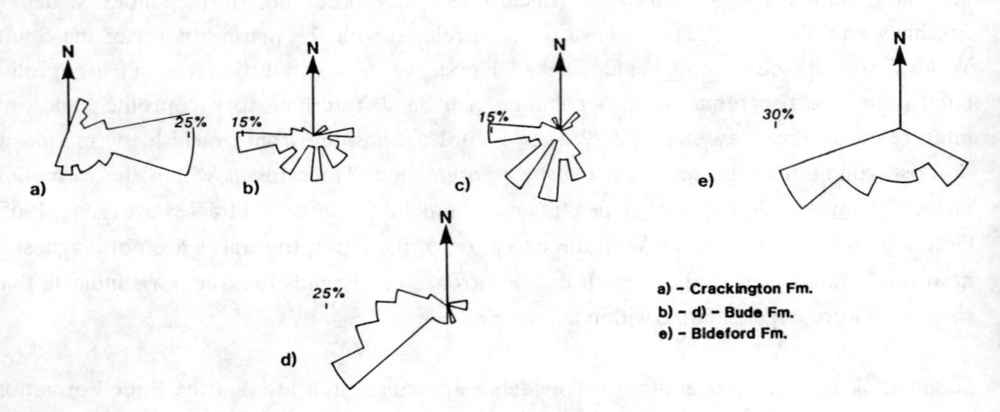

FIGURE 4
Representative lithological logs of the Crackington, Bideford and Bude Formations with palaeocurrent measurements. After Elliott (1976), Edmonds et al. (1979), Freshney et al. (1979) and Melvin, (1986). Note differeneces in scale. Bideford Formation from cycle 1 of Elliott (1976), see text for discussion of palaeocurrent varition within the Bude Formation.

Detailed examination of palaeocurrent data shows that palaeocurrent azimuths at different horizons (between index shales) vary considerably from NE to NW (Freshney et al. 1979 a; Melvin 1986; Fig. 4) and are thought to reflect point source switching. Studies of synsedimentary slumps also indicate a southerly dipping palaeoslope during most of Bude Formation desposition, however within the highest stratigraphical levels of the formation a northerly sense of vergence has been demonstrated (Enfield et al., 1985). Synsedimentary compressional deformation with a northward vergence has been documented by Whalley and Lloyd (1986).

Petrographically the Bude sandstones can be classified as subfels-and sublitharenites (Melvin, 1986), and are dominated by subangular to subrounded quartz grains. Lithic material usually comprises < 20 % of the rock and includes intrabasinal sandstone and siltstone clasts and extrabasinal chert, acidic and basic volcanics, acidic plutonics, schist, schistose quartzite and phyllite clasts (Freshney et al. 1979 a; Edmonds et al., 1979). Detrital feldspars consist of sodic plagioclase with rare perthitic K-feldspar grains occasionally present. Accessory minerals include tourmaline, zircon, sphene and rare epidote (Freshney et al. 1979 a; Edmonds et al. 1979). Melvin (1986) interpreted the composition of the sandstones to indicate derivation from a mature provenance consisting largely of pre-existing sediments and sedimentary rocks.

Bideford Formation

The principal feature of the 800 m thick Bideford Formation (see Fig. 3 b for age relationship to Bude and Crackington Formations) is the development of nine large-scale (50 - 100 m) coarsening upwards cycles (de Raaf et al., 1965; Elliot, 1976). The cycles commence with dark laminated shales which may contain thin (cms), normally graded, fine grained sandstones beds (Fig. 4). These are followed by ripple cross-laminated siltstones and sandstones with the amount of interbedded shale decreasing. The cycles are terminated by thick (10 - 30 m) medium-coarse grained sandstone bodies, which may be channelised or sharply based and commonly display low angle trough cross-stratification (indicating a northerly derivation; Fig. 4) or planar stratification. The thick sandstones are overlain by thin (5 - 15 cm) bioturbated siltstone beds and in one case a coal seam with associated seat-earth and roof containing plant fronds (Prentice, 1960). Marine faunas are sparse, with many of the shales containing brackish/freshwater assemblages.

The corarsening upwards cycles of the Bideford Formation have been attributed to the progradation of a fluvial dominated, elongate delta which extended southwards into a non-tidal, nearhore basin with limited wave activity (de Raaf et al., 1965; Elliott, 1976). The thick sandstones represent distributary channels which overlie delta front mudstones and siltstones. Abandonment facies are recorded by bioturbated siltstones and coals.

Basin History

Sedimentation in the Culm Basin commenced in the Namurian with deposition of the distal turbidites of the Crackington Formation. The Crackington Formation coarsens upwards and in the Westward Ho! area passes upwards into the deltaic Bideford Formation. The shallow marine/brackish sediments of the Bude formation are thought to pass laterally into the Bideford Formation (Freshney et al. 1979 a, b). All three formations contain the mid-Westphalian A G. amaliae horizon indicating that they are coeval (Fig. 3 b), and, notwithstanding any structural complications (discussed below), any palaeogeographic reconstruction must include all three formations.

Palaeocurrent and sedimentological data from the largely Namurian Crackington Formation are thought to reflect an E/W oriented axial drainage system (it is considered axial as the Bideford and Bude Formations are obviously more proximal and are largely derived from the north). Uplift of a northerly source area to supply the Bideford and Bude Formations had taken place by late Namurian/early Westphalian A times. Uplift and subaerial exposure is indicated by the presence of a probable in situ coal seam (Prentice, 1960) within the Bideford Formation.

A palaeogeographic reconstruction for all three formations during G. amaliae times is shown in Fig. 5, where the deltaic Bideford Formation passes laterally downslope (N to S) into the shallow marine/brackish sandstone dominated Bude Formation. The Bude Formation in turn passes downslope into the more basinal thin sheet sandstones of the Crackington Formation. A shallow marine/brackish environment with limited water circulation and no tidal influence is indicated by the biota and sedimentology. However, a problem with this scenario is the presence of the Crackington Formation to the north of the Bideford Formation during G. amaliae times in an area which apparently comprised the source for the Bideford Formation (Fig. 3 a, b). Edmonds et al. (1979) attributed this inconsistency to either local structural complexities or a rapid lateral facies change. Reading (1965) and Thomas (1988) also favour a structural solution to the problem possibly in terms of a low angle thrust fault juxtaposing a much more southerly derived thrust sheet containing the Bideford Formation adjacent to the temporally but not spatially equivalent Crackington Formation strata to the north of Bideford (Reading, 1965). However, structurally it is difficult to envisage a scenario in which a thrust sheet could be transported northwards yet not apparently be emplaced upon younger strata. Alternatively, an E-W oriented strike-slip fault would also explain this anomaly, however, as the fault would have to extend along the entire northern margin of the Bideford Formation outcrop (Fig. 3 a), it would form a major structural feature, a feature which has not been recognised. More detailed work in this area is required to clarify this problem, although, a sedimentological solution should not be discounted, including the possibility that the

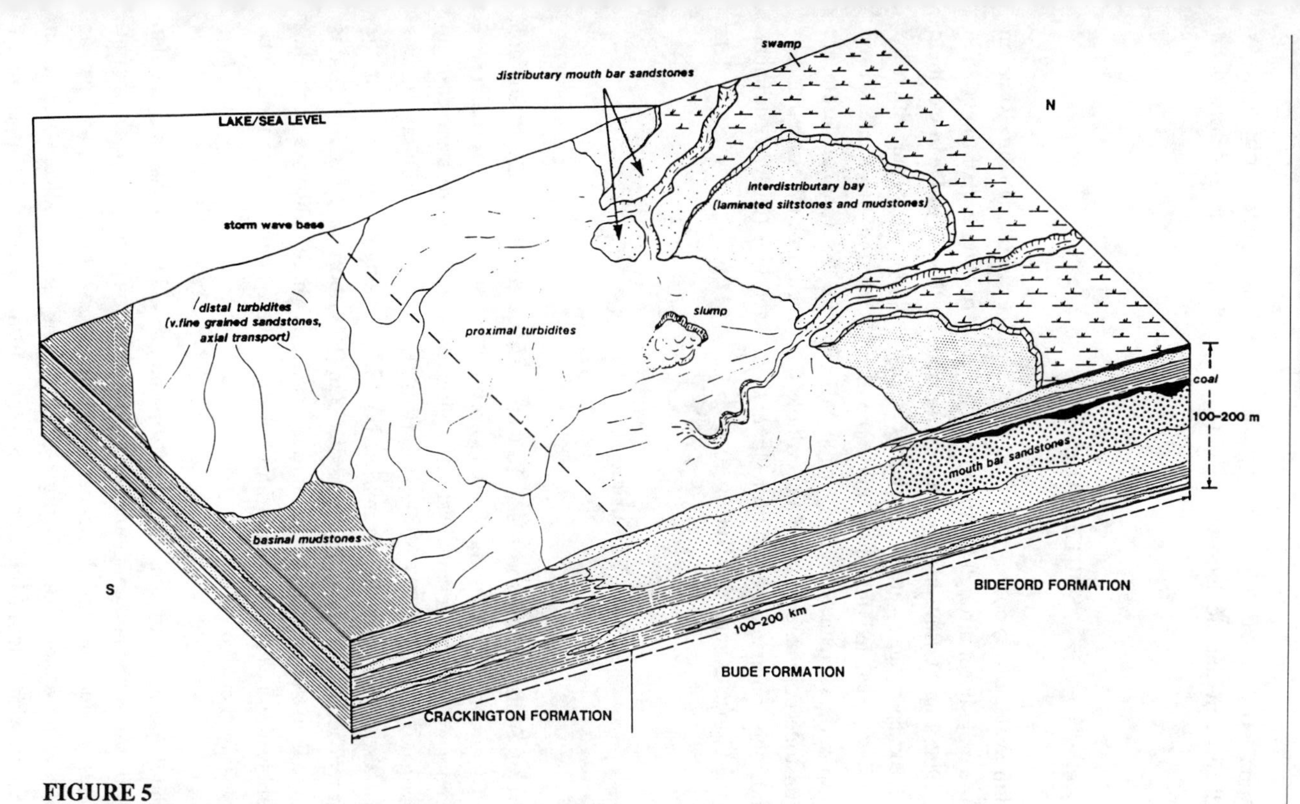

FIGURE 5
Schematic interpretation of the palaeogeography of the Culm Basin during mid-Westphalian A (G. amaliae) times. Inferred Lateral relationships derived from Walthers Law.

Crackington Formation north of Bideford comprises a lacustrine interdistributary bay sequence developed between the elongate channels (Elliott, 1965) of the Bideford delta system.

SOUTH WALES

Silesian rocks in South Wales are exposed in a broad E/W syncline stretching from Risca on the eastern edge of the main basin through Glamorgan to a westward extension of the basin in SW Dyfed (Jenkins, 1962; Fig. 1). Equivalently aged strata from the Forest of Dean Coalfield (located east of the Usk axis, Fig. 1) are also briefly described. Namurian strata within the South Wales Basin can be diveded into two units: the Basal Grit and the Shale Group separated by the Lower Marsdenian R. superbilingue MB (Fig. 6). Westphalian strata are composed of the mudstone dominated Lower and Middle Coal Measures (Westphalian A - early C), and the sandstone dominated Pennant (or Upper Coal) Measures (early Westphalian C - D) separated by the Upper Cwmgorse MB (Fig. 6).

Namurian

The Basal Grit
In the southern part of South Wales, the Basal Grit and associated Plastic Clay Beds are conformable on the Dinantian (T.N. George, 1970). However, northern outcrops reveal a sub-Namurian unconformity with a progressive eastwards directed overstep across increasingly older Dinantian corbonates (Ware, 1939; T.N. George, 1970). Deposition of the Basal Grit was largely restricted to the south of the basin in a marked NW-SW trending trough located in the Gower area where it reached its maximum thickness of 435 m (Ramsbottom, 1978; Fig. 2). It is thinly developed on the flanks of the trough and on the East Crop (where it displays its characteristic sandy and pebbly lithofacies) and at St. Brides Bay to the west, no deposition took place (Jones and Owen, 1967; Archer, 1965; Fig. 2).

The Basal Grit on the flanks of the trough, the East Crop and in SW Dyfed consists of tabular and lenticular quartzitic sandstone and conglomerate bodies often containing reworked marine shell debris. Sedimentary structures are dominated by trough cross-stratification and planar stratification. Palaeocurrent vectors (G.T. George, 1970; Kelling, 1974) indicate derivation from a variety of sources, with a dominantly northerly source from the North Crop, southerly source on the South Crop and easterly source from the East Crop (Fig. 5). Both coarsening and fining upwards sequences have been documented (Archer, 1965; G.T. George, 1970;

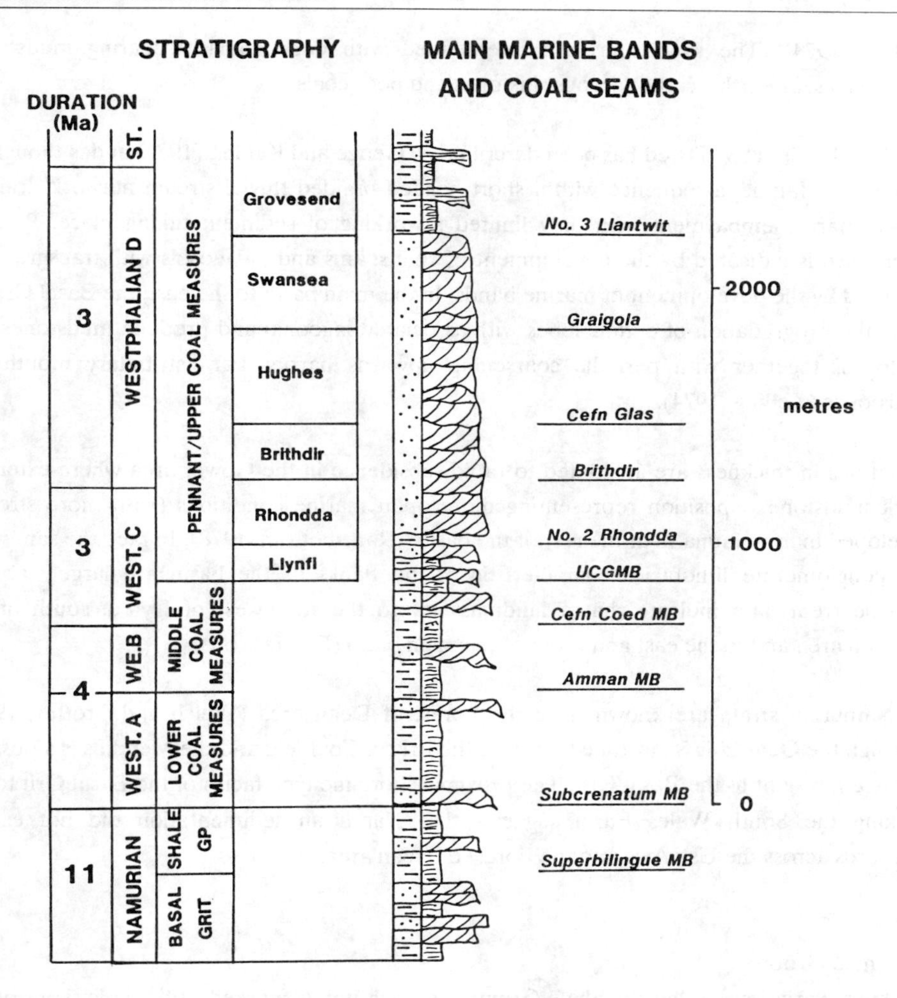

FIGURE 6
General stratigraphy of the Silesian of South Wales. Time scale based on Lippold et al. (1984) and Hess et al. (1985). ST. = Stephanian.

Kelling, 1974). The sandstones are interbedded with dark goniatite-bearing mudstones, occasional seat-earths, carbonaceous siltstones and poor coals.

The Basal Grit in SW Dyfed has been described by Geoge and Kelling (1982) and is thought to represent a fan-delta sequence with a short headed braided fluvial stream network draining into a marine embayment with only limited reworking of sediment taking place. Periodic emergence is indicated by the development of coal seams and palaeosols with transgressions recorded by the development of marine bands. In the main basin to the east, the Basal Grit records the progradation of deltaic lobes with associated lagoonal and prodelta mudstones and siltstones, together with periodic coarsening upwards barrier bar/distributary mouth bar sandstones (Kelling, 1974).

Variations in thickness are attributed to rapid subsidence in the Gower area where extensive black mudstone deposition representing more open marine conditions (with more strongly developed marine fauna than the rest of the basin - Ramsbottom, 1978). In fact, the sandstone and conglomerate lithofacies characteristic of the flanks of the basin are largely absent. Palaeocurrent data indicate that a landmass lay to the southwest of Dyfed, south of the Margam area and to the east and north of the main basin (Fig. 7a).

No Namurian strata are known from the Forest of Dean area (Welch and Trotter, 1961), although the Quartzitic Sandstone Group of the Bristol Coalfield and the Mendips is a possible lateral equivalent to the Basal Grit. The proximal shoreline type facies of the Basal Grit found flanking the South Wales Basin suggests that Namurian sedimentation did not extend eastwards across the Usk Axis into the Forest of Dean area.

The Shale Group

Thickness variations within the Shale Group, although not as marked, still mimic those of the underlying Basal Grit. Strata vary in thickness from 30 m in the east to 280 m in the Gower area to 120 m in the west. The Group is dominated by parallel laminated mudstones and siltstones (commonly containing a marine or brackish fauna), with occasionally developed laterally extensive, medium-coarse grained sheet sandstones. The sheet sandstones top coarsening upwards cycles and are locally rootleted with occasional coal smuts. Thin impure freshwater (Amroth, SW Dyfed) and marine limestones are rare (Archer, 1968).

Archer (1968) described the distribution of marine fauna in the Shale Group from the western part of the Coalfield. He noted that the Group contained marine fauna throughout, but that the distribution of fauna varied cyclically.

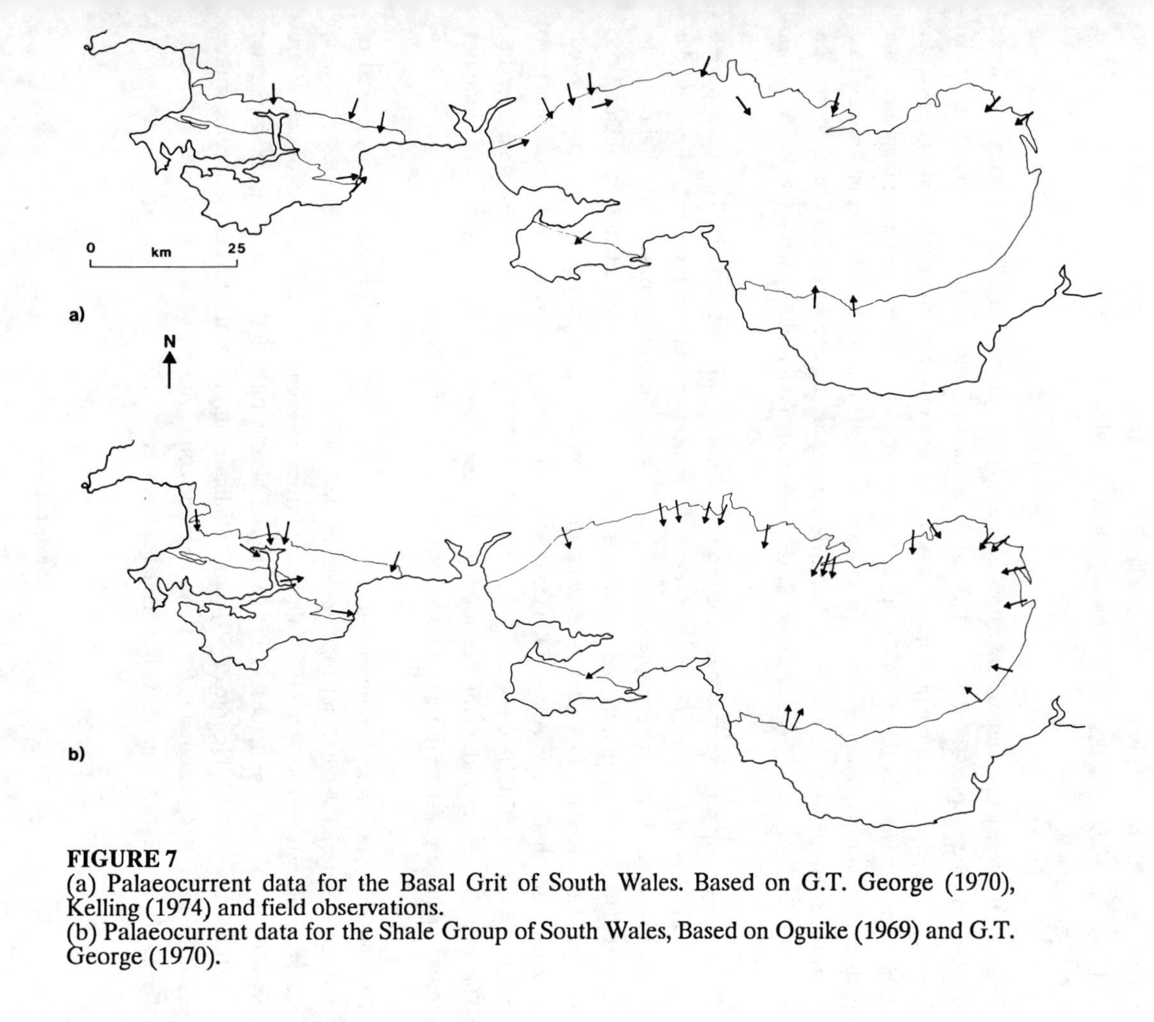

FIGURE 7
(a) Palaeocurrent data for the Basal Grit of South Wales. Based on G.T. George (1970), Kelling (1974) and field observations.
(b) Palaeocurrent data for the Shale Group of South Wales, Based on Oguike (1969) and G.T. George (1970).

In particular, in the vicinity of marine bands a much greater faunal diversity was found grading from a barren phase into Planolites, Lingula, mollusc spat, bivalves, Anthracoceras/Dimorphoce-ras to thicker shelled goniatites such as Gastrioceratids and a brachiopod phase.

Thick (5 to 10 m), channelised sandstone sequences are present at different horizons within the Group (see George 1982 for a detailed discription of one sequence, the Farewell rock, from SW Dyfed). On the north crop the sands were derived from the north and on the south crop from the west. In the main Coalfield Oguike (1969) identified three similar sandstone bodies. One in the Llandebie area derived from the north, one in the Aberkefnig area derived from the south and another in the Brynmawr area in the NE corner of the Coalfield derived from the east (Fig. 7b). These sandstones are developed in the middle part of the Shale Group.

The Shale Group represents a period of decreased energy within the basin. This is supported by the presence of thick sequences of black shales indicating a derease in clastic input into the basin following Basal Grit sedimentaion. The thick channelised sandstones are thought to represent distributary mouth bar sequences related to delta progradation (Oguike, 1969; George, 1970; Kelling, 1974) derived from a source area lying to the north and west of Dyfed, and to the north, east and south of the Coalfield. The laterally extensive sheet sandstones have been interpreted by G.T. George (1970) and Kelling (1974) as offshore barrier-bar deposits. The faunal cyclicity within the Group suggests that although salinity fluctuated conditions remained largely marine throughout deposition of the Shale Group

Petrographic data from Namurian sandstones indicates a source area largely comprised of quartzitic material yielding subangular-subrounded grains. Feldspar is largely absent. These features suggested to Kelling (1974) derivation from a sedimentary source. Igneous clasts are occasionally present and locally, in SW Dyfed, acidic tuffs and lavas form an important constituent, which G.T. George (1970) took to indicate supply from a Lower Palaeozoic source area. Kelling (1974), also noted that there was little compositional variation between sediments derived from the south and those derived from the north.

Westphalian

Lower and Middle Coal Measures

The Mudstone dominated Lower and Middle Measures comprise the economic part of the South Wales Coalfield. Thicknesses vary markedly from west to east across the basin (Fig. 2). The Coalfield of SW Dyfed forms an eastward extension to the main Coalfield (Jenkins, 1962; Fig. 1). Marine Bands (MB) occur in two main intervals within lower-mid Westphalian A and

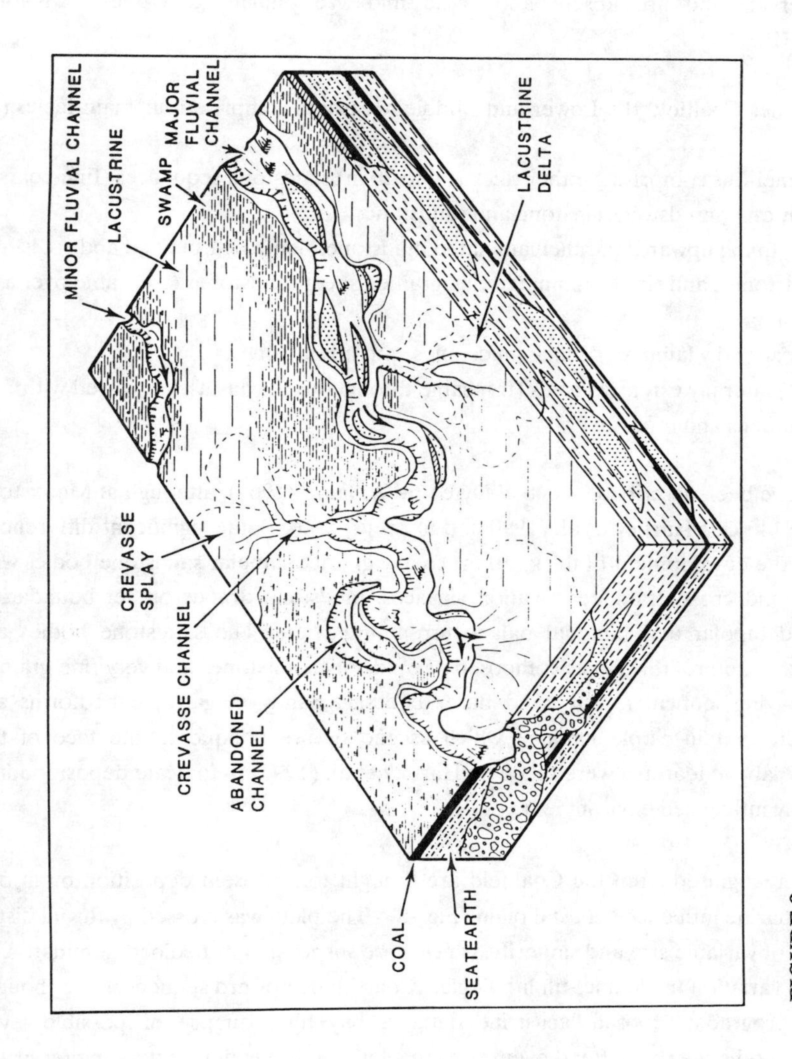

FIGURE 8
Reconstruction of Lower and Middle Coal Measures environments. Note that the major fluvial system may be of high or low sinuosity.

upper Westphalian B-lower Westphalian C strata. They are sporadically developed between these two intervals and are absent above the mid Westphalian C Upper Cwmgorse (Cambriense) MB.

In the South Wales Coalfield the Lower and Middle Measures comprise four main facies (cf. Fielding, 1984):

1) isolated channel-fills comprising either thick (up to 20 m) sandstone sequences (fine-coarse) or a combination of mudstone, siltstone and sandstone;
2) coarsening or fining upwards parallel laminated mudstones, lenticular bedded and cross-laminated siltstones, and ripple laminated fine grained sheet sandstones (traceable over a number of kms);
3) extensive horizontally laminated black mudstones with no siltstone;
4) thick (metres) laterally extensive coals (traceable throughout the basin), assosiated with rootleted siltstones and mudstones.

Similar facies have been described from SW Dyfed by Williams (1968), although at Monkstone Point (NGR SN146030) Hartley et al. (1990 and in prep.) noted some significant differences. These included the development of thick (10 m) coarse grained, tabular sandstone bodies with complex compound cross-strata, reactivation surfaces, mudstone drapes on set boundaries, pebble lags and bipolar and bimodal palaeocurrent indicators. The sandstone bodies are interbedded with a heterolithic facies composed of mudstones, siltstones and very fine grained sandstone displaying lenticular, wavy and flaser bedding. Sandy megaripple bedforms are developed which contain ripple laminae which ascend or are oblique to the face of the megaripple. The above features were taken by Hartley et al. (1990) to indicate deposition in a nearshore, tidally influenced environment.

The four facies recognised from the Coalfield are thought to represent deposition on a low-lying, swampy, marine influenced coastal plain (Fig. 8).- The plain was crossed by fluvial distributary channels of variable size and sinuosity, which were suspension or bedload dominated, as recorded by the variation in channel-fill lithologies. Coarsening upward sequences are thought to record the progradation of a lacustrine delta, a bay fill sequence or possible levee progradation. Periodic overbank flood events or breaching of levees during flooding (crevasse splay) resulted in deposition of fining upwards sheet sandstone and thin siltstone sequences. Thick sequences of mudstone were deposited in lakes and periodic emergence led to the development of peat swamps and associated palaeosols. Periodic glacioeustatically induced marine transgressions led to deposition of marine bands (Leeder, 1989).

Palaeocurrent and provenance data from channel-fills within the Lower and Middle Coal Measures reveals a complex drainage pattern (Fig. 9a), partially influenced by synsedimentary

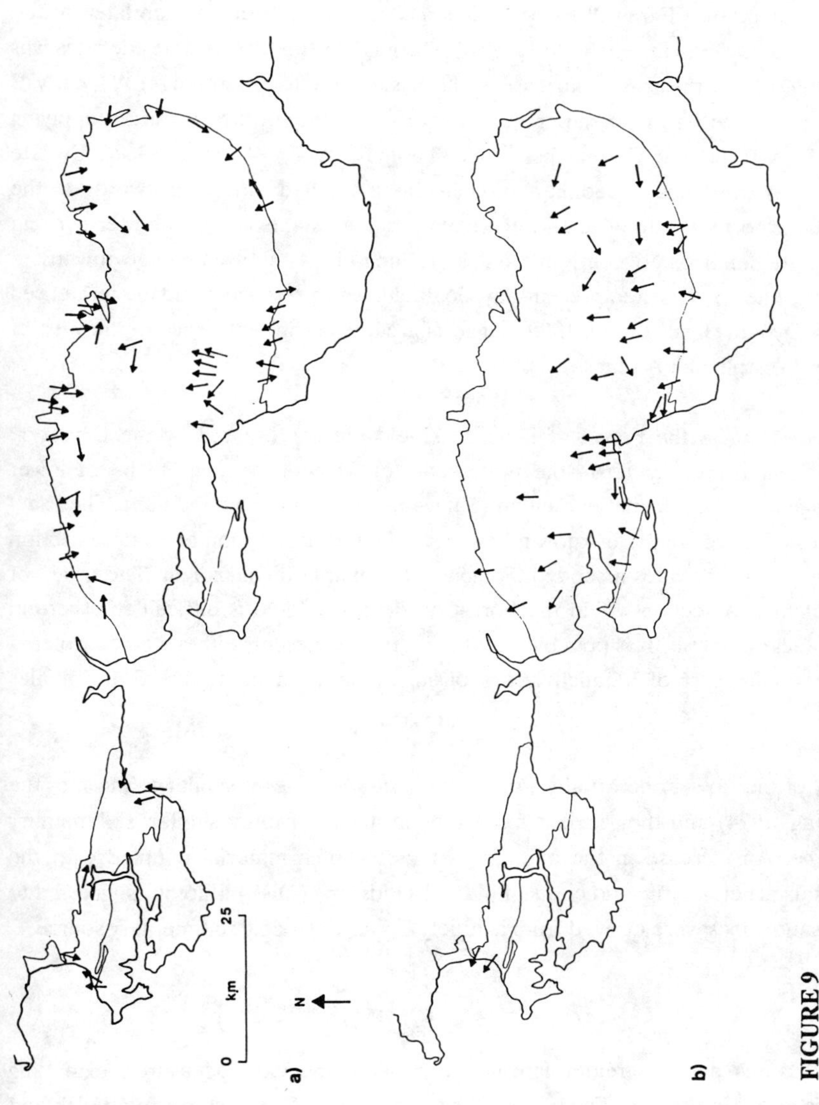

FIGURE 9
(a) Palaeocurrent data for the Lower and Middle Coal Measures (Westphalian A to early C) of South Wales. Based on Bluck (1961), Bluck and Kelling (1963), Thomas (1967), Williams (1968) and field observations.
(b) Palaeocurrent data for the Rhondda, Brithdir, Hughes and Swansea Beds of the Pennant Measures (Westphalian C/D) of South Wales. Based on Kelling (1968, 1974), Jones (1989) and field observations.

tectonic activity (Jones, 1989). Overlying the base of the Coal Measures is the northerly derived, lithologically mature Farewell Rock (which is either diachronous or unrelated to the Farewell Rock of SW Dyfed; cf. Leitch et al., 1958). During Westphalian A times detritus was increasingly supplied from the south and east of the basin. By late Westphalian A - early B times the northerly source had been superceded by a more lithic southerly source supplying detritus to SW Dyfed and the Carmarthen area (Thomas, 1967; Williams, 1968). By late Westphalian B times the locus of sedimentation in the south had moved eastwards to the Margam area, and a northerly supplied fluvial system became established in the Merthyr and Glynneath areas. Regional paleogeographic maps (Kelling, 1974) show drainage towards a coastline located to the south of the present day Coalfield. The presence of tidally influenced sediments in SW Dyfed (Hartley et al. 1990) suggests greater marine influence in the west of the basin in lower Westphalian A times.

Equivalently aged strata in the Forest of Dean Coalfield are represented by the Drybrook Sandstone (Welch and Trotter, 1961) the upper part of which is thought to be of lower Westphalian A age and unconformity-bounded (Sullivan, 1964; Wilson et al. 1988). The exact thickness and the sedimentology is unknown, however the brief description of 30 m of section given by Sullivan (1964) indicates a sandy lithofacies of probable fluvial origin. The origin of this small Westphalian A section within the Forest of Dean Coalfield is difficult to ascertain without detailed facies analysis. It is possible that the section represents either a small isolated basin or was originally part of a much larger basin, possibly linked to the South Wales Coalfield.

The petrography of the Lower and Middle Coal Measures sandstones is similar to those of the Namurian (Kelling, 1974) and they appear to have been derived from a similar, sedimentary dominated, source. An increase in the amount of metamorphic material coupled with the absence of detrital garnet, apatite and clasts of old red sandstone (ORS) lithology suggested to Kelling that the sandstones were derived from a largely Lower Paleozoic sedimentary source.

Pennant Measures

The Pennant Measures have a greater laterial extent than the Coal Measures, extending further eastwards into the Forest of Dean, north Oxfordshire and Gloucestershire (Welch and Trotter, 1961; Worssam, 1963; Poole, 1969). They reach a maximum thickness of 1830 m in the south of the Coalfield and thin laterally to 300 m on the East Crop (Fig. 1), where there is an intra-Pennant unconformity which omits the Swansea Beds (Fig. 2). The Measures commence with deposition of the Llynfi Beds which are transitional between the mudstone dominated Coal Measures and the overlying sandstone dominated Rhondda, Brithdir, Hughes and Swansea Beds (Fig. 6). The Grovesend Beds form the top unit of the Pennant Measures. They contain thick coals and record a return to mudstone dominated sedimentation. There is little variation in style of sandstone deposition between the Rhondda, Brithdir, Hughes and Swansea Beds. Typical sedimentary facies comprise erosive based channelised conglomerates containing log casts, rounded coal clasts, oxidised siderite nodules, siltstone pebbles and quartz granules. These pass upwards into immature, coarse grained sandstones which may be massive or display well developed trough and planar cross-stratification. Rootleted siltstones, thin coals and fine grained sheet sandstones are occasionally present above the coarse grained sandstones, but are more commonly eroded by the succeeding conglomerate.

The sandstones and conglomerates of the Pennant Measures record the development of an extensive low sinuosity fluvial system, with the basal Llynfi Beds reflecting a more sinuous fluvial system (Kelling, 1974; Jones, 1989). The finer grained sediments represent periods of channel abandonment and overbank flood events. Emergent areas were colonised by plants resulting in coal and palaeosol development. Palaeocurrent and provenance data indicate a dominantly southerly vegetated, coal bearing source area (Fig. 9b) with significant lithic detritus, and an occasional easterly quartzitic source (Kelling, 1968, 1974). A northerly quartzitic source supplied detritus to the basin during deposition of the Rhondda Beds (Kelling, 1968). Studies by Heard (1922), Heard and Davis (1924) and Kelling (1974) suggest that the southerly sourced lithic detritus has an Old Red Sandstone provenance. This supposition was based on the presence of abraded and recycled garnets, apatite and limonite cemented protoquartzite clasts. Although, the large amount of metamorphic clasts in the lithic fraction (metaquartzite, slate, quartz-chlorite-schist, quartz-sericite-schist) suggests that (unless reworked from the ORS) a Lower Palaeozoic source still remained exposed.

A 600 m thick, late Westphalian D to early Cantabrian section is preserved in thge Forest of Dean Coalfield (laterally equivalent to the top of the Swansea and Grovesend Beds, Fig. 2; Wagner and Spinner, 1972; Cleal, 1987). Three Groups have been identified: the lower Trenchard Group (which unconformably overlies the Westphalian A Drybrook Sandstones), Pennant and Supra-Pennant Group (Trotter, 1941). The Trenchard Group varies in thickness from 120 m in the north of the Coalfield where it consists of conglomerates, quartzites and rare shales to 60 m in the south where medium grained sandstones and shales dominate (Gayer and Stead, 1971). The rapid lateral facies and thickness changes suggest a local ? fault-bounded source located to the north of the Coalfield. The Pennant Group consists of coarse grained, thick-bedded sandstones and occasional coals similar in nature to the Pennant sandstones of the South Wales Coalfield. The Supra-Pennant Group is dominanted by shales and coals and is lithologically similar to the Govesend Beds of South Wales. The above observations suggest that the Trenchard Group records the initiation of a small basin in the Forest of Dean area in Westphalian D times and that by deposition of the Pennant Group the basin was probably linked to the South Wales Coalfield to the west.

Basin History

Following the break-up of the Dinantian carbonate ramp (Wright, 1986) a NE-SW trough centered on the Swansea Valley was initiated in the Namurian. The early Namurian Basal Grit was deposited in this trough and sourced mainly from the north and to a lesser extent from the south and east. A significant increase in basin width accompanied deposition of the overlying shallow marine Shale Group. Only limited clastic input related to delta progradation was developed in the Group. Source areas lay mainly to the north and south with limited easterly input in the main basin and to the north and west in Dyfed.

A gradual transition from a deltaic to a coastal through to an alluvial plain is recorded from late Namurian to early Westphalian C times decreasing marine influence, and an increase in channelisation and the fluvial nature of sandstones, in contrast to the sheet-like offshore and mouth bar sandstone sequences of the late Namurian. In Westphalian A times the basin saw a gradual decrease in marine activity from west to east, with tidal deposits in Dyfed and alluvial plain sedimentation in the Coalfield. Throughout the Lower and Middle Coal Measures clastic material was derived periodically from the north, south and east.

Deposition of the Pennant Measures in early Westphalian C times marked a change from a mudstone to sandstone dominated alluvial plain environment. The sudden influx of southerly derived coarse grained immature detritus, the large scale nature and decrease in sinuosity of Pennant fluvial systems, together with the lack of marine influence suggests rapid uplift of an immature source to the south of the Coalfield. A widespread feature of other coalfields south of the Wales-Brabant Massif is the development of thick sections of coarse immature sandstones of Westphalian C/D age, suggesting that the influx of the Pennant Measures was regional and not localised to South Wales. Although the exact timing of influx of sandstone varied along the margin of the Massif; it took place in early Westphalian C times (Bisson et al. 1967), in Berkshire in late Westphalian C times (Foster et al. 1989), in Oxfordshire in early Westphalian D times (Foster et al. 1989) and in Somerset in mid Westphalian C times (Kellaway, 1971).

A number of authors have demonstrated that synsedimentary tectonic activity had an important influence on Silesian sedimentation in South Wales. Within the main basin Jones (1989) documented contemporaneous tectonic activity along the E/W Pontypridd Anticline (Fig. 1) which resulted in drainage deflection, coal seam splits and marked thickness changes within the Coal and Pennant Measures. The presence of thickness changes and soft-sedimentary deformation structures were taken by Owen (1953, 1971), Weaver (1976) and Owen and Weaver (1983) to indicate that the three major NE/SW disturbances (Neath, Swansea Valley and Carreg Cennan; Fig. 1 were active during the Silesian. George (1956),

Squirrel and Downing (1964) and Barclay and Jones (1978) attributed the severe attenuation of the Silesian section in the east of the main basin (Fig. 2) to contemporaneous movement of the Usk Axis (Fig. 1) a hypothesis supported by the intra-Pennant unconformity in the eastern part of the Coalfield and the severely limited Silesian section in the Forest of Dean area. To the west of the main basin, Powell (1989) showed that the extensional E/W trending Druidston and Newgale Faults and compressional Ritec and Benton Faults (reactivated Devonian extensional faults) influenced Silesian sedimentation in SW Dyfed.

The above observations indicate that the South Wales Silesian Basin was being actively deformed during sedimentation by E/W, NE/SW and N/S oriented structures. It is generally believed that most of these structures are pre-Variscan in origin and that their differing orientations simply represent varying responses to NNW/N directed Variscan compression, rather than separate phases of deformation.

DISCUSSION

Palaeogeographic Reconstructions and Implications

A major problem in attempting palaeogeographic reconstructions for the Silesian of SW England and South Wales is restoring the true spatial relationship of the two areas prior to Variscan shortening (assuming as outlined previously, that no large scale strike-slip movement has taken place on the BCFZ). Styles of Variscan deformation and shortening estimates vary across SW Britain (eg. Shackleton et al., 1982; Sanderson, 1984; Coward and Smallwood, 1984; Smallwood, 1985; Powell, 1989; Jones 1989) making restoration of spatial relationship of different times between the two areas virtually impossible to estimate. Due to the above problems the palaeogeographic reconstructions presented here are purely schematic and only intended to give an impression of relative distances between SW England and South Wales.

Early Namurian E-R_1

In SW England and in limited parts of South Wales early Namurian sediments were deposited conformably upon Dinantian aged strata. In SW England early Namurian sediments are represented by the distal turbidites of the Crackington Formation, which contain bimodal E-W palaeocurrent indicators reflecting an axial drainage system. No evidence for a northerly source area present in Westphalian A - C times (see below) has been documented (Fig. 10 a).

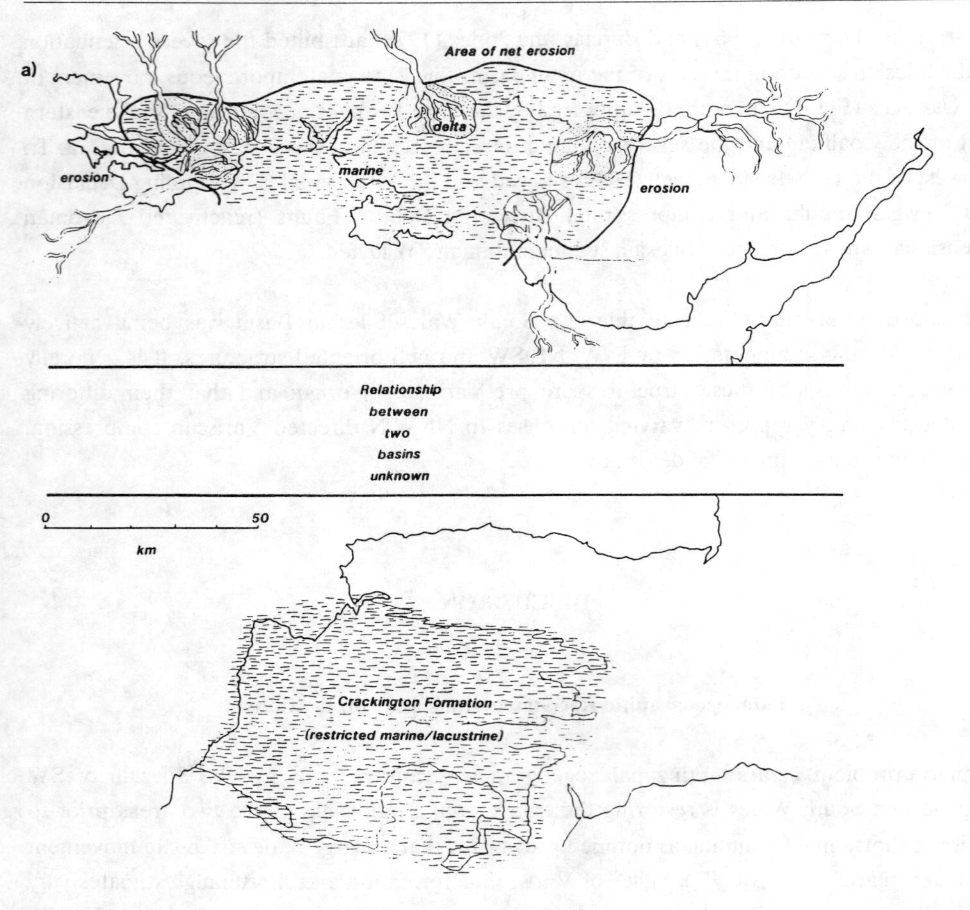

FIGURE 10
Schematic palaeogeographic maps of SW Britain: a) Namurian, b) Westphalian A and c) Westphalian C/D. Drainage patterns are based on palaeocurrent data, facies distributions and deduced from areas of maximum subsidence derived from thickness data. These reconstructions are based on present day topographies, whereas it should be noted that structural data estimate 20-50 % post-Westphalian shortening across SW Britain.

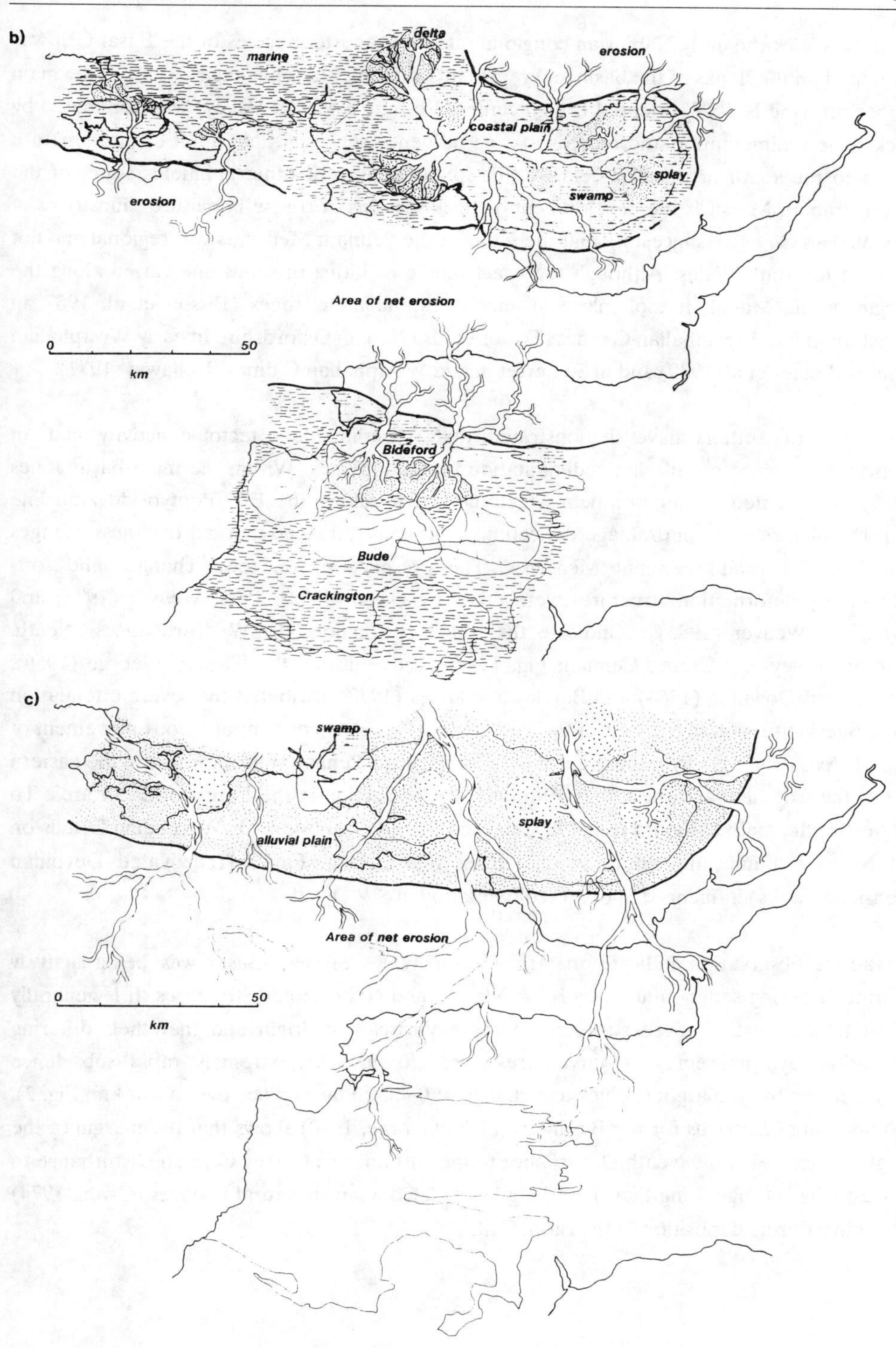
b)
delta
marine
erosion
coastal plain
splay
swamp
erosion
Area of net erosion
0
50
km
Bideford
Bude
Crackington
c)
swamp
alluvial plain
splay
Area of net erosion
0
50
km

In South Wales the early Namurian conglomerate and sandstone facies of the Basal Grit was developed on the flanks of the basin and represents a deltaic/shoreline facies whereas the main depocentre (the NE-SW oriented trough centered on the Swansea Valley) was dominated by black shale sedimentation and records marine influence suggests rapid uplift of an immature source to the south of the Coalfield. A widespreas feature of other coalfields south of the Wales-Brabant Massif is the development of thick sections of coarse immature sandstones of Westphalian C/D age, suggesting that the influx of the Pennant Measures was regional and not localised to South Wales. Although the exact timing of influx of sandstone varied along the margin of the Massif; it took place in early Westphalian C times (Bisson et al. 1967, in Berkshire in late Westphalian C times (Foster et al. 1989), in Oxfordshire in early Westphalian D times (Foster et al. 1989) and in Somerset in mid Westphalian C times (Kellaway, 1971).

A number of authors have demonstrated that synsedimentary tectonic activity had an important influence on Silesion sedimentation in South Wales. Within the main basin Jones (1989) documented contemporaneous tectonic activity along the E/E Pontypridd Anticline (Fig. 1) which resulted in drainage deflection, coal seam splits and marked thickness changes within the Coal and Pennant Measures. The presence of thickness changes and soft-sedimentary deformation structures were taken by Owen (1953), 1971, Weaver (1976) and Owen and Weaver (1983) to indicate that the three major NE/SW distrubances (Neath, Swansea Valley and Carreg Cennan; Fig. 1 were active during the Silesian. George (1956), Squirrel and Downing (1964) and Barclay and Jones (1978) attributed the severe attenuation of the Silesian section in the east of the main basin (Fig. 2) to contemporaneous movement of the Usk Axis (Fig. 1), a hypothesis supported by the intra-Pennant unconformity in the eastern part of the Coalfield and the severely limited Silesian section in the Forest of Dean area. To the west of the main basin, Powell (1989) showed that the extensional E/W trending Druidston and Newgale Faults and compressional Ritec and Benton Faults (reactivated Devonian extensional faults) influenced Silesian sedimentation in SW Dyfed.

The above observation indicate that the South Wales Silesian Basin was being actively deformend during sedimentation by E/W, NE/SW and N/S oriented structures. It is generally believed that most of these structures are pre-Variscan in origin and that their differing orientations symply represent varying responses to NNW/N extremely rapid subsidence relative to the basin margin (a thickness change of 0-435 m takes place over a few km; Fig. 2). Examination of isopachs for the Basal Grit (T.N. George, 1970) shows that the margin of the trough coincide with the Neath Disturbance to the east and the Carreg Cennan Disturbance to the west (Fig. 1) suggesting that these long-lived Caledonian structural features (Owen, 1974) were active during deposition of the Basal Grit.

Late Namurian R_2-G_1

Deposition of the Crackington Formation continuedduring late Namurian times still in SW England and axial E-W flow paleocurrents still to dominated. However, an increase in sandstone turbidite thickness suggests greater clastic input into the basin, possibly indicating closer proximity to the source area (Fig. 10 a). A feature supported by a probable upper Namurian age for the northerly derived Bideford Formation.

In South Wales the late Namurian Shale Group has a wider extent than the underlying Basal Grit and less pronounced thickness variations (Fig. 2), suggesting that a more regional subsidence pattern affected the basin. The quiet water, shallow marine conditions and decrease in coarse clastic input recorded by the Shale Group also supports an increase in the regional subsidence rate or alternatively, an increase in relative sea-level.

Variations in the diversity of fauna recorded by Archer (1968) from the Shale Group suggests variations in salinity within a dominantly marine basin. A gradual decrease in salinity is also apparently recorded from the Crackington Formation where fauna from being widespread in early-mid Namurian times becomes restricted to index shales by upper Namurian times.

Early to late Westphalian A

In SW England a change from the turbiditic/shallow marine sediments of the Crackington Formation (Westward Ho! section) to the deltaics of the Bideford Formation took place during early Westphalian A times. The Bideford Formation was sourced from an immature lithic source to the north, similar to that supplying detritus to the South and East Grops of the Coalfield respectively. At a slightly earlier time (basal Westphalian A) on the North Crop of the Coalfield a northerly source supplied quartzitic detritus to the fluvial systems. At the same time tidal sediments in SW Dyfed indicate an open marine influence in the western part of the South Wales Basin. These features suggest uplift of a source area between SW England and South Wales and reactivation of source areas to the north and east of the Coalfield (Fig. 10 b). The small outlier of lower Westphalian A strata in the Forest of Dean indicates that sedimentation obviously took place in the area, however, its relationship to the Coalfield is unclear without further work.

Late Westphalian A to Early Westphalian C (Upper Cwmgorse MB)

From late Westphalian A (G. amaliae) to early Westphalian C times, deposition of the lacustrine/shallow marine Bude Formation took place in SW England. The formation was sourced from the north and was deposited at the same time as a southerly source supplied lithic detritus to fluvial systems which were established in the Dyfed, Carmarthen and Margam areas. Whilst a northerly quartzitic source supplied fluvial systems in the Merthyr and Glynneath areas. Synsedimentary tectonic activity resulted in the deflection of drainage patterns within the Coalfield (Jones, 1989). Synsedimentary slumps within the Bude Formation (Enfield et al., 1985) indicate a change from a northerly to southerly dipping palaeoslope in late Westphalian B to early C times. Enfield et al., (1985) interpreted this change in vergence to represent tilting associated with the northerly propagation of the Variscan 'Front'. Also contemporaneous northerly verging compressive deformation has been documented from the Bude Formation by Whalley and Lloyd (1986).

The above features coupled with petrographic data suggest that a lithic source area largely composed of Lower Palaezoic sediments, metamorphics and ?ORS, was present between SW England and South Wales and that the area to the north of the Coalfield was subject to periodic uplift. Only limited easterly input is recorded on the East Crop of the Coalfield during this time period indicating relative source area quiescence, although continued attenuation of strata adjacent to the Usk Axis suggest that it remained a positive feature during sedimentation (Fig. 2). Synsedimentary slumps and compressional features in the Bude Formation show that the Culm Basin was being actively deformed during late Westphalian B to early C times. Synsedimentary tectonic activity in South Wales was largely confined to the reactivation of major long-lived structures. Thickness data (Fig. 2) suggest that subsidence was greater in the Culm Basin than in South Wales during Westphalian A to early C times.

Early Westphalian C to D

Post early Westphalian C sediments are not preserved in SW England, although, in South Wales deposition of the southerly derived Pennant measures commenced at this time (Fig. 2, 10 c). The change from the mudstone dominated coastal plain of the Middle Coal Measures to the sandstone dominated alluvial plain of the Pennant Measures was associated with a rapid increase in subsidence (Kelling, 1988; Fig. 2) and resulted in a sudden influx of coarse grained material from a rapidly uplifted, vegetated source area composed of coal, siltstone and siderite-bearing sediments as well as Devonian sandstones. It is unlikely that the coal clasts in the Pennant were sourced by the underlying Lower and Middle Coal Measures, as the Pennant Measures are conformable on the Coal Measures. The only presently exposed coal-bearing sediments south of the Pennant Measures are the Culm seams of the Bideford Formation (see Edmonds et. al., 1979 for a detailed review) which overlie Devonian quartzitic sandstones (Edmonds et al., 1979, 1985), possibly indicating uplift of the Culm Basin and associated Devonian rocks to source the Pennant Measures.

Alternatively, coal seams are present in the Asbian strata of the South Wales area (A.T.S. Ramsay, pers. comm. 1990) and uplift of a Dinantian source could equally explain the influx of coal, although, limestone clasts are largely absent from Silesian strata.

The large vertical thickness (1830 m), coarse grained nature, lack of fine grained sediments, the dominantly low sinuosity nature of Pennant fluvial systems and increased subsidence suggests that the Pennant source area was relatively closer to the Coalfield in early Westphalian C times than in Namurian or Westphalian A - B times, when finer grained detritus was supplied to shallow marine (Namurian) and coastal plain (Westphalian A - B) environments.

Tectonic Model

Sedimentological analysis of the Silesian Culm and South Wales Basins indicates that a southerly situated landmass intermittently supplied detritus to the Welsh Basin during the Namurian (Fig. 11 a). By Westphalian A and B times the landmass had become uplifted to separate and permanently source the Culm and Welsh basins. Early Westphalian C sedimentation suggests that this landmass (?together with the Culm Basin) was uplifted and deformed to form a source for the Pennant Measures of the Coalfield. Thickness data (Fig. 2) indicate that during Westphalian A to early Westphalian C times subsidence was greater in the Culm than in the South Wales Basin. A feature which may reflect the greater proximity of the Culm Basin to any lithospheric load related to the northward migration of the Variscan Orogeny.

Within the Variscan tectonic context of a northerly propagating orogen, it is suggested that the landmass which divided the Culm and Welsh Basins was uplifted in response to fault movement within the basin-fill sequence. It is probable that the BCFZ was originally a Devonian extensional fault (Gayer and Jones, 1989) and following the development of the compressional tectonic regime inversion along the fault resulted in uplift of the source area which divided and supplied the Culm and South Wales Basins. Post-Culm Basin deformation resulted in the development of a thrust-system which linked with the BCFZ resulting in detachment of the Culm Basin to form a thrust-sheet-top basin (Ori and Friend, 1984). This hypothesis is supported by the fact that the BCFZ is the only major structural lineament between the two basins (Fig. 1) and that it is known to be a major Variscan thrust (Brooks et al., 1988).

The proximal nature of the Westphalian C-D Pennant Measures suggests that by early Westphalian C times the Variscan deformation front had migrated through the foreland to

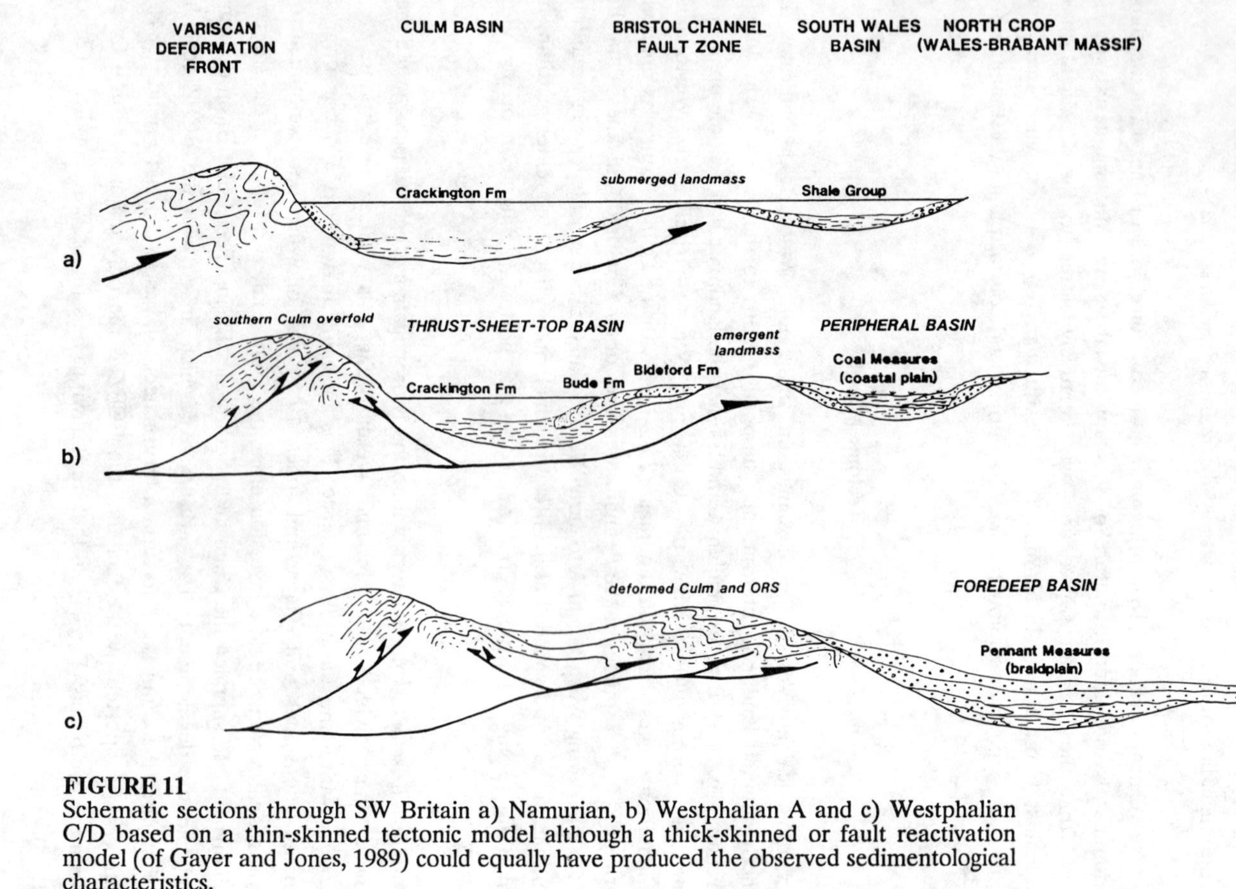

FIGURE 11
Schematic sections through SW Britain a) Namurian, b) Westphalian A and c) Westphalian C/D based on a thin-skinned tectonic model although a thick-skinned or fault reactivation model (of Gayer and Jones, 1989) could equally have produced the observed sedimentological characteristics.

directly source the Pennant sandstones in a foredeep type basin (Allen et al., 1986; Fig. 11 c). As such, the Pennant Measures represent the final and most proximal infill related to the Variscan Orogeny.

The periodic uplift of a source area to the north of the South Wales Basin throughout Namurian to early Westphalian C times is thought to result from inversion of pre-existing Devonian extensional faults during Variscan compression. The sporadic easterly supply of material to the South Wales Basin and isolation of the Forest of Dean Coalfield during the Silesian was probably due to periodic movement on the Usk Axis, although the nature and timing of movement is poorly constrained.

Acknowledgements

Receipt of a NERC funded post-doctoral post within the Special Topic; BAsin Dynamics is gratefully acknowledged, L. Warr, B. Besly, G. Kelling and an anonymous reviewer are thanked for critical reviews of the manuscript.

REFERENCES

Allen, P.A., Homewood, P and Williams, G.D. 1986, Foreland basins: an introduction. In: Allen, P.A. and Homewood, P. (eds), Foreland Basins. Int. Assoc., Sediment. Spec. Publ. No. 8: 3-12.

Allen, P.A. and Underhill, J.R. 1989. Swalex cross-stratification produced by unidirectional flows, Bencliff Grit (Upper Jurassic), Dorset, U.K. J. Geol. Soc. London, 146: 241-252.

Archer, A.A. 1965. Notes on the Millstone Grit of the North Crop of the Pembrokeshire Coalfield. Proc. Geol. Assoc., 76: 137-150.

Archer, A.A. 1968. Geology of the South Wales Coalfield, Special Memoir. Gwendraeth Valley and adjoining areas. Mem. Brit. Geol. Surv., 216pp.

Ashwin, D.P. 1957. The coastal outcrop of the Culm Measures of southwest England, Abs. Proc. Conf. Geol. and Geomorph. in southwest England, 2: 2-3.

Barclay, W. J. and Jones, D.G. 1978, Recent boreholes on the attenuated Corboniferous strata of the Blaenavon-Pontypool area, Gwent. Bull. Geol. Surv. G.B., No. 67: 17pp.

Barclay, W.J., Taylor, K. and Thomas, L.P. 1988. Geology of the South Wales Coalfield, part. V. The country around Merthyr Tydfil (Third Edition). Mem. Brit. Geol. Surv. Sheet 231, 53pp.

Beaumont, C. 1981. Foreland basins. Geophys. J.R. astron. Soc., 65: 291-329.

Bisson, G., Lam., R.K. and Calver, M.A. 1967. Boreholes in the concealed Kent Coalfield between 1948 and 1959. Bull. Geol. Surv. G.B., No. 26: 99-166.

Bluck, B.J. 1961. The sedimentary history of the rocks between the horizon of G. subcrenatum and the Garw Coal in the South Wales Coalfield. Unpublished Ph.D. thesis, Univ. of Wales (Swansea), 130pp.

Bluck, B.J. and Kelling, G. 1963. Channels from the Upper Carboniferous Coal Measures of South Wales. Sedimentology, 2: 29-53.

Brooks, M. Trayner, P.M. and Trimble, T.J. 1988. Mesozoic reactivation of Variscan thrusting in the Bristol Channel area. J. Geol. Soc. London, 145: 439-444,

Cleal, C.J. 1987. Macrofloral biostratigraphy of the Newent Coalfield, Gloucestershire. Geol. J., 22: 207-217.

Coward, M.P. and Smallwood, S. 1984. An interpretation of the Variscan tectonics of SW Britain. In: Hutton, D.W.H. and Sanderson, D.J. (eds). Variscan Tectonics of the North Atlanctic Region. Spec. Publ. Geol. Soc. London, 14: 89-101.

Edmonds, E.A. 1974. Classification of the Carboniferous rocks of south-west England. Rep. Inst. Geol. Sci., 74/13, 7pp.

Edmonds, E.A., Williams, B. J. and Taylor, R.T. 1979. Geology of Bideford and Lundy Island. Mem. Geol. Surv. G.B.

Edmonds, E.A. Whitaker, A. and Williams, B.J. 1985. Geology of the country around Ilfracombe and Barnstaple. Mem. Geol. Surv. G.B.

Elliott, T. 1976. Upper Carboniferous sedimentary cycles produced by river-dominated, elongate deltas. J. Geol. Soc. London, 132: 199-208.

Enfield, M.A., Gillcrist, J.R., Palmer, S.N. and Whalley, J.S. 1985. Structural and sedimentary evidence for the early deformation of the Bude and Crackington Formations of north Cornwall and Devon. Proc. Ussher Soc., 6: 165-172.

Fielding, C.R: 1984. A coal depositional model for the Durham Coal Measures of northeast England. J. Geol. Soc. London, 141: 919-939.

Foster, D., Holliday, D.W., Jones, C.M., Owens, B. and Welsh, A. The concealed Upper Palaeozoic rocks of Berkshire and South Oxfordshire. Proc. Geol. Ass., 100: 395-407.

Freshney, E.C., Edmonds, E.A., Taylor, R.T. and Williams, B.J. 1979a. Geology of the country around Bude and Bradworthy. Mem. Geol. Surv. G.B.

Freshney, E.C., Beer, K.E. and Wright, J.E. 1979b. Geology of the Country around Chumleigh. Mem. Brit. Geol. Surv.

Freshney, E.C. and Taylor, R.T. 1972. The Upper Carboniferous stratigraphy of north Cornwall and west Devon. Proc. Ussher Soc. 5: 464-471.

Gayer, R.A. and Jones, J. 1989. The Variscan foreland in South Wales. Proc. Ussher Soc., 8: 177-179.

Gayer, R.A. and Stead, J.T.G. 1971. The Forest of Dean Coal and Iron Ore fields. In: Basset, D.A. and Basset, M.G. (eds). Geological excursions in South Wales and the Forest of Dean. Roberts, Cardiff. pp. 20-36.

George, G.T. 1970. The sedimentology of Namurian sequences in South Pembrokeshire. Unpubl. Ph.D thesis, Univ. of Wales (Swansea), 202pp.

George, G.T. and Kelling, G. 1982. Stratigraphy and sedimentology of Upper Carboniferous sequences in the coalfield of southwest Dyfed. In: Bassett, M.G. (ed). Geological excursions in Dyfed, southwest Wales. National Museum of Wales, Cardiff, 175-202.

George, G.T. 1982. Sedimentology of the Upper Sandstone Group (Namurian G_1) in southwest Dyfed: a case study. In: Bassett, M.G. (ed). Geological Excursions in Dyfed southwest Wales. Nat. Museum Wales, Cardiff, pp.203-214.

George, T.N. 1956. The Namurian Usk Anticline. Proc. Geol. Assoc., 66: 297-316.

George, T.N. 1970. British Regional Geology: South Wales. H.M.S.O. London (3rd edition), 152pp.

Goldring, R. and Seilacher, A. 1971. Limulid undertracks and their sedimentological implications. N. Jahrb. Geol. Palaont. Abh., 137: 422-442.

Hartley, A.J., Bamford, M. and Alexander. J. 1990. Tidal sedimentation from the Westphalian of SW Dyfed. Abs. Conf. Int. Ass. Sed. Nottingham.

Hartley, A.J. and Warr, L.N. 1990. Upper Carboniferous foreland basin evolution in SW Britain. Proc. Ussher Soc., 7, 212-216.

Heard, A. 1922. The petrology of the Pennant Series. Geol. Mag., 59: 83-92.

Heard, A. and Davies, R. 1924. The Old Red Sandstone of the Cardiff district. Q. J. Geol. Soc. London, 80: 489-519.

Hess, J.C., Lippolt, H.J., Holub, V.M. and Pesek, J. 1985. Isotopic ages of two Westphalian C tuffs-a contribution to the Upper Carboniferous time scale. Abstr. Eur. Geophys. Assoc. Cambridge.

Higgs, R. 1983. The possible influence of storms in the deposition of the Bude Formation (Westphalian), north Cornwall and north Devon. Proc. Ussher Soc., 5: 477-478.

Higgs, R. 1984. Possible wave influenced sedimentary structures in the Bude Formation (Lower Westphalian, southwest England), and their environmental implications. Proc. Ussher Soc., 6: 88-94.

Higgs, R. 1986. 'Lake Bude' (early Westphalian, SW England): storm dominated siliciclastic shelf sedimentation in an equatorial lake. Proc. Ussher Soc., 8: 417-418.

Holder, M.T. and Leveridge, B.E. 1986. Correlation of the Rhenohercynian Variscides. J. Geol. Soc. London, 143: 143-147.

Jenkins, T.B.H. 1962. The sequence and correlation of the Coal Measures of Pembrokeshire. Q. J. Geol. Soc. London, 118: 65-101.

Jones, D.G. 1974. The Namurian Series in South Wales. In: Owen, T.R. (ed.), The Upper Palaeozoic and Post-Palaeozoic Rocks of Wales. University of Wales Press, Cardiff, pp. 117-132.

Jones, D.G. and Owen, T.R. 1967. The Millstone Grit succession between Brynmawr and Blorenge. Proc. Geol. Assoc., 77: 187-198.

Jones, J. 1989. The influence of contemporaneous tectonic activity on Westphalian sedimentation in the South Wales Coalfield. In: Atherton, R., Gutteridge, P. and Nolan, S. (eds). Devonian and Carboniferous Tectonics and Sedimentation. Proc. Yorks. Geol. Soc. Spec. Publ. (in press).

Kellaway, G.A. 1971. The Upper Coal Measures of Southwest England, compared with those of South Wales and the southern Midlands. C.R. 6eme Congr. Int. Strat. Geol.. Carb. Sheffield, 1967, 3: 1039-1055.

Kellaway, G.A. and Hancock, P.L. 1983. Structure of the Bristol District, the Forest of Dean and the Malvern Fault Zone, In: Hancock, P.L. (ed.). The Variscan Fold Beld in the British Isles. Hilger, Bristol, pp. 88-107.

Kelling, G. 1968. Patterns of sedimentation in the Rhondda Beds of South Wales. Bull. Amer. Assoc. Petrol. Geol., 52: 2369-2386.

Kelling, G. 1974. Upper Carboniferous sedimentation in South Wales. In: Owen, T.R. (ed). The Upper Palaeozoic and post-Palaeozoic Rocks of Wales. University of Wales Press, Cardiff pp. 185-224.

Kelling, G. 1988. Silesian sedimentation and tectonics in the South Wales Basin: a brief review. In: Besly, B. and Kelling, G. (eds), Sedimentation in a Synorogenic Basin Complex: the Upper Carboniferous of Northwest Europe. Blackie, Glasgow and London, p. 38-42.

King, A.F. 1965. Xiphosurid trails from the Upper Carboniferous of Bude, north Cornwall. Proc. Geol. Soc. London, 1626: 162-165.

Leeder, M.R. 1989. Recent developments in Carboniferous geology: a critical review with implications for the British Isles and NW Europe. Proc. Geol. Assoc., (in press).

Leitch, D., Owen, T.R. and Jones, D.G. 1958. The basal Coal Measures of the South Wales Coalfield from Llandebie to Brynmawr. Q. J. Geol. Soc. London, 113: 461-486.

Lippolt, H.J., Hess, J.C. and Burger, K. 1984. Isotopische Alter pyroklastischer Sanidine aus Kaolin-Kohlentonstein als Korrelationsmarken für das Mitteleuropäisch. Oberkarbon. Fortschr. geol. Rheinld. Westf., 32: 119-150.

McKeown, M.C., Edmonds, E.A., Williams, M., Freshney, E.C. and Masson Smith, D.J. 1973. Geology of the country around Boscastle and Holdsworthy. Mem. Geol. Surv. G.B.

Melvin, J. 1986. Upper Corboniferous fine grained turbiditic sandstones from southwest England: a model for growth in an ancient, delta-fed subsea fan. J. Sediment. Petrol., 56: 19-34

Oguike, R.O. 1969. Sedimentation of the Middle Shales (Upper Namurian of the South Wales Coalfield. Unpublished Ph. D Thesis, Univ. of Wales (Swansea), 274 pp.

Ori, G.G. and Friend. P.F. 1984. Sedimentary basins formed and carried piggyback on active thrust sheets. Geology, 12: 475-478.

Owen, T.R. 1953. The structure of the Neath Disturbance between Bryniau Gleision and Glynneath, South Wales. Q. J. Geol. Soc. London, 109: 333-365.

Owen, T.R. 1971. The relationship of Carboniferous sedimentation to structure in South Wales. C.R. 6eme Congr. Strat. Carb. Sheffield, 1967, 3: 1305-1316.

Owen, T.R. 1974. The Variscan Orogeny in Wales. In: Owen, T.R. (ed). The Upper Palaeozoic and post-Palaeozoic Rocks of Wales. University of Wales Press, Cardiff, pp. 285-294.

Owen, T.R. and Weaver, J.D. 1983. The structure of the main South Wales in the British Isles. Hilger, Bristol. pp. 74-87.

Poole, E.G. 1969. The stratigraphy of the Geological Survey Apley Barn borehole, Witney, Oxfordshire. Bull. Geol. Surv. G.B., 29: 1-104.

Postma, D. 1982. Pyrite and siderite formation in brackish and freshwater swamp sediments. Am. J. Sci., 282: 1151-1183.

Powell, C.M. 1989. Structural controls on Palaeozoic basin evolution and inversion in SW Wales. J. Geol. Soc. London, 146: 439-446.

Prentice, J.E. 1960. The stratigraphy of the Upper Carboniferous rocks of the Bideford region, north Devon. Q. J. Geol. Soc. London, 116: 397-408.

Ramsbottom, W.H.C. 1978. Namurian mesothems in South Wales and northern France. J. Geol. Soc. London, 135: 307-312.

Ramsbottom, W.H.C., Calver, M.A., Eager, R.M.C., Hodson, F. Holliday, D.W. Stubblefield, C.J. and Wilson, R.B. 1978. A correlation of the Silesian rocks in the British Isles. Spec. Rep. No. 10, Geol. Soc. London.

de Raaf, J.F.M., Reading, H.G. and Walker, R.G. 1965. Cyclic sedimentation in the Lower Westphalian of North Devon, England. Sedimentology, 4: 1-52.

Reading, H.G. 1965. Recent finds in the Upper Carboniferous of southwest England and their significance. Nature, 208: 745-747.

Sanderson, D.J. 1984. Structural variations across the northern margin of theVariscides in NW Europe. In: Hutton, D.W.H. and Sanderson, D.J. (eds). Variscan Tectonics of the North Atlantic Region. Spec. Publ. Geol. Soc. London, 14: 149-165.

Selwood, E.B. and Thomas, J.M. 1986. Variscan facies and structure in central SW England. J. Geol. Soc. London, 143: 199-207.

Shackleton, R.M. Ries, A.C. and Coward, M.P. 1982. An interpretation of the Variscan structures in SW England. J. Geol. Soc. London, 139; 543-554.

Smallwood, S.S. A thin-skinned model for Pembrokeshire. J. Struct. Geol., 7: 583-595.

Squirrel, H.C. and Downing, R.A. 1964. The attenuation of the Coal Measures in the southeast part of the South Wales Coalfield. Bull. Geol. Surv. G.B., No. 21, pp. 119-132.

Sullivan, H.J. 1964. Miospores from the Drybrook Sandstone and associated measures in the Forest of Dean Basin, Gloucestershire. Paleontology, 7: 351-392.

Thomas, J.M. 1988. Basin history of the Culm Trough of Southwest England. In: Besly B. and Kelling, G. (eds). Sedimentation in a Synorogenic Basin Complex: the Upper Carboniferous of Northwest Europe. Blackie, Glasgow and London, pp. 24-37.

Thomas, L.P. 1967. A sedimentary study of the sandstones between the horizons of the Four-Foot Coal and the Gorllwyn Coal of the Middle Coal Measures of the South Wales Coalfield. Unpublished Ph.D thesis, Univ. of Wales (Swansea) 176pp.

Trotter, F.M. 1941. Geology of the Forest of Dean coal and iron ore field. Mem. Brit. Geol. Surv. 95pp.

Wagner, R.H. and Spinner, E. 1972. The stratigraphic implication of the Westphalian D macro- and microfauna of the Forest of Dean Coalfield (Gloucestershire), England. Proc. Int. Geol. Conr. Montreal, 1972, 7: 428-431.

Walker, R.G. 1970. Deposition of turbidites and agitated-water siltstones: a study of the Upper Carboniferous Westward Ho! Formation, north Devon. Proc. Geol. Ass., 81: 43-67.

Ware, W.D. 1939. The Millstone Grit of Carmarthenshire. Proc. Geol. Assoc., 50: 168-204.

Weaver, J.D. 1976. Seismically induced load structures in the basal Coal Measures, South Wales. Geol. Mag., 113: 535-543.

Welch, F.B.A. and Trotter, F.M. 1961. Geology of the country around Monmouth and Chepstow. Me. Brit. Geol. Surv. 164pp.

Whalley, J.S. and Lloyd, G.E. 1986. Tectonics of the Bude Formation, north Cornwall-the recognition of northerly directed decollement. J. Geol. Soc. London, 143: 83-89.

Williams, P.F. 1968. The sedimentation of Westphalian (Ammanian) Measures in the Little Haven-Amroth Coalfield, Pembrokeshire. J. Sediment. Petrol., 38: 332-362.

Wilson, D., Davies, J.R. Smith, M. and Waters, R.A. 1988. Structural controls on Upper Palaeozoic sedimentation in southeast Wales. J. Geol. Soc. London, 145: 901-914.

Basin Inversion and Foreland Basin Development in the Rhenohercynian Zone of South-West England

L. N. Warr
Department of Geology, University of Exeter, EX4 4QE, UK
Present address: Geologisch-Paläontologisches Institut, Ruprecht-Karls-Universität, Im Neuenheimer Feld 234, D-6900 Heidelberg 1, Germany

ABSTRACT

Geological mapping and structural analysis undertaken in north Cornwall reveals that basin inversion influenced the structural development of the central Rhenohercynian zone of SW England. A model for the extensional geometry of the Middle to Upper Devonian Trevone Basin shows sequence thinning northward across a fault controlled margin, onto a northern shelf area. Lower Carboniferous footwall uplift lying north of this shelf provided the sediment source for adjacent paralic facies (Boscastle Formation). These E-W striking basin-controlling faults continued their influence by acting as buttresses to the NNW-directed D1 compression during Namurian times. This resulted in a prominent phase of D1 backthrusting and the formation of a confrontation zone within the basin during its inversion.

Upper Carboniferous foreland basin style sedimentation of the Culm Basin is proposed to have occurred synchronously with regional D1 inversion of the Trevone Basin. Basin inversion controls are also suggested to have influenced the site of foreland basin development; areas of thick syn-rift sequences (e.g. Trevone Basin) undergoing uplift and thrust loading, while areas of thinner syn-rift (beneath the Upper Carboniferous Culm) becoming the site of subsidence. The second regional (D2) phase added further complications to the orogen, with footwall cleavage development, lateral and oblique ramp geometries and out-of-sequence thrusting. An important phase of D2 underthrusting beneath the southern part of the Culm Basin initiated backthrusting within the Upper Carboniferous rocks. This resulted in elevated temperatures (M2) and high coaxial shear strains within the underthrusting wedge.

INTRODUCTION

The Rhenohercynian zone of SW England is considered more complex in its deformational history than laterally equivalent areas, for example the Rheinische Schiefergebirge of mainland Europe (compare Sanderson & Dearman 1973; Shackleton et al. 1982; Sanderson 1984; Coward & Smallwood 1984 with Weber 1981 and Weber & Behr 1983; Behr 1983; Meissner et al. 1984; Behr et al. 1984). These complexities include:

- multiple cleavage development (up to four phases), complex in-teractions of folding and thrusting involving lateral ramps, oblique thrust traces, out-of-sequence movements (Selwood et al. 1984, Fig. 3, Turner 1982; Andrews et al. 1988; Warr 1988) and early phases of backthrusting (Shackleton et al. 1982; Coward & Smallwood 1984; Andrews et al. 1988; Seago & Chapman 1988).

- confrontation zones across which transport directions of the same deformation event oppose each other, examples being the Padstow Confrontation Zone (Gauss 1966; 1967; 1973; Roberts & Sanderson 1971; Andrews et al. 1988; Pamplin & Andrews 1988; Selwood & Thomas 1988; Durning 1989) and the Rusey Fault Zone (Selwood & Thomas 1984; Selwood & Thomas 1986a; Warr 1989).

- localized zones of high ductile strain and multiple deformations with coaxial strain histories, e.g. the Tintagel High Strain Zone (Sanderson 1972, 1973, 1982; Andrews et al. 1988; Warr 1989).

- localized areas of transpression, e.g. the Start-Perranporth line (Holdsworth 1989)

In addition to these structural complications in the regional pattern, the stratigraphic record of the pre-Upper Carboniferous rocks of central SW England and north Cornwall show rapid facies variations (House 1971; Isaac et al. 1982; Selwood & Thomas 1986b, 1987; Bluck et al. 1988). These are recognized within a number of structural successions, commonly bounded by low-angle thrusts. Complex facies relationships indicate localized areas of differential subsidence, varying both along and across the regional strike.

Many of the previous regional structural models for the Rhenohercynian of SW England highlighted thin-skinned thrust processes without considering the influence of a pre-existing extensional basin development (e.g. Shackleton et al. 1982; Coward & McClay 1983; Coward & Smallwood 1984; Andrews et al. 1988). A model is here presented based on the excellent coastal exposure of north Cornwall and adjacent parts of north Devon (Fig. 1a). This model proposes that the stratigraphic, structural and low temperature metamorphic history was

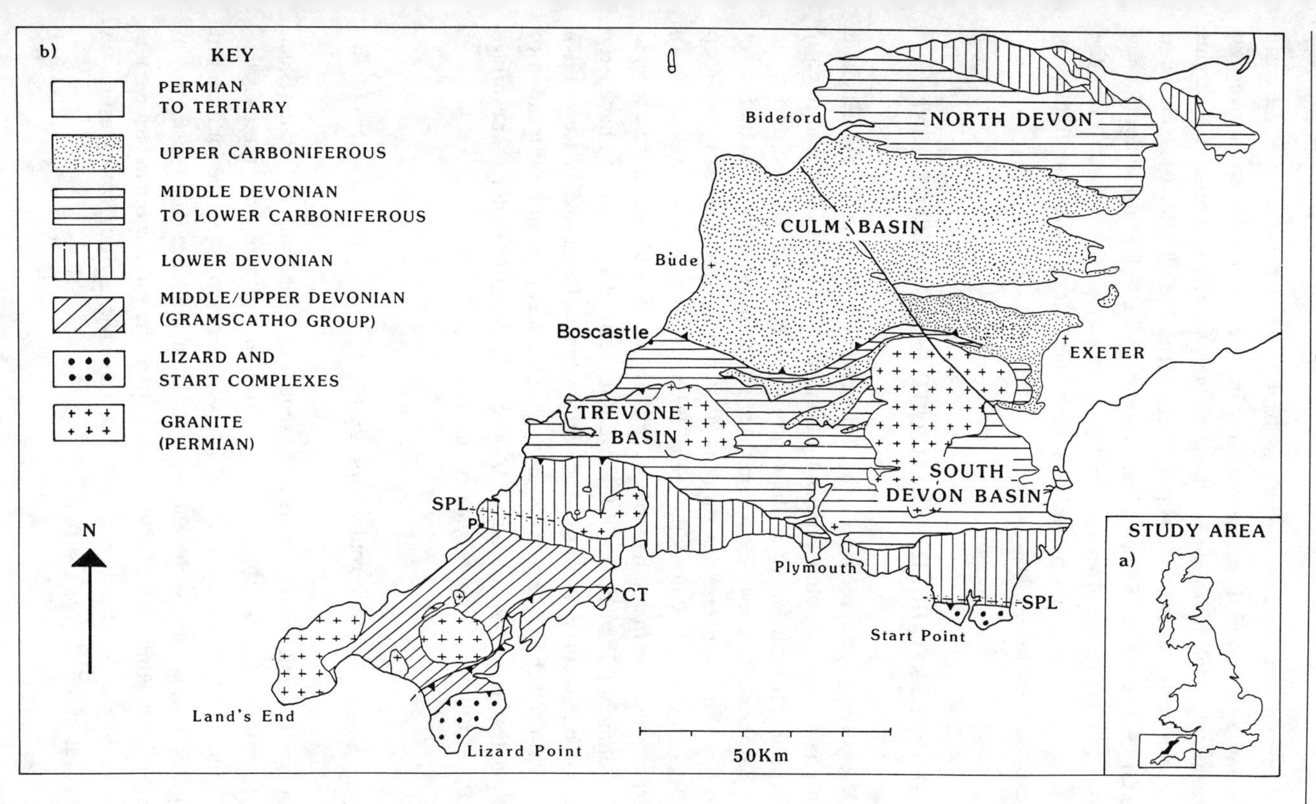

FIGURE 1
a) Location of the study area in SW England (Darkened area)
b) Simplified geological map of SW England. Abbreviations used; P- Perranporth, CT- Carrick Thrust, SPL- Start-Perranporth Line.

controlled by the interacting processes of crustal extension, basin inversion, foreland basin development and continuing fold and thrust movements. In recent years the basin inversion concept has been increasingly applied to the restacking of highly stretched continental margins within orogenic zones, with examples from the Alps (Gillcrist et al. 1987; Butler 1989; Hayward & Graham 1989; De Graciansky et al. 1989), the Subvariscan foldbelt of South Wales (Hanna & Graham 1989; Powell 1989) and southern Ireland (Williams et al. this volume). It is the theme of this paper to apply such a concept to the problematical Rhenohercynian foldbelt of SW England and to explore resulting implications.

THE STRATIGRAPHIC RECORD IN RELATION TO BASIN DEVELOPMENT

The Upper Palaeozoic stratigraphic record of the Rhenohercynian zone of SW England indicates that a main phase of north-south extension occurred during mid Devonian to early Carboniferous times (although some rifting in the south may have been initiated during the early Devonian, see Barnes & Andrews 1986; Smith & Humphreys 1989). This period of extension appears to have been synchronous with the deposition of the Gramscatho Group and the assembly of the Lizard ophiolite to the south (Davis 1984; Styles & Rundle 1984). These units, which are represented in south Cornwall (Fig. 1b) are considered to have formed within an oceanic environment (Kirby 1979, 1984; Floyd & Leveridge 1987) and to have been tectonically emplaced probably during Lower Carboniferous times (Barnes & Andrews 1986; Wilkinson & Knight 1989). As the plate tectonic setting and palaeogeographical location of these rocks remains unclear (compare Barnes & Andrews 1986 with Holder & Leveridge 1986a) they are excluded from the Rhenohercynian discussed here.

Lower Devonian stratigraphy

Lower Devonian, dominantly shelf sequences are represented by the Meadfoot Group (Bluck et al. 1988) and the Dartmouth Slates, which crop-out along an E-W belt from the coastline between Perranporth and the Plymouth area (Fig. 1b). There is growing evidence for localized subsidence occurring during this time, with sequences of lacustrine facies amid distal fluvial and playa sediments (Smith & Humphreys 1989). The nature of the pre-Devonian basement is poorly known, but probably includes Lower Palaeozoic (Ordovician) quartzites resting on Cadomian basement (Freshney & Taylor 1980).

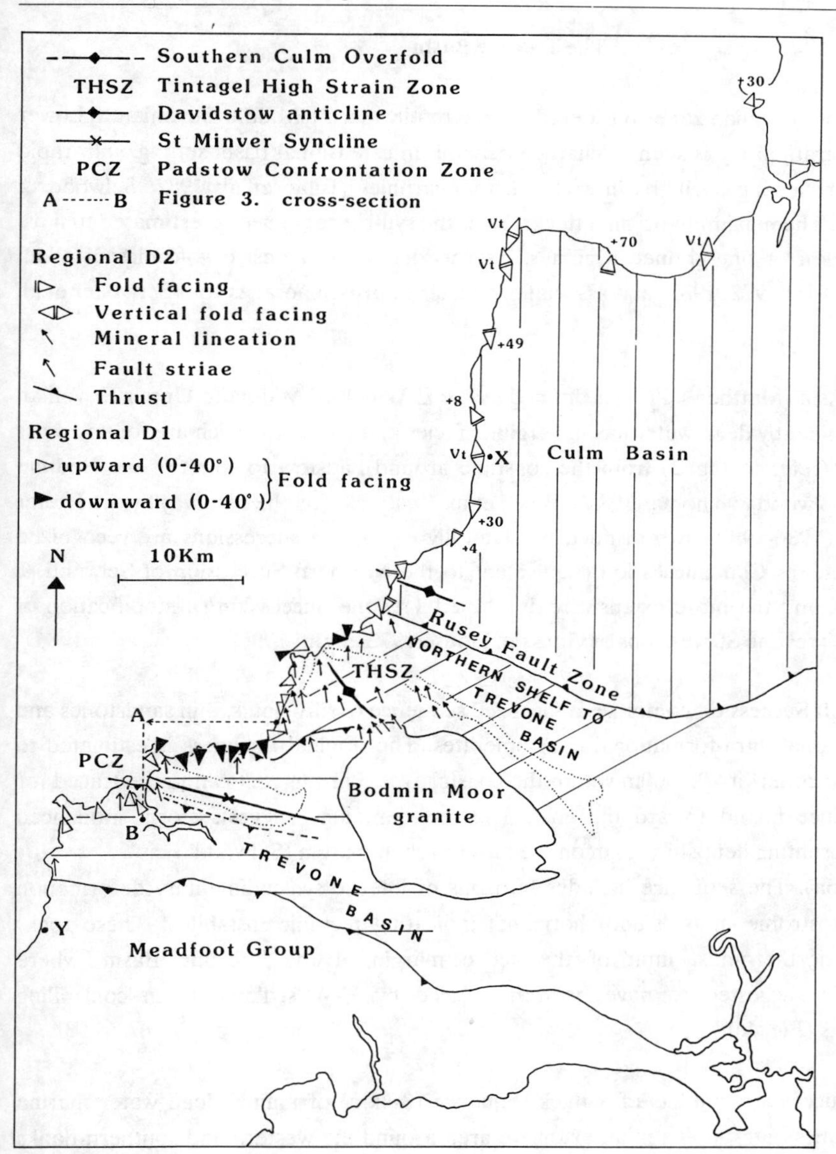

FIGURE 2
Structural map of north Cornwall (including parts of north Devon) showing the transport indicators of regional D_1 and D_2 deformations. A-B refers to Fig. 3 and X-Y to Fig. 4.

The Trevone Basin

Within the Rhenohercynian zone considered here, a totally marine Middle Devonian to Lower Carboniferous stratigraphy is seen as characteristic of an extensional basin setting, with rapid facies changes reflecting shelf, basin and rise topographies (Isaac et al. 1982; Selwood & Thomas 1986b). The maximum original thickness of the syn-rift sequences is estimated to have approached 4-5km of fine-grained argillites, interbedded with extensive volcanics of alkali basalt affinity (Floyd 1972; 1982) and associated with stratiform mineralization (Scrivener et al. 1989).

The Trevone Basin (Matthews 1977) is defined as the E-W belt of Middle to Upper Devonian rocks of predominantly deep water basinal argillite facies and volcanics which are found across north Cornwall (refer to Fig. 2) from the coastline around Padstow to south of the Bodmin Moor granite (Selwood & Thomas 1986b; Bluck et al. 1988). During the resurvey by the Exeter contract group (1986-1990), two distinct, but laterally equivalent successions are recognized (Fig. 3), the Bound's Cliff Succession (equivalent to the Northern Succession of Selwood & Thomas 1986a), and the more extensively developed Trevone Succession (a modification of the Pentire and Trevone Successions of Gauss & House 1972; Smith 1990).

The Bound's Cliff Succession comprises a localized sequence of silty slates, thin sandstones and tuffs with occasional intra-formational conglomerates. The original thickness is estimated to have been in the order of 200-300m within the coastal area, but this unit cannot be traced for any great distance inland toward the east. The sedimentology suggests storm influenced conditions representing deposition high on the basin to shelf margin (Selwood & Thomas 1986a). The sequence includes horizons of intense syn-sedimentary deformation and thick olistrostrome units (slump horizons) indicating tectonic instability. These rocks represent the northernmost limit of the active margin of the Trevone Basin, where sedimentation is suggested to have been influenced by E-W striking, basin controlling extensional faults (Fig. 4c).

The Trevone Succession consists of a thick sequence (3-4km) of mainly deep water marine slates and volcanics outcropping in an extensive area around the western and southern flanks of the Bodmin Moor granite. In the north of this area conglomerates and volcanic horizons are found toward the northern active margin of the basin. These features are less abundant in the southern part of the Trevone Succession, south of the Camel Estuary. The southern margin of the basin is not seen, as a gently dipping north-directed thrust boundary emplaces Lower Devonian Meadfoot Group upon Trevose Slates of deep water basinal argillite facies (Fig. 2).

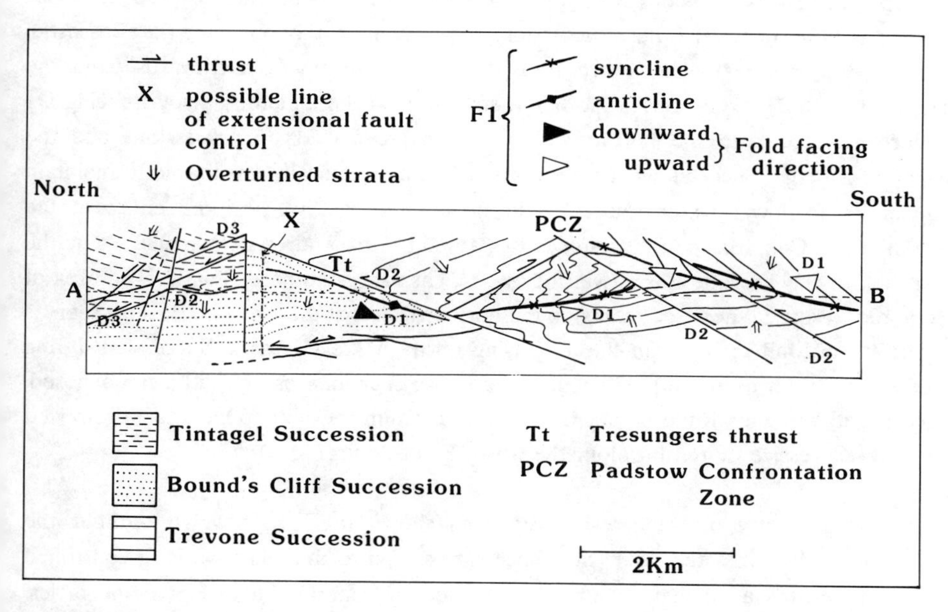

FIGURE 3
Generalised across strike north to south coastal section showing the structure between Tregardock Beach and Padstow). The position of A-B is marked on Fig. 2.

To the north of the Trevone Basin lie two contrasting and younger successions; the Boscastle Succession which consists of shallow water marine clastics and limestones of Upper Devonian to earliest Namurian age (Selwood et al. 1985) and the Tintagel Succession (Freshney et al. 1972), a sequence of marine Upper Devonian to Lower Carboniferous argillaceous slates and volcanics of outer shelf and argillaceous basinal facies (Selwood & Thomas 1986a). The palaeogeography of these successions is a subject of debate. Selwood & Thomas (1986a, 1986b) and Isaac et al. (1982) envisage these successions as highly allochthonous units (i.e. Boscastle, and Tredorn fold nappes) which were transported large distances northwards from the southern shelf of the Trevone Basin. However, structural workers identify a south-facing D_1 event which affected both the northern part of the Trevone Basin successions and the Boscastle and Tintagel Successions. These workers find it structurally favourable to maintain these generally shallower water sequences in their same relative position, either at the southern shelf of a Culm Basin (Sanderson 1984; Pamplin 1988; Andrews et al. 1988) or on the northern shelf of the Trevone Basin (Matthews 1977) as suggested here (Fig. 2). The area of uplift required to source these sediments on the northern shelf of the Trevone Basin is seen as resulting from footwall uplift in an the area lying north of the Boscastle Succession during crustal extension (refer to Fig. 4c). Although evidence for such an area of uplift is not exposed, it is suggested to lie beneath the Upper Carboniferous Culm rocks or to have been removed laterally by transpressive shortening along the Rusey Fault Zone (Fig. 2).

The predicted geometry of the Trevone Basin is shown in Fig. 4c which outlines the distribution of facies across the basin. Sequences are shown to thin across a fault-controlled active margin towards a northern shelf, accompanied by general shallowing of the facies. Footwall uplift to the north of this shelf provides the source for the paralic sediments of the Boscastle Formation during the late Devonian to early Carboniferous period. Restricted basinal facies of Lower Carboniferous age are found along the margins of the present day Upper Carboniferous Culm Basin. These thin sequences were deposited during a time of continuing extension during the early Carboniferous period accompanied by a general rise in the sea level (Selwood & Thomas 1986b.

The Culm Basin

Lying conformably on thinly developed Dinantian basinal facies, the Upper Carboniferous Culm sequence (2-3km thick) is preserved in an E-W trending basin, which crosses north Cornwall and Devon. It has an anticlinorial structure with a maximum width of 45km and an along strike length extending over 75km, before passing beneath the Permo-Triassic cover in the Exeter region (Fig. 1b).

The Upper Carboniferous rocks comprise a thick basal sequence of monotonous distal turbidites (the Crackington Formation) of Namurian to early Westphalian age, passing northwards into Westphalian strata of more variable turbiditic sandstones (the Bude Formation: interpretation of Melvin 1986) and shallower marine Deltaics (Bideford Formation: De Raaf 1965; Elliott 1976). These deposits are characterized by fine-grained orthoquartzite sandstones, pointing to a mature continental source (Melvin 1986; Thomas 1988). It is difficult to pinpoint the areas of uplift during the Namurian times, especially as the palaeocurrents of the distal turbidites demonstrate E-W axial flow within the basin (Edmonds 1974), but a northerly source of mature sediment seems most likely. They lack the coarser tongues of clastics which would be expected from direct derivation from the south, although removal of such deposits along the Rusey Fault Zone is a possibility. There is, however, good evidence for a topographic high emerging in the Bristol Channel area during Westphalian times which provided a northerly source to the Culm basin and a southerly source to the South Wales Coalfield. Hartley (this volume) suggests that the emergence of a Variscan thrust (Mechie & Brooks 1984; Brooks et al. 1988), culminating in the Bristol Channel area controlled the sediment dispersal patterns across this area. It is also suggested that along these lines earlier, but unrelated uplift initiated by late Caledonian strike-slip movements in the Bristol Channel influenced sedimentation during Lower to Mid-Devonian times in north Devon (Tunbridge 1986).

In the model proposed here (Fig. 4) the Upper Carboniferous Culm Basin is suggested to be a foreland basin, which developed within a progressively thickening orogen, as opposed to an extensional origin (Sanderson 1984; Pamplin 1988). Sedimentation is considered as been synchronous with deformation to the south, including D_1 inversion of the Trevone Basin. Support for a foreland basin model is presented as follows:

- the stratigraphy lacks the rapid facies variations, volcanic activity and associated mineralization which characterize the Mid-Devonian to early Carboniferous sequences.

- the style of sedimentation is seen as characteristic of basin infilling in an environment of northward-migrating tectonism (Melvin 1986; Thomas 1988; Hartley this volume).

- estimates of the geothermal gradients (40½C/km) from vitrinite reflectance data in the Upper Carboniferous Culm (Cornford et al. 1987) are comparable with gradients recorded from foreland basins of the Eastern Interior of North America (Beaumont et al. 1987).

- K-Ar dating in SW England (Dodson & Rex 1971) suggests the first phase of deformation and associated metamorphism occurred during Namurian times (M_1 event

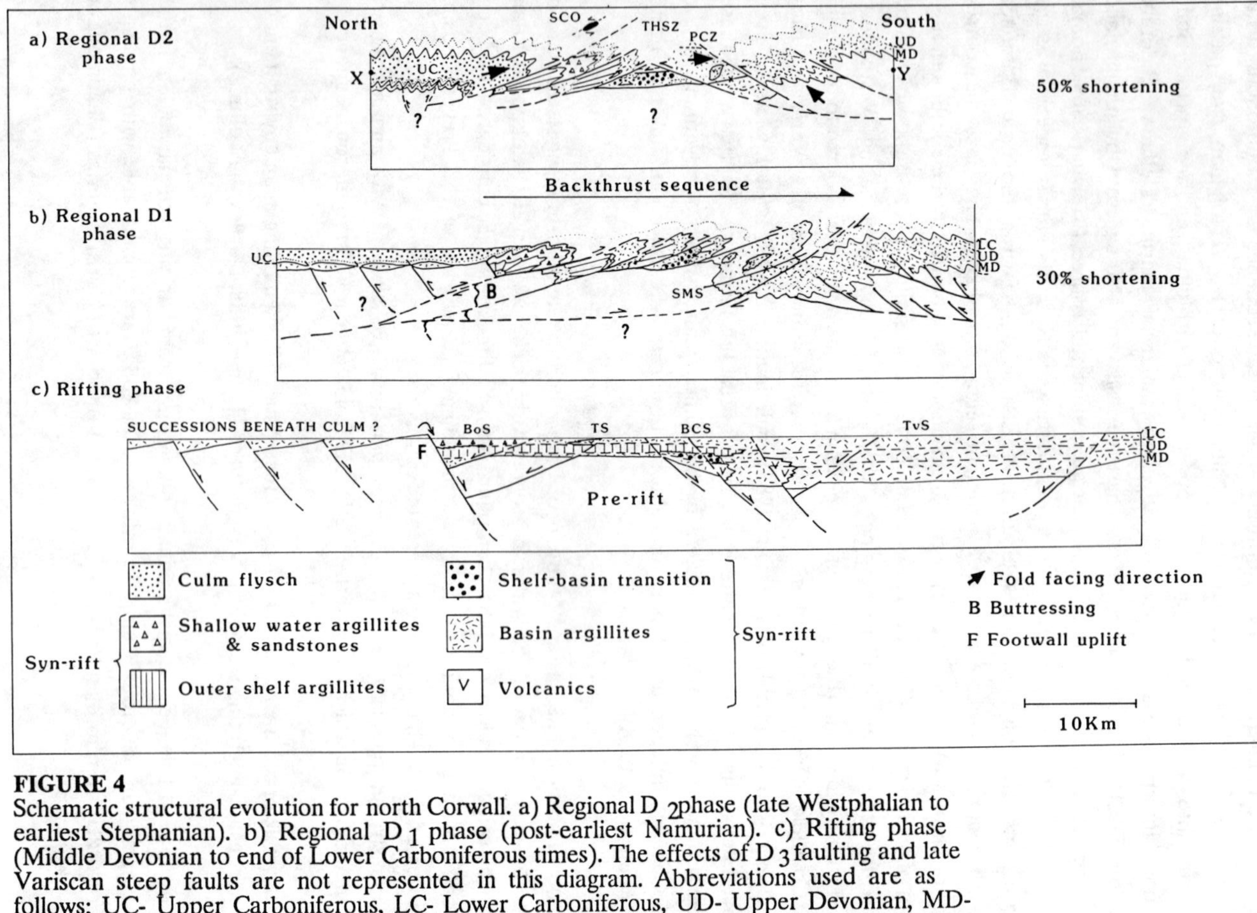

FIGURE 4
Schematic structural evolution for north Corwall. a) Regional D_2 phase (late Westphalian to earliest Stephanian). b) Regional D_1 phase (post-earliest Namurian). c) Rifting phase (Middle Devonian to end of Lower Carboniferous times). The effects of D_3 faulting and late Variscan steep faults are not represented in this diagram. Abbreviations used are as follows; UC- Upper Carboniferous, LC- Lower Carboniferous, UD- Upper Devonian, MD- Middle Devonian, SCO- Southern Culm Overfold, THSZ- Tintagel High Strain Zone, PCZ- Padstow Confrontation Zone, SMS- St Minver Syncline, BoS- Boscastle Succession, TS- Tintagel Succession, BCS- Bound'sCliff Succession, TvS- Trevone Succession.

in Fig. 8). These dates therefore overlap with sedimentation during the Upper Carboniferous.

the structural style of deformation is related to stratigraphic depth of burial which is markedly different from that of the Trevone Basin (Warr 1991). This type of stockwork tectonics is characteristic of the Upper Carboniferous Ruhr coalfield of Germany (Brix et al. 1988).

REGIONAL (D_1) SHORTENING

Inversion of the Trevone Basin

ards the end of the Lower Carboniferous and the beginning of the Upper Carboniferous a sition from crustal extension to compression occurred which led to the restacking of the nal contents of the Trevone Basin and the successions on its northern shelf. The youngest s deformed during this phase are of earliest Namurian age (Freshney et al. 1972), which ocated immediately south of the Rusey Fault Zone (Fig. 2).
structural zones can be recognized within north Cornwall based on the kinematics of first e deformation structures; a zone lying north of the Camel Estuary in which D_1 transport in a southerly direction and a zone to the south of the estuary where D_1 transport was rds the NNW (Fig. 2 & 3). These two opposing transport directions confront each other at well known "Padstow Confrontation Zone", but the relative timing of events across this is still a subject of debate (see Andrews et al. 1988; Selwood & Thomas 1988; Warr & ning 1990; Andrews et al. 1990).

Padstow Confrontation Zone lies within the Trevone Basin and is here attributed to the amental control exerted by inferred pre-existing extensional structures (Fig. 3 & 4). The ence of such extensional faults would be expected to affect the compressive deformational , influencing the scale and geometry of folding, the direction of structural transport (facing matics) and the sequence of deformation. Fig. 5 shows an analogous experiment of sion tectonics using sand and mica grains (McClay 1989) in which thrusting is onstrated to be influenced by the initial syn-rift geometry.

ping across the northern margin of the Trevone Basin has revealed regional scale tight to inal south-facing folds, dismembered by out-of-sequence (D_2) thrusts (Fig. 3). Primary ges in the facing kinematics and downward facing folds have been identified which are dered to represent original D_1 inverted geometries. Folds with wavelengths of over 3km

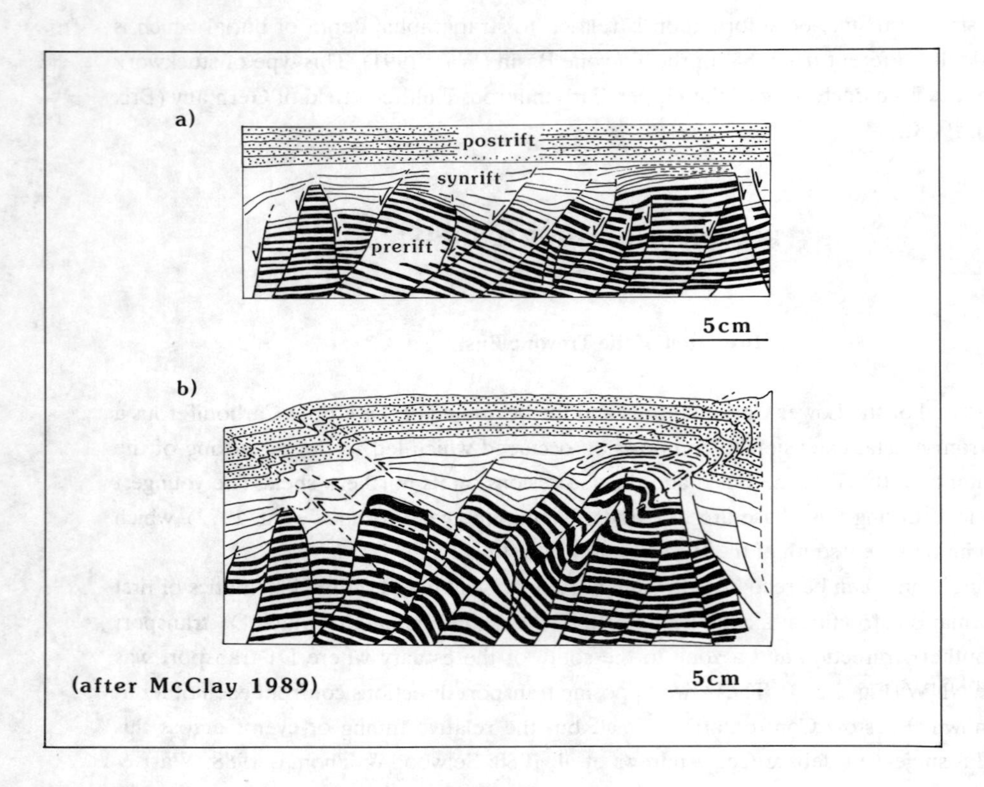

FIGURE 5
Analogous model of inversion tectonics (McClay 1989) using a sand-mica and mica layered model with homogeneous extension and subsequent contraction above a horizontal detachment. Extended 50 % and then contracted 19 %. This figure illustrates the influence of syn-rift geometry on compressive structures.

have been recognized (e.g. the St Minver Syncline, Gauss & House 1972), which show long overturned limbs and shorter rightway-up limbs. More than 70% of the strata represented within the coastal section is estimated to be upside down. A reinterpretation of the K-Ar dates (Dodson & Rex 1971) for north Cornwall (see section 6) suggests Namurian cooling ages for the regional D_1 phase which can be stratigraphically constrained to post- earliest Namurian times in the Boscastle area (Warr 1989).

In the model for the (D_1) inversion of the Trevone Basin (Fig. 4b) it is proposed that the backthrust sequence was initiated by buttressing against extensional faults to the north of the basin. The existence of such extensional faults has already been discussed in section 2.2 in relation to basin development. The reactivation of antithetic faults may have influenced the location of backthrusts within the basin, although this is not established from direct field observation.

Foreland basin development and early deformation

It is proposed that the Upper Carboniferous Culm was deposited in advance of a structurally inverting part of the orogen and represents syn-orogenic sedimentation. Foreland basins usually develop in response to lithospheric loading due to nappe emplacement (Beaumont 1981) in the external parts of orogenic belts, between the front of a mountain chain and an adjacent craton (Allen et al. 1986). Such mechanisms of basin development (Fig. 6a) have recently been applied to the South Wales Coalfield and the Upper Carboniferous Culm of SW England (Kelling 1988; Gayer & Jones 1989). In their general model, Gayer & Jones (1989, Fig. 2) suggest the Culm was deposited within a thrust-sheet-top basin in the hanging wall of a major Variscan thrust which lies beneath the Bristol Channel fault zone (Brooks et al. 1988). Such a nappe loading model is here considered applicable to the formation of the Culm Basin (Fig. 6a), although controls exerted by basin inversion are also believed to be important (Fig. 6b). It is proposed that there is a basic relationship between the thickness of Middle Devonian to Lower Carboniferous syn-rift sequences and the position of subsequent Upper Carboniferous sedimentation in SW England. Areas which suffered greater amounts of fault controlled subsidence during the earlier extensional phase responded by incurring larger amounts of uplift during inversion (related to the inversion ratio of Williams et al. 1989). The Culm basin is suggested to have formed in an area of thin syn-rift relative to the Trevone Basin and north Devon area (including Bristol Channel area). The inverted areas of thick syn-rift sedimentation emerged as structural highs and acted as thrust loads, while the area of thinner syn-rift sequences (beneath the Upper Carboniferous Culm) became relative lows and sites of subsidence.

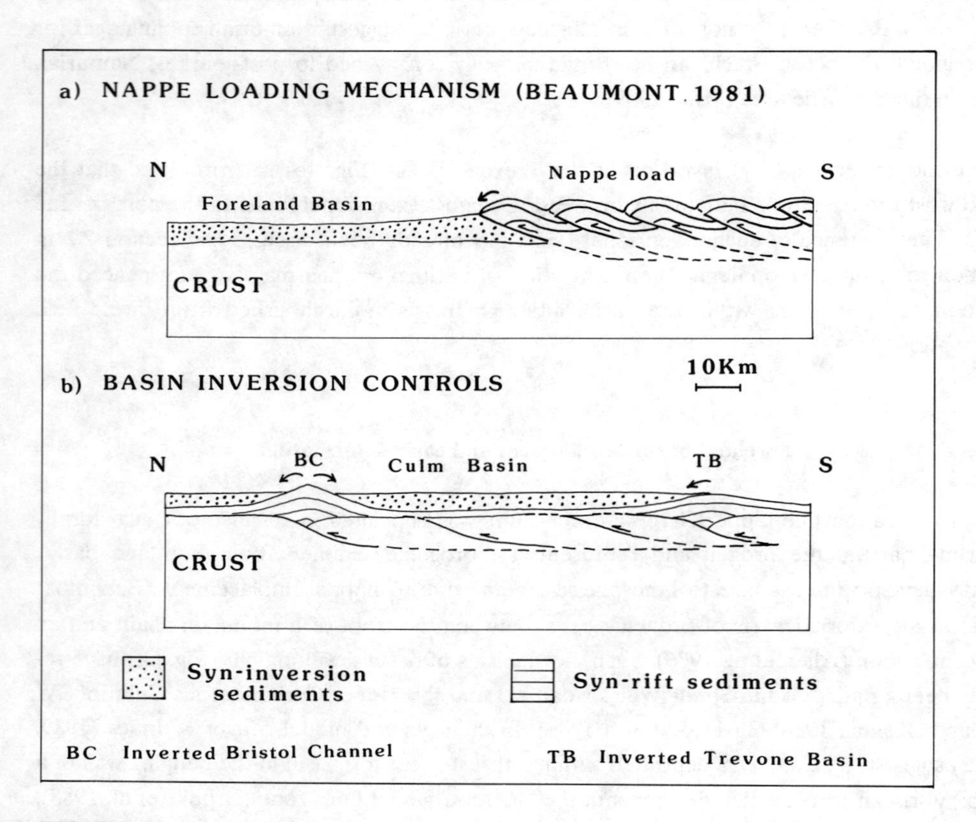

FIGURE 6
Controlling influences on foreland basin evolution. a) Nappe loading controls (Beaumont 1981). b) Basin inversion controls.

The pattern of deformation within the Culm sequence is one of increasing intensity and metamorphism with burial depth. Toward the top of the sequence, Whalley & Lloyd (1986) present evidence for early deformation within the Westphalian Bude Formation. They describe a progression from syn-sedimentary deformation within slump horizons which then act as detachment horizons for north-directed thrusts during early tectonic shortening. This activity is here suggested to be a response of deformation and uplift in the north Devon and Bristol channel area (Fig. 6b).

REGIONAL D_2 SHORTENING

D_2 thrusting in the Trevone Basin

Following D_1 backthrusting, a phase of out-of-sequence D_2 thrusting towards the NNW occurred across the Trevone Basin. These D_2 thrusts dismembered the F_1 regional folds into a series of thrust slices (Fig. 3) which have considerable influence on the present distribution of stratigraphic units (Isaac et al. 1982; Selwood et al. 1985; Selwood & Thomas 1986b; Warr 1988). Lateral ramp geometries and oblique thrust traces are evident (Turner 1982; Andrews et al. 1988; Warr 1988) features which are commonly found in orogenic regions characterized by inversion (Hayward & Graham 1989). Localized non-coaxial overprinting of deformations within the footwall of D_2 thrusts led to complex zones of refolding and multiple cleavage development. Two examples serve to illustrate these features, the first being the Padstow Confrontation Zone, where a D_2 thrust emplaces F_1 north-facing folds upon F_1 south-facing folds (Durning 1989; Warr & Durning 1990; see Fig. 4a). A strong S_2 crenulation cleavage is associated with refolding the S_1 (south-facing) cleavage in the footwall and is most intensely developed between Gravel Caverns and Trebetherick Point. The second example is in the footwall of a D_2 thrust which emplaces the Lower Devonian Staddon Grits over the southern limit of the Trevone Basin Succession, where the S_2 cleavage overprints the S_1 (north-facing) cleavage (e.g. SX 051680).

Underthrusting of the Culm Basin

D_2 underthrusting of the southern margin of the Culm Basin occurred sometime around late Westphalian to earliest Stephanian times (Warr 1989). Within the Tintagel High Strain Zone, early D_2 ductile shear directed towards the NNW was coaxial with F_1 fold structures giving rise to a composite S_1/S_2 slaty cleavage, increasing the sheathing of F_1 fold axes and variations in the facing kinematics (Warr 1989). This NNW-ward movement initiated southward simple shear in the southern part of the structurally higher Culm Basin, causing modification of initially upright chevron folds in the Bude Formation (Lloyd & Whalley 1986) and formation of the Southern Culm Overfold (Freshney et al. 1972; Sanderson 1974). The ENE-WSW trending Rusey Fault Zone therefore appears as an inter-basinal Confrontation Zone, across which oblique underthrusting towards the NNW, initiated backthrusting towards the SSW of the overlying basin (Warr 1989; see Fig. 2). Such obliquity of opposed transport (i.e. dextral transpression) is characteristic of NNW underthrusting of a regionally striking E-W basin (Sanderson 1984).

The same general geometrical relationship of transport directions is found across the Padstow Confrontation Zone (an intra-basinal confrontation; Fig. 2) which is attributed to the interaction of the regional D1 thrusting direction (toward the NNW) with the pre-existing extensional architecture of the Trevone Basin (striking E-W).

THE PATTERN OF LOW TEMPERATURE METAMORPHISM ACROSS NORTH CORNWALL AND DEVON.

The regional pattern of low-temperature metamorphism from north Devon to north Cornwall is best illustrated by illite "crystallinity" studies (Fig. 7; Primmer 1985a 1985b; Kelm 1986; Warr 1991). The transition of metamorphism from diagenesis to epizone conditions is broadly related to stratigraphy (M_1) and a diastathermal or extensional control (Robinson & Bevins 1989; Robinson 1987) is suggested. However, an anomaly in the regional pattern is the Tintagel High Strain Zone with its protracted heating (M_2) to epizone conditions (Primmer 1985a, 1985b). This metamorphism is not related to stratigraphic age as a transition to anchizone conditions occurs southwards into older slates (Primmer 1985a; Pamplin 1988; Warr 1991). Petrological evidence from the Tintagel Volcanic Formation shows greenschist facies to be synchronous with early D_2 ductile strain (e.g. Andrews et al. 1988). However, a recent isocryst survey of this area (Warr & Robinson 1990) shows the isocryst contour representative of the anchizone/epizone boundary to run oblique to mapped regional D_2 thrusts in the Camelford area. This trend in grade is noted to be roughly parallel with the structurally overlying D_2 backthrusting of the southern Culm Basin. Epizone metamorphic conditions and high strains are therefore considered to have been caused by increasing temperatures and confining pressures during underthrusting of the Culm Basin.

INTERPRETATION OF THE REGIONAL K-AR DATES.

A regional survey of K-Ar dating on slates from SW England was conducted by Dodson & Rex (1971), the results of which are summarized in Fig. 8. The apparent diachroneity in the ages

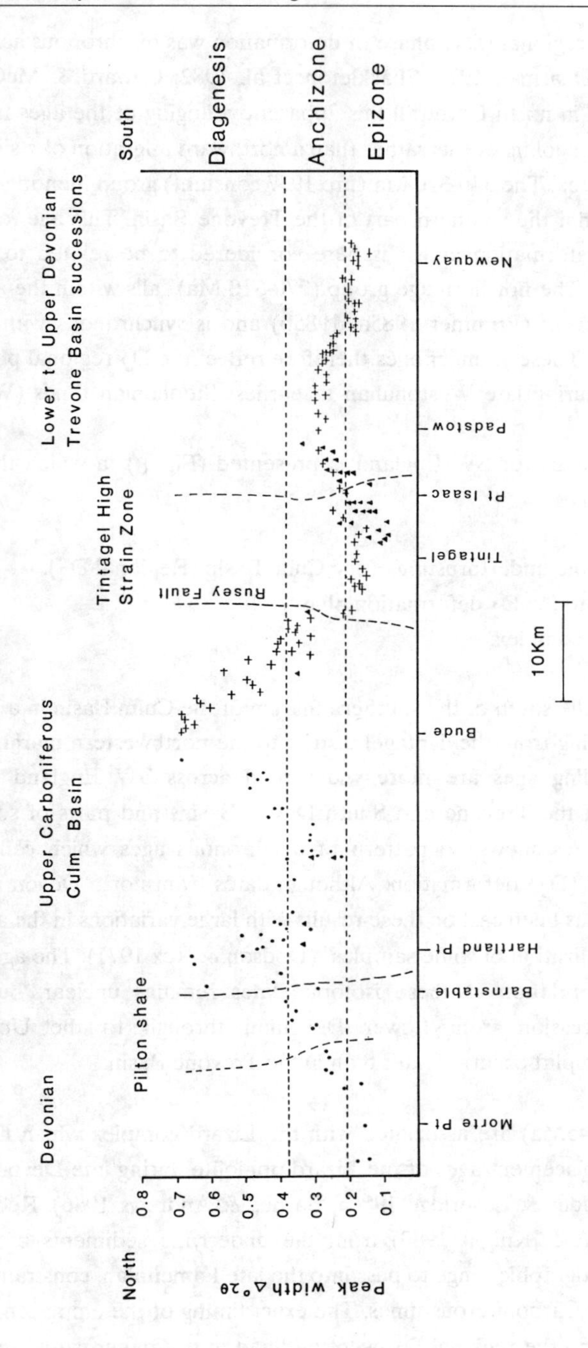

FIGURE 7
Summary of illite "crystallinity" data across north Devon and Cornwall, compiled from Ph.D studies conducted at the University of Bristol. All measurements were run under consistent machine conditions. Reference to data points are as follows; dots- Kelm (1986), crosses- Primmer (1985a) and triangles- Warr & Robinson (1990).

has been used to suggest that the regional (D_1) phase of deformation was diachronous across SW England (e.g. Sanderson & Dearman 1973; Shackleton et al. 1982; Coward & McClay 1983; Sanderson 1984). However, in north Cornwall this apparent younging of the ages from south to north represents separate cooling events rather than a northward migration of a single phase of deformation across the area. The 340-320 Ma (pre 1977 constant) group identified by Dodson & Rex (1971) occurs within the southern part of the Trevone Basin. This age range they attributed to a Namurian deformation and it is here considered to be related to the regional (D_1) deformation phase. The northern age group (270-310 Ma) falls within the area affected by a M_2 metamorphic event (Primmer 1985a, 1985b) and is synchronous with D_2 underthrusting of the Culm Basin. These younger ages therefore reflect the D_2 regional phase of deformation which occurred during late Westphalian to earliest Stephanian times (Warr 1989).

A reinterpretation of the K-Ar dates for SW England is presented (Fig. 8) in which three populations of ages are recognized:

- M_2 cooling ages, following the underthrusting of the Culm Basin (Regional D_2).
- M_1 cooling ages attributed to the D_1 deformation phase.
- emplacement of the Lizard complex.

M_2 cooling ages are restricted to the south of the southern margin of the Culm Basin in a belt approximately 10km wide, extending from the Tintagel district to the northwestern margin of the Dartmoor Granite. M_1 cooling ages are more widespread across SW England and incorporate the southern parts of the Trevone and South Devon Basins and parts of south Cornwall. This distribution of dates shows no pattern of diachronous ages which can be attributed to a migrating regional (D_1) deformation. Although dates from north Devon also fall within this population, doubt has been cast on these results with large variations in the ages attributed to "incomplete recrystallization of some samples" (Dodson & Rex 1971). The age of deformation in north Devon in relation to these isotopic dates remains unclear, but a conformable stratigraphical succession from Lower Devonian through to the Upper Carboniferous times suggests that uplift occurred later than in the Trevone Basin.

The oldest group of dates (365-345Ma) are associated with the Lizard complex which have been considered to represent emplacement ages of the Lizard ophiolite during late Devonian or early Carboniferous times (Holder & Leveridge 1986a, Barnes & Andrews 1986). Recent palynological evidence (Wilkinson & Knight 1989) from the underlying sediments of the Gramscatho Group shows the stratigraphic range to pass into the late Famennian, constraining deformation to latest Devonian or Carboniferous times. The exact timing of the emplacement of the Lizard ophiolite in relation to the regional D_1 event outlined in this paper would seem to remain uncertain.

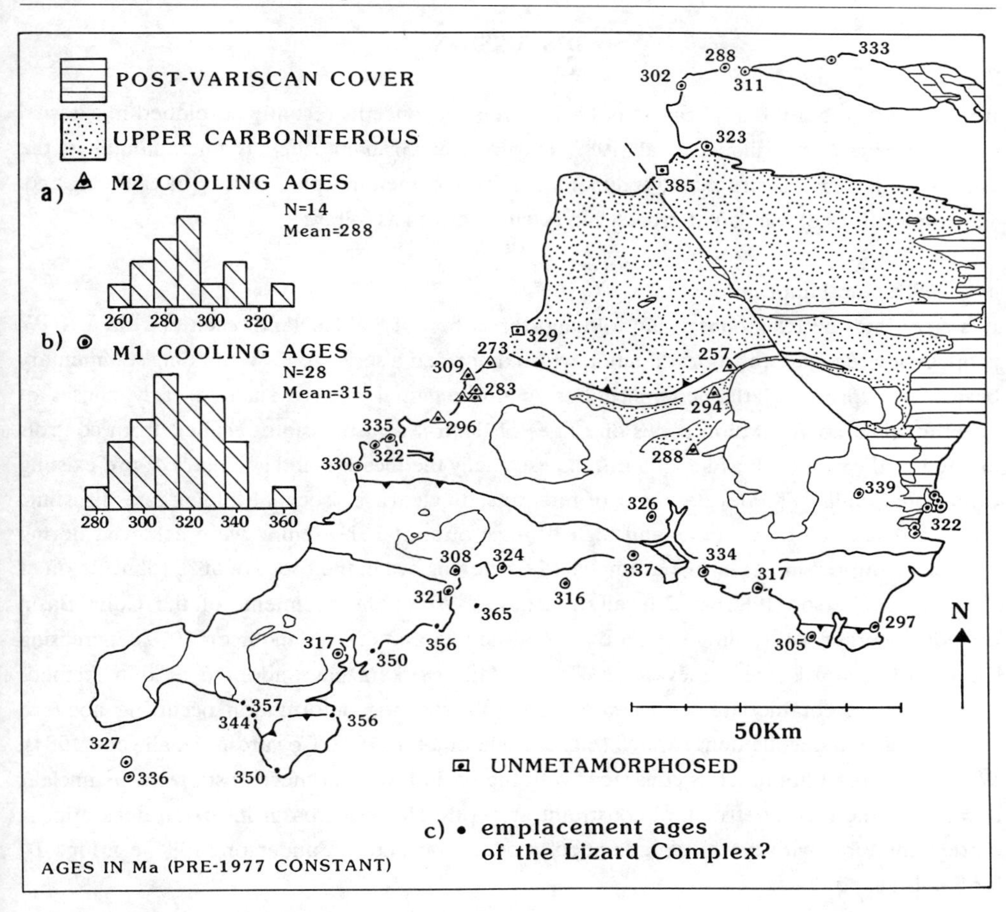

FIGURE 8
K-Ar dates for SW England (Dodson & Rex 1971). Three age group are outlined a) M2 cooling ages. b) M1 cooling ages. c) emplacement ages for the Lizard Complex.

DISCUSSION

It is concluded that the application of basin inversion concepts recently developed in external orogenic belts (e.g. Gillcrist et al. 1987, Hayward & Graham 1989) is fundamental to the understanding of the structural, sedimentological and metamorphic history of SW England. The implications of such an approach can be summarized as follows:

a) Structural implications

It is clear that the Rhenohercynian fold and thrust belt of SW England resulted from a NNW propagating regional transport direction which deformed a series of E-W trending sedimentary basins. It is proposed that complications in this regional pattern such as early phases of backthrusting, confrontation zones and lines of dextral transpression, resulted from controls exerted by the original basin configuration, especially the location and geometry of pre-existing extensional faults. The development of intense slaty cleavage, isoclinal folding and thrusting, indicate that significant strains and high degrees of stratal shortening were achieved during Variscan compression. Average estimates of shortening are in the order of 50% (Shackleton et al. 1982; Sanderson 1984) and locally reaches 60-70%. The sediments of the Culm Basin provide accurate shortening estimates of between 35 and 60% (Sanderson 1979), increasing southwards towards the Rusey Fault Zone. Many workers have adopted a "thin-skinned" approach to accommodate such strains at depth, showing deformation occurring above a shallow regional decollement surface (e.g. Shackleton et al. 1982, Coward & Smallwood 1984). Whether such an approach is consistent with the basin inversion mechanism remains unclear because of the lack of structural constraint at depth. However, basin inversion does appear more compatible with the "thicker skinned" model proposed by Sanderson (1984, e.g. Fig. 10) for SW England.

Basin inversion can accommodate considerable shortening (by more internal strain in the syn-rift strata), without necessarily involving major displacements between individual nappe units. This is an important point to be borne in mind when attempting to reconstruct the pre-deformational distribution of units by employing facies analysis (e.g. Selwood & Thomas 1986a, 1986b; Isaac et al. 1982). Throughout north Cornwall, the syn-rift sequences, although complexly folded and faulted and suffering high amounts of internal strain are overall suggested to have maintained their relative regional pre-deformational positions (refer to Fig. 3).

b) Sedimentological implications

The concept that sedimentation occurred synchronously with inversion during Upper Carboniferous times leads to the question of what mechanisms controlled the site of sedimentation in SW England. It is suggested here that relative amounts of uplift associated with basin inversion determined the areas of differential thrust loading and downflexuring within the Rhenohercynian belt. The principle controls as to whether uplift or subsidence results would be determined by the initial thickness of syn-rift sequences that were inverted and the timing of inversion. Such a mechanism conforms to the use of the term foreland basin as defined by Allen et al. (1986).

c) Implication for low-temperature metamorphism

One of the main problems facing workers in low-temperature metamorphic terrains of the external parts of orogenic belts is distinguishing between burial and deformation-related controls of metamorphism (e.g. Bevins & Robinson 1988; Roberts et al. 1990; Bevins & Robinson 1989). A diastathermal (extensional) model (Robinson 1987) is suggested for the M_1 (burial) metamorphic event of north Cornwall and Devon, in which a high thermal flux developed during crustal thinning and was maintained during regional (D_1) deformation. The M_2 event found in the Tintagel High Strain Zone is attributed to deformation related metamorphism with increasing temperatures and confining pressures occurring during underthrusting of the southern Culm Basin. The controls of heat flow within the Culm basin are less well understood, although a low geothermal gradient would be expected for a foreland basin setting (Beaumont 1981). The regional pattern of metamorphism and implications are explored fully in Warr et al. (in press).

d) General implications to the Rhenohercynian of Europe.

The controls exerted by basin inversion provides a fundamental key to the understanding of tectono/stratigraphic variations along the strike of the orogen. Differences in the amount and direction of initial fault-controlled crustal stretching and in the timing, magnitude and direction of subsequent orogenic compression can inevitably result in structural variations along the strike. In SW England, basin inversion appears to have occurred shortly after extension. The structural complications in the fold and thrust belt described here result from the influence of the pre-existing extensional basin architecture on the compressive deformation. The relative obliquity of the pre-existing extensional framework (E-W) to the NNW regional compression adds further complications to the pattern of deformation. The pattern of basin inversion is also considered to have influenced the position and nature of Upper Carboniferous foreland basin development, determining the location of thrust-loading, erosion and subsidence.

Acknowlegdements

The author thanks Oonagh Smith, Robin Clayton and Bridget Durning as members of the NERC/University of Exeter mapping group and the staff of the BGS Exeter office (especially

Brian Leveridge) for their help and support. Thanks also to Jim Andrews and Mary Ford for reviewing the manuscript and Brian Selwood, Mike Thomas and Doug Robinson for their guidance. The fieldwork for this study was undertaken as part of a NERC/University of Exeter mapping group (1986-1990) contracted to revise the BGS 1:50,000 Geological Sheets 336 & 335 (Camelford & Trevose Head sheets).

REFERENCES

Allen, P.A., Homewood, P. & Williams, G.D. 1986. Foreland basins: an introduction. In: ALLEN, P.A.& WILLIAMS, G.D. (eds.) Foreland Basins. Special Publication of the International Association of Sedimentologists, 8-12

Andrews, J.R., Barker, A.J. & Pamplin, C.F. 1988. A reappraisal of the facing confrontation in north Cornwall; fold- or thrust-dominated tectonics? Journal of the Geological Society, London, 145, 777-788.

Andrews, J.R., Barker, A:J. & Pamplin, C.F. 1990. Reply to: Discussion on a reappraisal of the facing confrontation in north Cornwall; fold- or thrust-dominated tectonics? Journal of the Geological Society, London, 147, 408-410.

Barnes, R.P. & Andrews, J.R. 1986. Upper Palaeozoic ophiolite generation and obduction in south Cornwall. Journal of the Geological Society, London, 143, 117-124.

Beaumont, C. 1981. Foreland Basins. Geophysical Journal of the Royal Astronomical Society, 65, 291-329.

Beaumont, C., Quinlain, G.M. & Hamilton, J. 1987. The Alleghanian orogeny and its relationship to the evolution of the Eastern Interior, North America. In: BEAUMONT, C. & TANKARD, A.J. (eds.) Sedimentary Basins and Basin-forming Mechanisms. Canadian Society of Petroleum Geologists, Memoir 12, 425-445.

Behr, H.J. 1983. Intracrustal thrust-tectonics at the northern margin of the Bohemian Massif. In: MARTIN, H. & EDER, F.W. (eds.) Intracontinental Fold Belts. Springer-Verlag, Berlin, 365-404.

Behr, H.J., Engel, W., Franke, W., Giese, P. & Weber, K. 1984. The Variscan belt in Central Europe: main structures, geodynamic implications, open questions. Tectonophysics, 109, 15-40.

Bevins, R.E. & Robinson, D. 1988. Low grade metamorphism of the Welsh Basin Lower Palaeozoic succession: an example of diastathermal metamorphism. Journal of the Geological So-ciety, London, 145, 363-366.

Bevins, R.E. & Robinson, D. 1989. Reply to: Discussion on low grade metamorphism of the Welsh Basin Lower Palaeozoic succession: an example of diastathermal metamorphism? Journal of the Geological Society, London, 146, 885-890.

Bluck, B.J., Haughton, P.D.W., House, M.R., Selwood, E.B. & Tunbridge, I.P. 1988. Devonian of England, Wales and Scot-land. In: MCMILLAN, N.J., EMBRAY, A.F. & GLASS, D.J. (eds) Devonian of the World. Canadian Society of Petroleum Geolo-gist's, Memoir 14, (Vol. 1).

Brix, M.R., Drozdzewski, G., Greiling, R.O., Wolf, R. & Wrede, V. 1988. The N. Variscan margin of the Ruhr coal district (Western Germany): structural style of a buried thrust front? Geologische Rundschau, 77/1, 115-116.

Brooks, M., Trayner, P.M. & Trimble, T.J. 1988. Mesozoic reactivation of Variscan thrusting in the Bristol Channel area U.K. Journal of the Geological Society, London, 140, 257-265.

Butler, R.W.H. 1989. The influence of pre-existing basin struc-ture on thrust system evolution in the Western Alps. In: COOPER, M.A. & WILLIAMS, G.D. (eds.), Inversion Tectonics, Geological Society Special Publication No 44, 105-122.

Cornford, C., Yarnell, L. & Murchinson, D.G. 1987. Initial vitri-nite reflectance results from the Carboniferous of north Devon and Cornwall. Proceedings of the Ussher Society, 6, 461-467.

Coward, M.P. & McClay, K.R. 1983. Thrust tectonics of S Devon. Journal of the Geological Society, London, 140, 215-228.

Coward, M.P. & Smallwood, S. 1984. An interpretation of the Varis-can tectonics of SW Britain. In: HUTTON, D.H. & SANDERSON, D.J. (eds) Variscan tectonics of the North Atlantic region. Special publication of the Geological Society, London, 14, 89-102.

Davis, G.R. 1984. Isotopic evolution of the Lizard Complex. Journal of the Geological Society, London, 141, 3-14.

De Graciansky, P.C., Dardeau, G., Lemoine, M. & Tricart, P. 1989. The inverted margin of the French Alps and foreland basin inversion. In: COOPER, M.A. & WILLIAMS, G.D. (eds.), Inversion Tectonics, Geological Society Special Publication Nr. 44, 87-104.

De Raaf, J.F.M., Reading, H.G., & Walker, R.G. 1965. Cyclic sedi-mentation in the Lower Westphalian of north Devon, England. Sedimentology, 4, 1-52.

Dodson, M.H. & Rex, D.C. 1971. Potassium-argon ages of slates and phyllites from south-west England. Quarterly Journal of the Geological Society of London, 126, 469-499.

Durning, B. 1989. A new model for the development of the Variscan Facing confrontation at Padstow, north Cornwall. Proceedings of the Ussher Society, 7, 141-145.

Edmonds, E.A. 1974. Classification of the Carboniferous rocks of south-west England. Report of the Institute of Geological Sciences, 74/13. (7pp).

Elliot, T. 1976. Upper Carboniferous sedimentary cycles produced by river-dominated, elongated deltas. Journal of the Geological Society, London, 132, 199-208.

Floyd, P.A. 1972. Geochemical, origin and tectonic environment of the basic and acidic rocks of Cornubia, England. Proceedings of the Geologists' Association, 83, 385-404.

Floyd, P.A. 1982. Chemical variation in Hercynian basalts relative to plate tectonics. Journal of the Geological Society, London, 139, 505-520.

Floyd, P.A. & Leveridge, B.E. 1987. Tectonic environment of the Devonian Gramscatho basin, South Cornwall: framework mode and geochemical evidence from turbidite sandstones. Journal of the Geological Society, 144, 531-542.

Freshney, E.C., McKeown, M.C. & Williams, M. 1972. Geology of the coast between Tintagel and Bude. Memoir of the Geological Survey of Great Britain.

Freshney, E.C. & Taylor, R.T. 1980. The Variscides of southwest Britain. In: OWEN, T.R. (ed.). United Kingdom: Introduction to general Geology and Guide to excursions 002, 055, 093, and 151. International Geological Congress, Paris 1980. Institute of Geological Sciences, London.

Gauss, G.A. 1966. Some aspects of slaty cleavage in the Padstow area of north Cornwall. Proceedings of the Ussher Society, 1, 221-224.

Gauss, G.A. 1967. Structural aspects of the Padstow area, north Cornwall. Proceedings of the Ussher Society, 1, 284-285.

Gauss, G.A. 1973. The structure of the Padstow area, north Cornwall. Proceedings of the Geologists' Association, 84, 283-313.

Gauss, G.A. & House, D.M. 1972. The Devonian successions in the Padstow area, North Cornwall. Journal of the Geological Society, London, 128, 151-172.

Gayer, R. & Jones, J. 1989. The Variscan foreland in South Wales. Proceeding of the Ussher Society, 7, 177-179.

Gillcrist, R., Coward, M.P. & Mugnier, J.L. 1987. Structural in-version and its controls: examples from the Alpine foreland and the French Alps. Geodinamica Acta (Paris), 1, 5-34.

Hanna, S.S. & Graham, R.H. 1988. A structural context of strain measurements on reduction spots in the Alpes Maritimes and the Hercynian fold belt of southern Britain. Annales Tectonicae, 2, 71-83.

Hartley, A. this volume Silesian sedimentation in southwest Britain: sedimentary responses to the developing Variscan Orogeny.

Hayward, A.B. & Graham, R.H. 1989. Some geometrical characteristics of inversion. In: COOPER, M.A. & WILLIAMS, G.D. (eds.), Inversion Tectonics, Geological Society Special Publications No. 44, 17-39.

Holder, M.T. & Leveridge, B.E. 1986. A model for the tectonic evolution of south Cornwall. Journal of the Geological Society, London, 143, 125-134.

Holdsworth, R.E. 1989. Short Paper: The Start-Perranporth line: a Devonian terrane boundary in the Variscan orogen of SW England? Journal of the Geological Society, London, 146, 419-421.

House, M.R. 1971. Devonian faunal distributions. In: MIDDLEMISS, F.A., RAWSON, F.A. & NEWALL, G. (eds.), Faunal provinces in Space and Time. Geological Journal special issue no. 4, 77-94.

Isaac, K.P., Turner, P.J. & Stewart, I.J. 1982. The evolution of the Hercynides of central SW England. Journal of the Geological Society of London, 139, 521-531.

Kelling, G. 1988. Silesian sedimentation and tectonics in the South Wales Basin: a brief review. In: BESLEY; B. & KELLING, G. (eds.), Sedimentation in a synorogenic basin complex: Upper Carboniferous of northwest Europe. Blackie , Glasgow and London, 38-42.

Kelm, U. 1986. Mineralogy and illite crystallinity of the pelitic Devonian and Carboniferous strata of north Devon and western Somerset. Proceedings of the Ussher Society, 6, 338-343.

Kirby, G.A. 1979. The Lizard Complex as an ophiolite. Nature, 282, 58-60.

Kirby, G.A. 1984. The petrology and geochemistry of dykes of the Lizard ophiolite Complex, Cornwall and Devon. Journal of Geological Society of London, 141, 53-9.

Lloyd, G.E. & Whalley, J.S. 1986. The modification of chevron folds by simple shear: examples from north Cornwall and Devon. The Journal of the Geological Society, London, 143, 89-94.

Matthews, S.C. 1977. The Variscan fold belt in southwest England. Neues Jahrbuch für Geologie und Palaeontologie Abhandlungen, 154, 94-127.

Mechie, J. & Brooks, M. 1984. A seismic study of deep geological structure in the Bristol Channel. Geophysical Journal of the Royal Astronomical Society, 78, 661-689.

Melvin, J. 1986. Upper Carboniferous fine-grained turbiditic sandstones from southwest England: a model for growth in an ancient, delta-fed subsea fan. Journal of sedimentary Petrology, 56, 19-34.

McClay, K.R. 1989. Analogue models of inversion tectonics. In: COOPER, M.A. & WILLIAMS, G.D. (eds.), Inversion Tectonics, Geological Society Special Publication No 44, 41-59.

Meissner, R., Springer, M. & Fluh, E 1984. Tectonics of the Variscides in north-western Germany based on seismic reflec-tion measurements. In: HUTTON, D.H.W. & SANDERSON, D.J. (eds) Variscan Tectonics of the North Atlantic Region. Special Publication of the Geological Society, London, 14, 215-228.

Pamplin, C.F. 1988. A re-examination of the Tintagel High Strain Zone and the Padstow Confrontation, north Cornwall. Unpubli-shed Ph.D thesis, University of Southampton.

Pamplin, C.F. & Andrews, J.R. 1988. Timing and sense of shear in the Padstow Facing Confrontation, north Cornwall. Proceedings of the Ussher Society, 7, 73-76.

Powell, C.M. 1989. Structural controls on Palaeozoic basin evolution and inversion in southwest Wales. Journal of the Geological Society, London, 146, 439-446.

Primmer, T.J. 1985a. A transition from diagenesis to greenschist facies within a major Variscan fold/thrust complex in south-west England. Mineralogical Magazine, 49, 365-374.

Primmer, T.J. 1985b. The pressure-temperature history of the Tintagel district, Cornwall: metamorphic evidence on the evolution of the area. Proceedings of the Ussher Society, 6, 218-223.

Roberts, B., Evans, J.A., Merriman, R.J. & Smith, M. 1989. Discussion on low grade metamorphism of the Welsh Basin Lower Palaeozoic succession: an example of diastathermal metamor-phism? Journal of the Geological Society, London, 146, 885-890.

Roberts, R.L. & Sanderson, D.J. 1971. Polyphase development of slaty cleavage and the confrontation of facing directions in the Devonian rocks of north Cornwall. Nature: Physical Science, 230, 87-89.

Robinson, D. 1987. The transition from diagenesis to metamorphism in extensional and collisional settings. Geology, 15, 966-969.

Robinson, D. & Bevins, R.E. 1986. Incipient metamorphism in the Lower Palaeozoic marginal basin of Wales. Journal of Metamor-phic Petrology, 4, 101-113.

Sanderson, D.J. 1972. Oblique folding in southwest England. Proceedings of the Ussher Society, 2, 438-442.

Sanderson, D.J. 1973. The development of fold axes oblique to the regional trend. Tectonophysics, 16, 55-70.

Sanderson, D.J. 1974. Chevron folding in the Upper Carboniferous rocks of north Cornwall. Proceedings of the Ussher Society, 3, 96-103.

Sanderson, D.J. 1979. The transition from up-right to recumbent folding in the Variscan fold belt of southwest England: a model based on the kinematics of simple shear. Journal of Structural Geology, 1, 171-180.

Sanderson, D.J. 1982. Models of strain variation in nappes and thrust sheets- a review. In: G.D. WILLIAMS (ed), Strain within thrust belts, Tectonophysics, 88, 201-233.

Sanderson, D.J. 1984. Structural variations across the northern margin of the Variscan in NW Europe. In: HUTTON, D.H. & SANDERSON, D.J. (eds.) Variscan Tectonics of the North Atlantic Region. Special Publication of the Geological Society, London, 14, 149-165.

Sanderson, D.J. & Dearman, W.R. 1973. Structural zones of the Variscan fold belt in SW England, their location and development. Journal of the Geological Society, London, 129, 527-536.

Scrivener, R.C., Leake, R.C., Leveridge, B.E. & Shepherd, T.J. 1989. Volcanic- exhalative mineralization in the Variscan province of south-west England. Terra abstracts, 1(1), 125.

Seago, R.D. & Chapman, T.J. 1988. The confrontation of structural styles and the evolution of a foreland basin in central SW England. Journal of the Geological Society, London, 145, 789-800.

Selwood, E.B, Edwards, R.A., Simpson, S., Chesher, J.A., Hamblin, R.J.O., Henson, M.R., Riddolls, B.W. & Waters, R.A. 1984. The geology of the country around Newton Abbot. Memoir of the Geological Survey of Great Britain.

Selwood, E.B., Stewart, I.J. & Thomas, J.M. 1985. Upper Palaeo-zoic sediments and structure in north Cornwall- a reinter-pretation. Proceedings of the Geologists' Association, 96, 129-141.

Selwood, E.B. & Thomas, J.M. 1984. Structural models of the geology of the north Cornwall coast: a reappraisal. Proceedings of the Ussher Society, 6, 134-136.

Selwood, E.B. & Thomas, J.M. 1986a. Upper Palaeozoic successions and nappe structures in north Cornwall. Journal of the Geological Society, London, 143, 75-82.

Selwood, E.B. & Thomas, J.M. 1986b. Variscan facies and structure in central SW England. Journal of the Geological Society, London, 143, 199-207.

Selwood, E.B. & Thomas, J.M. 1987. The tectonic framework of Upper Carboniferous sedimentation in central southwest Eng-land. In: Sedimentation in a Synorogenic Basin Complex: the Upper Carboniferous of northwest Europe (BESLEY, B.M. & KELLING, G.) Blackie, Glasgow & London.

Selwood, E.B. & Thomas, J.M. 1988. The Padstow Confrontation, north Cornwall: a reappraisal. Journal of the Geological Society, London, 145, 801-808.

Shackleton, R.M., Ries, A.C. & Coward, M.P. 1982. An interpreta-tion of the Variscan structures in SW England. Journal of the Geological Society, London, 139, 533-541.

Smith, O. 1990. A stratigraphical revision of the Trevone Basin, north Cornwall and its structural implications. Proceedings of the Ussher Society, 7, 307.

Smith, S.A. & Humphreys, B. 1989. Lakes and alluvial sandflat- playas in the Dartmouth Group, south-west England. Proceedings of the Ussher Society, 7, 118-124.

Styles, M.T. & Rundle, C.C. 1984. The Rb-Sr isochron age of the Kennack Gneiss and its bearing on the age of the Lizard Complex, Cornwall. Journal of the Geological Society, London, 141, 15-19.

Thomas, J.M. 1988. Basin history of the Culm Trough of southwest England. In: Sedimentation in a synorogenic basin complex: the Upper Carboniferous of northwest Europe. (eds. BESLEY, B.M. & KELLING, G.) Blackie, Glasgow & London, 24-37.

Tunbridge, I.P. 1986. Mid-Devonian tectonics and sedimentation in the Bristol Channel area. Journal of the Geological Society, London, 143, 107-115.

Turner, P.J. 1982. The anatomy of a thrust: a study of the Greystone thrust Complex, East Cornwall. Proceedings of the Ussher Society, 5, 270-278.

Warr, L.N. 1988. The deformational history of the area northwest of the Bodmin Moor granite. Proceedings of the Ussher Society, 7, 67-72.

Warr, L.N. 1989. The structural evolution of the Davidstow Anticline and its relationship to the Southern Culm Overfold, north Cornwall. Proceedings of the Ussher Society, 7, 136-140.

Warr, L.N. 1991. Basin inversion controls in the Variscan of SW England. Unpublished Ph.D thesis, University of Exeter.

Warr, L.N. & Durning, B. 1990. Discussion on the reappraisal of the facing confrontation in north Cornwall; fold- or thrust- dominated tectonics? Journal of the Geological Society, London, 147, 408-410.

Warr, L.N. & Robinson, D. 1990. The application of the illite "crystallinity" technique to geological interpretation: a case study from north Cornwall. Proceeding of the Ussher Society, 7, 223-227..

Warr, L.N., Primmer, T.J. & Robinson, D. in press. Variscan low grade Metamorphism in southwest England: a diastathermal and thrust-related origin. Journal of Metamorphic Geology.

Weber, K. 1981. The structural development of the Rheinische Schie-fergebirge. Geologie en Mijnbouw, 60, 149-159.

Weber, K. & Behr, M.J. 1983. Geodynamic interpretation of the mid-European Variscides. In: MARTIN, H., EDER, F.W. (eds.) Intra-continental Fold belts. Springer-Verlag, Berlin, 427-469.

Whalley, J.S. & Lloyd, G.E. 1986. Tectonics of the Bude Forma-tion, north Cornwall- the recognition of northerly directed decollement. Journal of the Geological Society, London, 143 (1), 83-88.

Williams, E.A., Ford, M., Edwards, H.E. & O'Sullivan, M.J. this volume. An outline evolution of the Late Devonian Munster Basin, south-west Ireland.

Williams, G.D., Powell, C.M. & Cooper, M.A. 1989. Geometries and kinematics of inversion tectonics. In: COOPER, M.A. & WILLIAMS, G.D., (eds), Inversion Tectonics, Geological Society Special Publication No. 44, 3-1.

Wilkinson, J.J. & Knight, R.W. 1989. Short Paper: Palynological evidence from the Portleven area, south Cornwall: implications for Devonian stratigraphy and Hercynian structural evolution. Journal of the Geological Society, London, 146, 739-742.

Structural Geology

Magnetic Fabric Relationship between Crystalline and Variscan Sedimentary Complexes in Eastern Bohemian Massif

F. Hrouda
Geofyzika, State Company, P.O. Box 62, 61246 Brno, CSFR

ABSTRACT

Magnetic anisotropy of both crystalline and Lower Carboniferous sedimentary rocks of the E part of the Bohemian massif was measured in order to investigate the fabric relationship between crystalline and sedimentary complexes. The magnetic fabric is shown to be deformational in origin not only in the crystalline rocks, but also in most sedimentary rocks. Moreover, the orientations of the deformational magnetic fabric elements are similar in many sedimentary formations and crystalline complexes. This implies that at least one Variscan deformation phase affected both crystalline and sedimentary rocks. This deformation was probably associated with the creation of the nappe structure in the eastern part of the Bohemian massif.

INTRODUCTION

Anisotropy of magnetic susceptibitily (AMS) is a geophysical technique for the investigation of preferred orientation of magnetic minerals in rocks. Its main advantage is rapidity - the measurements of one specimen takes 2 to 5 minutes. In strongly magnetic rocks it indicates the preferred orientation of ferromagnetic minerals either by grain shape (magnetite) or by crystal lattice (hematite, pyrrhotite), while in weakly magnetic rocks it often indicates the preferred orientation of dark silicates (micas, amphiboles, pyroxenes) by crystal lattice. Using modern sensitive instruments (e.g. KLY - 2 Kappabridge), the AMS can be measured on almost all rock types and, therefore, it can be used in the investigation of the fabric relationship between crystalline and sedimentary rocks, which is tedious when using other methods.

During the past twenty years the AMS has been investigated in various geological units in the E part of the Bohemian massif. Large parts of the results have been published by Hrouda (1970, 1971, 1976, 1978, 1979, 1985, 1989), Hrouda and Janák (1971) and Hrouda et al. (1970, 1971). In these publications detailed data including maps of sampling sites are presented. A smaller part of the AMS data obtained in cooperation with Drs. Cháb, Chlupácová, Kos, Prichystal, Raclavská, Rejl and Vasek have remained unpublished. The colleagues are thanked for permission to use their data in the present paper.

In the present paper the measured data are re-evaluated using the package of data base oriented statistical programs developed by Dr. V. Jelínek and interpreted in terms of the fabric relationship between crystalline and sedimentary complexes with the aim to discover the geological processes which gave rise to this relationship.

GEOLOGICAL SETTING

This section is based on the reviews by Svoboda et al. (1968) and Mísar et al. (1983) to which the reader is referred for further details.

The easternmost part of the Bohemian masif is built up of the Bruno-Vistulic crystalline complex consisting of various gneisses and granitoids, basically pre-Cambrian in age, though in places deformed and partially recrystallized during the Variscan orogeny. This complex extends under Tertiary sediments of the West Carpathian Foredeep and Flysch Belt, where it has been drilled in numerous boreholes, and perhaps also under Palaeozoic sediments in Moravia where it also crops out as two granitoid massif: the Brno massif and the Dyje massif (Fig. 1). It evidently represents the basement of Variscan nappes and it is often incorporated into the Variscan nappe structure.

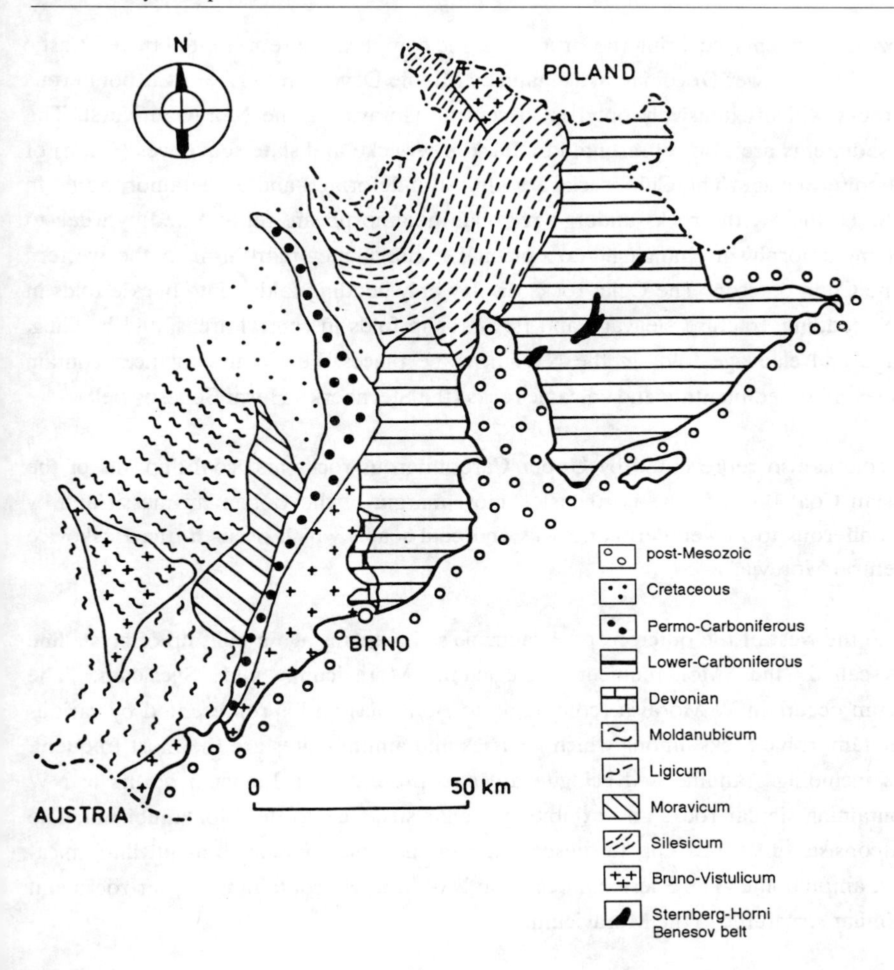

FIGURE 1
Geological sketch map of the E part of the Bohemian massif. Adapted from Svoboda et al. (1966).

The Palaeozoic sediments, covering the Bruno-Vistulic complex, are represented by red clastic sediments, probably Lower Devonian in age, and by Middle Devonian to Lower Carboniferous carbonate rocks which extensively outcrop in central Moravia as the Moravian Karst. The Palaeozoic sediments are also represented by thick greywacke and slate sequences (Culm) of Lower Carboniferous age. The Culm rocks are almost undeformed and unmetamorphosed in the E and S. To the W, the rocks undergo regional metamorphism, represented by weak to strong anchimetamorphism changing into epizonal regional metamorphism in the western margin of the Culm outcrop. The Culm rocks display only bedding folded into buckle folds in the E and S, bedding, fracture cleavage and flexural slip folds in central areas, and bedding, slaty cleavage and cleavage folds in the NW. In some places the Culm sequences contain klippen of Devonian sedimentary and volcanic rocks (the Sternberk - Horní Benesov belt).

The Palaeozoic is also represented by Upper Carboniferous rocks and coal deposits of the Upper Silesian Coal Basin in the northeasternmost margin of the Bohemian massif and by upper Carboniferous to Lower Permian rocks and coal seams which occur in the Boskovice graben in central Moravia.

The region to the west of the outcrops of Palaeozoic sediments is again built up of crystalline complexes, called the Moldanubicum, Lugicum, Moravicum, and Silesicum. The Moldanubicum occurs in W Moravia (continuing to Bohemia) and is represented by various catazonal metamorphic rocks among which gneisses and amphibolites are the most frequent. Other rocks including granulite and eclogite are also present. The Lugicum occurs in NW Moravia containing similar rocks and exhibiting similar structure to the Moldanubicum. The Moravicum consists of various epi- to mesozonal metamorphic rocks such as phyllite, mica-schist, gneiss, amphibolite. The Silesicum occurs in NW Moravia containing similar rocks and displaying similar structures to the Moravicum.

As early as in the beginning of this century, nappe structures were revealed in the E Bohemian massif by Suess (for review see Suess 1926) who hypothesized that the Moldanubicum and Lugicum, as large nappes, have overriden the Moravicum and Silesicum which appear as windows (or half-windows). After world war II Suess' ideas were in general rejected (for review see Swoboda et al. 1968). Recently, they have not only been accepted again, but imbricate and duplex structures have been revealed also in the Bruno-Vistulicum on the basis of reflection seismic survey (Ibrmajer and Tomek, pers. comm.) and in the Culm sequences (Cháb 1986, Tomek, pers. comm.).

DATA PRESENTATION

The AMS data are presented in Table 1 and Figs. 2 to 4. The table presents the arithmetical means of the anisotropy degree, $P = k_1/k_3$ ($k_1>k_2>k_3$ are principal susceptibilities) and of the shape parameter, $T = 2 \ln (k_2/k_3)/\ln (k_1/k_3)-1$ ($T>O$ indicates planar magnetic fabric, $T<O$ indicates linear magnetic fabric), their standard deviation (s(P), s(T), respectively), numbers of specimens (N) and numbers of localities (n) investigated in each geological unit. The figures present contour diagrams of magnetic foliation poles and of magnetic lineations in the common geographic coordinate system and in the so-called palaeogeographic coordinate system defined by the horizontal main mesoscopic foliation (bedding or metamorphic schistosity) and north.

BRNO-VISTULIC CRYSTALLINE COMPLEX

In this complex, the AMS has been investigated only in the Brno massif, which consists mostly of granitoid rocks (granodiorite, diorite) creating two large bodies: the Eastern and Western Granitoid Zones separated by the Central Metabasite Zone) consisting originally of extrusive and intrusive basic rocks.

In the Eastern Granitoid Zone the AMS is very variable both in the anisotropy degree and the shape parameter (Table 1). The magnetic foliations and magnetic lineations are widely scattered in space (Fig. 2 a, b). Nevertheless, one of the main maxima in magnetic lineation is oriented NE - SW and the magnetic foliation poles tend to create an irregular girdle oriented NW - SE (Fig. 2 a, b). As a transition from massive granodiorites to those exhibiting cataclastic foliation has been found and the magnetic foliation and magnetic lineation shows the same orientations in massive and cataclazed granodiorites, it has been concluded that the magnetic fabric is deformational in origin in the Eastern Granitoid Zone (for details see Hrouda 1985).

In the Western Granitoid Zone the anisotropy degree in slightly lower, but still variable as well as the magnetic fabric shape (Table 1). The magnetic foliations and magnetic lineations are even more scattered than in the Eastern Granotoid Zone (Fig. 2c, d). However, the magnetic foliation poles tend to create a girdle oriented NE - SW (Fig. 2 c, d). Similar relationship between massive and foliated granodiorites to that in the Eastern Granitoid Zone lead us to the conclusion that the fabric in these rocks is also mostly deformational in origin (Hrouda 1985).

In extrusive rocks of the Metabasite Zone the anisotropy degree is variable, but relatively high in some localities (Table 1). The magnetic foliation is steep and oriented NE - SW, strongly

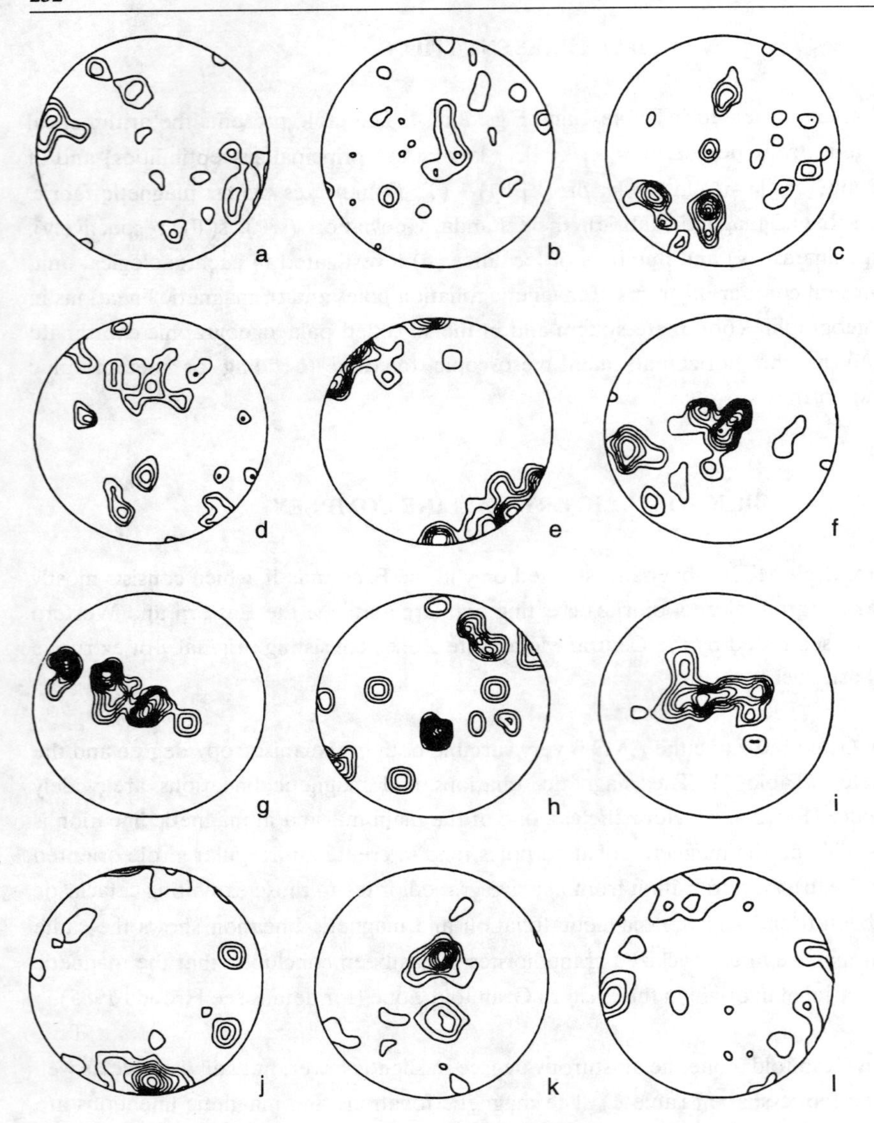

FIGURE 2
Contour diagrams of magnetic foliation poles (a, c, e, g, i, k) and magnetic lineations (b, d, f, h, j, l) in the Brno massif (a, b - Eastern Granitoid zone, c, d - Western Granitoid Zone, e, f - Central Metabazite Zone), Devonian red sediments (g, h), Moravicum (i, j) and Moldanubicum (k, l).
The most external contour corresponds to the density of 1%, the others increase by 1% (a, b, e, f) or 2% (the other figures). Geographical coordinate system, equal-area projection on lower hemisphere.

deviating from the N - S oriented mesoscopic foliation which is parallel to the course of the zone (Fig. 2 e, f). This magnetic fabric has been interpreted as deformational in origin and generated during sinistral simple shear along the metabasite zone (Hrouda 1985).

RED CLASTIC SEDIMENTS OF THE MORAVIAN KARST

The AMS indicates the preferred orientation of hematite in these sediments. The magnetic foliation poles create a narrow partial girdle oriented NW - SE, while the magnetic lineations are widely scattered NE - SW (Fig. 2 g, h). The AMS shows a close relationship to the bedding and to the palaeo-current directions derived from the orientation of pebbles and is, therefore, regarded as sedimentary in origin (Hrouda and Janák 1971).

CULM FORMATION

The Culm rocks form two large outcrops, the Nízky Jeseník Mts. and the Drahanská vrchovina upland, separated by the Tertiary and Quartenary (Fig. 1). In the Nízky Jeseník Mts., the AMS is very different in the E part (east of the Devonian Sternberk - Horní Benesov belt) and in the W part (west of this belt). In the E part, anisotropy degree is low and the magnetic fabric is mostly planar (Tab. 1). The magnetic foliation is near the bedding and the magnetic lineation is often parallel to the sedimentary structures indicating the palaeocurrent directions, but sometimes it deviates from these sedimentary structures, being near the intersection lines between cleavage and bedding. Both the magnetic foliation poles and the bedding poles create girdles oriented NW - SE and the magnetic lineation is mostly oriented NE - SW, parallel to the bedding/cleavage intersection lines (Fig. 3 a - d). This magnetic fabric indicates an effect of a weak lateral shortening, probably accompanying the cleavage formation (Hrouda 1976, 1979).
In the W part, the anisotropy degreee is much higher and the magnetic fabric is planar (Tab. 1). The magnetic foliation is parallel to either the slaty cleavage or the metamorphic schistosity and the magnetic lineation is parallel to the cleavage/ bedding intersection lines. The magnetic foliation create a wide W-E girdle, while the magnetic lineation is mostly oriented NE-SW (Fig. 3 e, f). The magnetic fabric is undoubtedly deformational in origin and the main deformation is associated with the formation of slaty cleavage perpendiculr to the strong rock shortening (Hrouda 1979).

In the Drahanská Vrchovina upland, the sedimentary magnetic fabric has been more or less affected by ductile deformation which was in general much weaker than that in the Nízky Jeseník Mts. The influence of the deformation was smaller in the younger Myslejovice

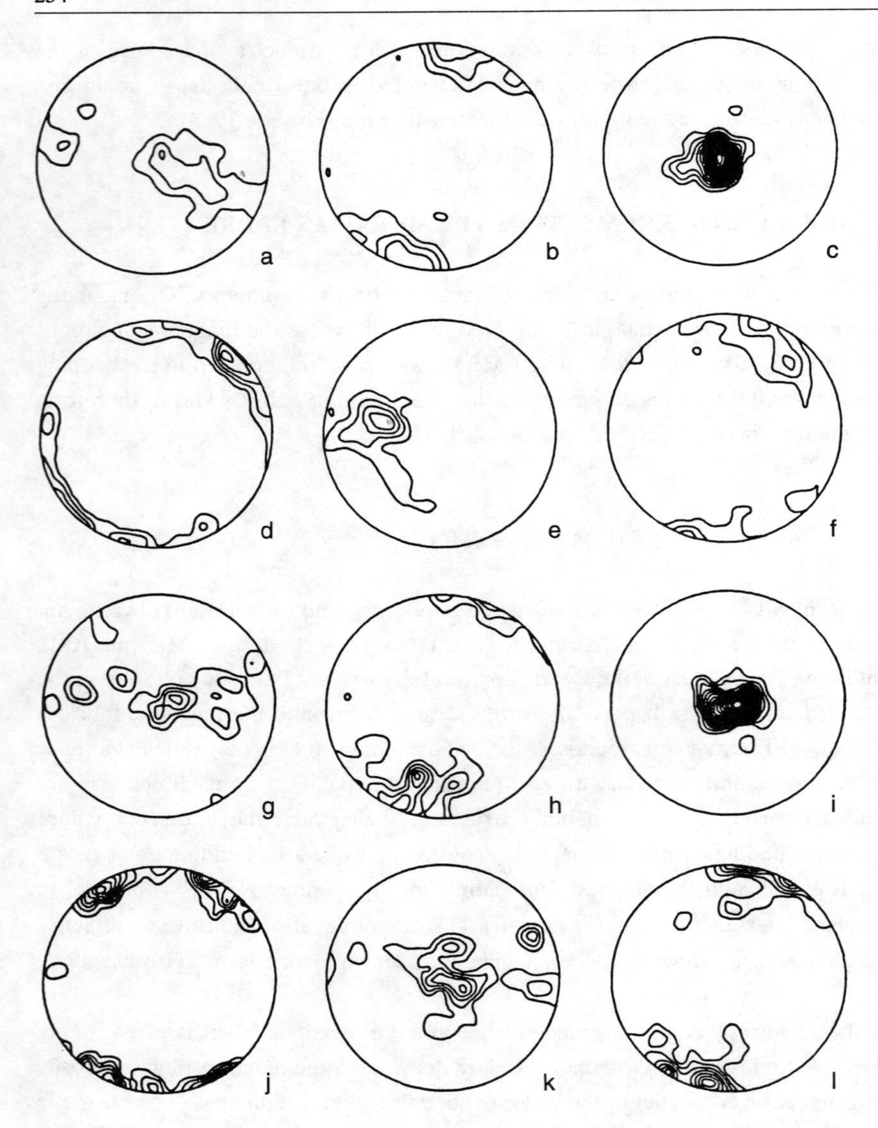

FIGURE 3
Contour diagrams of magnetic foliation poles (a, c, e, g, i, k) and magnetic lineations (b, d, f, h, j, l) in the eastern part of the Nízky Jeneník Mts. (a, b - geographic coordinate system, c, d - palaeogeographic coordinate system), in the western part of the Nízky Jeseník Mts. (e, f - geographic coordinate system), and in the Drahanská vrchovina upland (g, h - all specimens, geographic coordinate system, i, j - Myslejovice Formation, palaeogeographic coordinate system, k, l - Protivanov Formation, palaeogeographic coordinate system).
The most external contour corresponds to 2% density, the other increase by 2%. Equal area projection on lower hemisphere.

Formation, where the magnetic foliation is mostly parallel to the bedding (Fig. 3 i, j) and the effect of deformation manifests in the deviation of the magnetic lineation from the palaeo-current directions. In the older Protivanow Formation (Fig. 3 k, l) the effect of strain was stronger, giving rise to significant deviations of the magnetic foliation from the bedding. The magnetic foliation poles create a girdle oriented W - E and the magnetic lineations are oriented from NNW - SSE to NNE - SSW (Fig. 3 g, h, Kos 1987).

VOLCANIC ROCKS OF THE STERNBERK - HORNI BENESOV BELT

In these rocks, the anisotropy degree is low and the magnetic fabric shape is very variable (Tab. 1). The magnetic foliations mostly strike N - S and dip E, the magnetic lineations mostly plunge SE (Fig. 4 k, l). The magnetic fabrics of the various volcanic bodies are similar to one another. The orientations of magnetic foliations resemble those in the surrounding deformed sedimentary rocks: one can conclude that the magnetic fabric is partially deformational in origin.

SILESICUM

In the Silesicum the AMS has been investigated in various rock types: Devonian phyllites, pre-Cambrian gneisses, amphibolites of unknown (but considered Devonian) age. The anisotropy degree in these rocks is moderate to very high and the magnetic fabric is mostly planar (Table 1) The magnetic foliation is often parallel to the metamorphic schistosity and the magnetic lineation to the mesoscopic lineation. However, the magnetic foliation also often deviates significantly from the metamorphic schistosity, probably indicating the effects of a younger deformation (Fig. 4 i, j).

In all the units investigated, the AMS is deformational and complex in origin, reflecting at least two deformations phases. Nevertheless, the orientations of the principal susceptibilities in all the units are similar. The magnetic foliation poles create a wide girdle oriented NW - SE, and the magnetic lineations are oriented mostly NE - SW (Fig. 4 a, b).

MORAVICUM

In the Moravicum, the AMS was investigated in the Bítes gneiss, Svratka gneiss and in various amphibolites. The anisotropy degree in these rocks is moderate to high (Table 1). The magnetic foliation poles create an imperfect girdle oriented W - E and the magnetic lineation

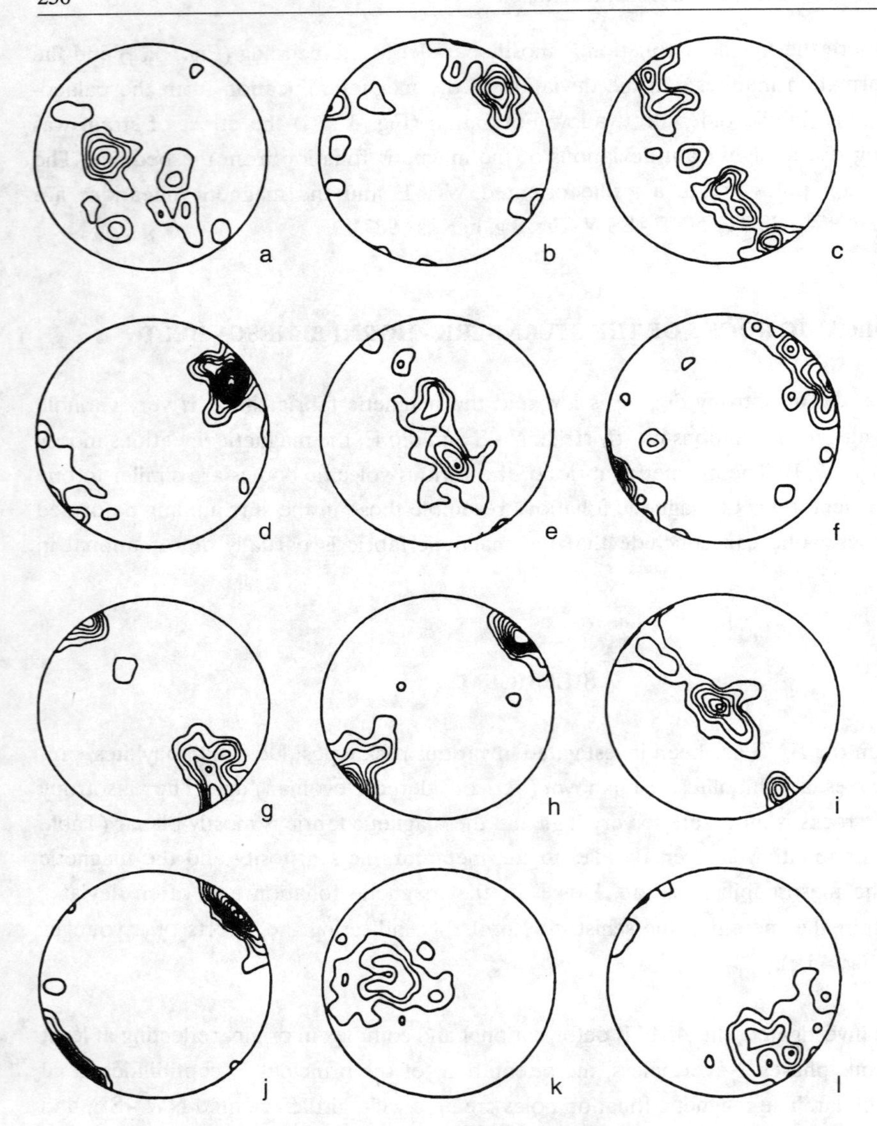

FIGURE 4
Contour diagrams of magnetic foliation poles (a, c, e, g, i, k) and magnetic lineations (b, d, f, h, j, l) in the Silesicum (a, b - Vrbno Formation, c, d - plagioclase phyllite, e, f - Desná dome gneiss, g, h - amphibolites in geographic coordinate system, i, j - amphibolites in palaeogeographic coordinate system) and in volcanic rocks of the Sternberk - Horní Benesov belt (k, l).
The most external contour corresponds to 1%, the others increase by 2%. Equal-area projection on lower hemispere.

Table 1 Magnetic anisotropy parameters for the investigated geological units of the eastern Bohemian massif

Geological unit	*P*	*s(P)*	*T*	*s(T)*	*N*	*n*
Bruno-Vistulicum of the Brno massif						
Eastern Zone	1.077	0.048	0.14	0.25	151	24
Western Zone	1.059	0.073	0.15	0.24	104	21
Central Zone	1.052	0.042	0.14	0.27	40	4
Devonian of the Moravian Karst						
red sediments	1.035	0.029	0.49	0.30	24	6
Devonian of the Šternberk - Horní Benešov belt						
volcanic rocks	1.062	0.038	0.25	0.32	172	19
Culm Formations						
E Nízký Jeseník	1.064	0.020	0.33	0.35	505	44
W Nízký Jeseník	1.115	0.047	0.34	0.26	498	57
Drahanská vrch.	1.056	0.089	0.33	0.34	178	17
Silesicum						
phyllite	1.125	0.056	0.31	0.35	107	7
plagioclase phyll.	1.162	0.048	0.02	0.40	73	7
gneiss	1.167	0.105	0.30	0.44	174	87
magnetic amph.	1.329	0.237	0.13	0.44	73	22
non-magnetic amph.	1.075	0.040	0.41	0.36	92	22
Moravicum						
Bíteš gneiss	1.101	0.070	0.67	0.14	12	3
Svratka gneiss	1.079	0.021	-0.25	0.17	9	3
magnetic amph.	1.426	0.216	-0.12	0.14	5	5
non-magnetic amph.	1.103	0.082	0.21	0.61	39	5
Moldanubicum						
magnetic amph.	1.429	0.285	0.20	0.25	16	5
non-magnetic amph.	1.125	0.032	0.72	0.31	52	5

is mostly oriented N - S (Fig. 2 i, j). The AMS here originated either during ductile deformation or during a rock recrystallization in an anisotropic stress field (Hrouda et al. 1971).

MOLDANUBICUM

In the Moldanubicum, the AMS has been investigated only in amphibolites. The anisotropy degree in these rocks varies from low to very high and the AMS origin is variable in different areas (Table 1). In some places, the magnetic fabric is closely related to the mesoscopic rock fabric, being probably associated with the metamorphism of the rocks. In other places, the magnetic fabric is unrelated to the rock fabric and the AMS may be younger than the rock origin. In some other places, the amphibolites were enriched in young magnetite and their magnetic fabric is mimetic (Hrouda et al. 1971). Both the magnetic foliation and lineation show large scatters (Fig. 2 k, l).

DISCUSSION AND CONCLUSION

Although the rocks building up the eastern part of the Bohemian massif are very different in age (ranging from pre-Cambrian to Upper Palaeozoic) and origin (metamorphic, intrusive, effusive and sedimentary rocks), their AMS patterns are not very different. The principal AMS feature is a girdle pattern in magnetic foliation poles, oriented W - E to NW - SE and the orientation of magnetic lineation N - S to NE - SW. This magnetic fabric feature is exhibited by all the geological units investigated, except the Moldanubicum and the Western Granitoid Zone of the Brno massif where the girdle in the magnetic foliation poles is NE - SW and the magnetic lineation is mostly oriented NW - SE.

The origin of magnetic fabric in sedimentary rocks is only rarely sedimentary in origin; mostly the magnetic fabric shows signs of moderate to strong ductile deformation. In volcanic rocks the magnetic fabric is also deformational in origin. In the granitoid rocks investigated the magnetic fabric has often been strongly affected by ductile deformation. In metamorphic rocks, the magnetic fabric is either due to recrystallization in an anisotropic stress field or due to younger deformation.

If the magnetic fabrics of the majority of the investigated rocks, regardless of their origin, is deformational and the AMS patterns are more or less uniform in the majority of the geological units of the E Bohemian massif, one can conclude that the deformation that formed the magnetic fabric observable today was younger than Lower Visean. This deformation strongly overprinted the older metamorphic and deformational magnetic fabrics in metamorphic rocks and the depositional, intrusive and flow magnetic fabrics in sedimentary, intrusive and effusive rocks, respectively.

If one accepts the concept of nappe structures in the E Bohemian massif as introduced by Suess and extended by Tomek and Cháb, it seems very likely that the deformation that overprinted the older magnetic fabrics was associated with the generation of the nappes. In the initial stages the Bruno-Vistulicum, Palaeozoic, Moravicum and Silesicum were imbricated, while in the terminal stages they were overriden by the Moldanubicum and Lugicum which have their own different fabrics.

Finally, it should be emphasized that in this paper only one phase of the very complex deformational development of the E. Bohemian massif has been interpreted. The interpreted phase was selected because it is recorded in the magnetic fabric most conspicuously. There have been other deformations (e.g. development of shear zones, etc.) which may have been reflected in the magnetic anisotropy as well. As their magnetic fabric imprint can be separated from the complex magnetic fabric only with great difficulties, these possible deformations are not interpreted in the present paper.

REFERENCES

Cháb, J. 1986: Tectonics of the Morava-Silesian brach of the European Late Palaeozoic orogeny (a working hypothesis) (in Czech). Véstník Ustr. úst. geol., 61, 113-120

Hrouda, F., 1970: The relation between the fabric and anisotropy of magnetic susceptibility for some West Moravian geneisses. Véstník Ustr. úst. geol., 45, 147-156.

Hrouda, F., 1971: The magnetite frabric of some massive and mylonitized granodiorites of the most northern part of the Brno massif. Cas. miner. geol., 16, 37-45.

Hrouda, F., 1976. The origin of the cleavage in the light of magnetic anisotropy invenstigations. Phy. Earth Planet. Inter., 13, 132-142.

Hrouda, F., 1978: The magnetic fabric in some folds. Phys. Earth Planet. Inter., 17, 89-97.

Hrouda, F., 1979: The strain interpretation of magnetic anisotropy in rocks of the Nízky Jeseník Mountains (Czechoslovakia). Sbor. geol. ved, UG, 16, 27-62.

Hrouda, F., 1981: On the superposition of regional slaty cleavage on folded strata and its reflection in magnetic anisotropy. Cas. miner. geol. 26, 341-348.

Hrouda, F., 1985: The magnetic fabric of the Brno massif. Sbor. geol. ved, UG, 19, 89-112.

Hrouda, F. and F. Janák, 1971. A study of the hematite fabric of some red sediments on the basis of their magnetic susceptibility anisotropy. Sediment. Geol., 6, 187-199.

Hrouda, F., F. Janák and J. Stelcl, 1970; Über die Verwendbarkeit der Anisotropie der magnetischen Suszeptibilitaet zur Bestimmung der Magnetitorientierung. In: P. Paulisch (editor) - Experimental and natural rock deformation, Springer Verlag Berlin, 253-262.

Hrouda, F., F. Janák, L. Rejl and J. Weiss, 1971: The use of magnetic susceptibility anisotropy for estimating the ferromagnetic mineral fabric of metamorphic rocks. Geol. Rundsch., 60, 1124-1142.

Kos, J., 1987: The investigation of the anisotropy of magnetic susceptibiligy of the Lower Carboniferous sedimentary complex of the Drahany upland (in Czech). Thesis Charles University in Prague, 76 pp.

Mísar, Z., A. Dudek, V. Havlena, J. Weiss, 1983. Geology of CSSR I. Bohemian massif (in Czech). SPN Publishing House Praha, 333 pp.

Suess, F., 1926: Das Grossgefüge der böhmischen Masse. Zbl. Mineral. Geol. Palaeont., Abt. B, 97-1o9, Leizig.

Svoboda, J., et al. (1966): Regional Geology of Czechoslovakia. The Bohemian massif. Academia Praha, 668 pp.

Some Aspects of Interactivity between Folding and Thrusting in the Ruhr Carboniferous

V.Wrede
Geologisches Landesamt Nordrhein-Westfalen, De-Greiff-Str.195, D-4150 Krefeld, Germany

ABSTRACT

The main feature of the tectonics of the Ruhr Carboniferous are disharmonic folds, which decrease gradually in intensity from south to north. In a vertical sense it is possible to distinguish 3 or 4 tectonic levels ("stockwerks").

The orogenic shortening is equal in all these stockwerks. The stockwerks are neither stratabound nor controlled by the rock material. Within this stockwerk arrangement, thrusts in the central stockwerk accommodate the differences in fold styles between upper and lower levels. Another factor generating thrusts is the accommodation of volume surpluses in the center of chevron folds. The distribution of northward and southward directed thrusts being almost random in the Ruhr Carboniferous a gross horizontal mass transport cannot be observed. Many thrusts are substituted by folds either along strike or downwards, giving evidence of the contemporary activity of both tectonic elements.

In general, the thrusts can be interpreted as part of the mechanism of folding.

The angle between thrust and bedding depends on the dip of strata and is subject to changes of angle that can be described by a numerical model. This relation between bedding-dip and angle between thrust and bedding-plane can be computed easily into a relation between bedding-dip and thrust-dip. These relationships are valid not only for thrusts that have been present before folding, but also for thrusts coming into existence during the folding process.

INTRODUCTION

The Ruhr Carboniferous is the best documented part of the subvariscan fold belt in Cental Europe. This area of more than 120 km length and 50 km across strike is nearly completely mined to a depth of more than 1000 m. These mining activities have supplied a large amount of geological data, analysed by the "Tiefentektonik" research programme at the Geological Survey of Northrhine-Westphalia in Krefeld during the last 15 years (Drozdzewski et al. 1980, 1981, 1985; Kunz et al. 1988). The results of these analyses are important not only for mine-planning but put a new face on the details of tectonic processes of the Variscan orogeny.

STOCKWERK-TECTONICS AND THRUSTS

The tectonics of the Ruhr Carboniferous are mainly determined by folds, decreasing in intensity from south to north. In the southern part of the area orogenic shortening by folding and thrusting is more than 50% gradually decreasing towards the north, where shortening is less than 10% (Wrede 1987). About another 10% of shortening resulted from the internal strain of the rock. Folding in the Ruhr Carboniferous is strongly disharmonic: in vertical direction it is possible to distinguish 4 tectonic levels ("stockwerks"): the uppermost stockwerk is defined by large synclinoria reaching wavelengths of some 10 km displaying horizontal bedding and separated by narrow, intensively folded anticlinoria (Fig. 1). In the second stockwerk minor folds are common within the synclinoria, combined with intense thrust tectonics. The third stockwerk is determined by intensified minor folding, but in general thrusts are missing in this level. Parallel to the intensification of minor folding within the synclinoria, the intensity and amplitude of minor folds in the anticlinoria diminishes downwards. As a result of this development, the tectonic style of synclinoria and anticlinoria becomes more and more equalized and in a lowermost, fourth, stockwerk it is no longer possible to distinguish these main elements seen in the upper levels (Wrede 1988). This stockwerk-arrangement is not stratabound nor dependent on facies. In the entire Ruhr district, however, a marked interdependance of stockwerk arrangement and fold axis undulation can be observed. In axial depressions the upper stockwerks penetrate down to relatively old strata, while in axial culminations relatively young beds are involved in lower stockwerks (Fig. 2).

Orogenic shortening is equal in all these stockwerks; the large number of small-scale folds in the lower stockwerks is compensated by the great amplitudes of the anticlinoria in the upper stockwerks. Thrusts in the Ruhr Carboniferous generally are bound to the second stockwerk. They accommodate the differences in tectonic style between the uppermost and third stockwerk. They die out both upwards and downwards (Fig. 3, 4). The distribution of thrusts being dependent on stockwerk arrangement, thrusts root in different stratigraphic levels and no regional detachment is developed in the Ruhr Carboniferous (Brix et al. 1988). Likewise, fault propagation folds or fault bend folds as typically described in the haging wall of ramps (Rich 1934) cannot be observed in the Ruhr Carboniferous. In most cases both footwall and hangingwall of thrust are deformed in similar style, so axial planes can be traced across the thrusts.

In general, thrust planes do not run into bedding planes, which might be regarded as flats in a "flat-ramp" arrangement. On the contrary, thrusts often die out dipping in the opposite direction to the bedding dip (Fig. 4, 5). Since northward and southward directed thrusts are distributed nearly randomly in the Ruhr Carboniferous (cf. section 4), a bulk horizontal displacement of the complete unit by thrusting is neglectable. The mechanics of orogenic shortening by thrusting without horizontal displacement has been illustrated by so-called "fish-tail-structures" (Drozdzewski 1979),with pairs or groups of opposite dipping thrust in a step-like-arrangement (Fig. 6). These thrusts are shortening bed-lengths but cause no larger horizontal displacement.

The concentration of thrusts in a specific tectonic stockwerk explains why thrusts outcrop mainly in the central part of the Ruhr Carboniferous: Large thrusts are missing in the southern part where the third and fourth stockwerks are exposed and likewise they are absent in the northern part, where only the uppermost stockwerk is developed.

TEMPORAL STATE OF THRUSTS

The close relations between the distribution of thrusts and stockwerk tectonics in the Ruhr Carboniferous indicates that the thrusts are elements of the folding process, generated by and during orogenic shortening by folds. This view is supported by numerous observations of folds replacing thrusts along the strike or in a vertical direction (Fig. 7). Another typical example of the combination of thrusting and folding is the frequent observation of thrusted hinges of box folds (Fig. 8). The contemporaneous activity of folding and thrusting and the causal dependency of thrusts on folding is also seen in a diagram displaying frequency of thrusts in relation to dip of bedding (Drozdzewski & Wrede 1989). In a first approach Fig. 9 indicates this relation for the Ruhr Carboniferous. Surprisingly, in general the number of thrusts seem to be decreasing with increasing dip of strata (i.e. increasing intensity of folding). This coincides with the observation that thrusts are absent in the intensively folded lower stockwerks of the Ruhr Carboniferous and supports the idea that thrusts might substitute folds.

However, apart from this general trend, the relation between the number of thrusts and bedding-dip is not continuous. There is a distinct minimum in the graph at 35½ and a maximum at about 65½. So it is possible to distinguish two populations of thrusts: when bedding dips exceed 30 - 35½. Their number increases with progressive folding. These thrust might be interpreted as accommodation thrusts developing in locking chevron folds.

However, this diagram faces the objection that it might be rather a function of the frequency of bed dips and not a function of thrust mechanics. According to the results of Büttner et al. (1985) more than 50% of the bedding-planes in the Ruhr Carboniferous are dipping with less than 20½, and, on the other hand, only 15-25% of the dips exceeding 35½. For this reason a determination of frequency of bed dips in the Ruhr Carboniferous has been applied to the above graph. As result of this computation a second graph displays the distribution of thrusts normalized for the distribution of bed dips (Fig. 10). There still are two populations of thrusts, but in this graph the total number of thrusts in general increases with the bed dips. This relation gives a strong argument for a causal dependency between folds and thrusts: An increase in folding causes an increase in thrusting. So there is no reason to regard thrusts which are deformed in the folding-process, to be "early thrusts" that were flat-lying at the onset of folding (Bradley, 1989): The processes of thrusting and folding took place simultaneously and are closely connected. For the Ruhr Carboniferous the age of "folded thrusts" in relation to the folding has been a matter of long and controversial discussion (e. g. Stahl 1949, Scholz 1956, Seidel 1957), which finally led to a model of contemporaenous folding and thrusting.

A CLASSIFICATION OF THRUSTS

To classify different types of thrusts, we can distinguish "synthetic" and "antithetic" thrusts. In this paper the terms "synthetic" and antithetic" are used in relation to the bedding dip, i.e. thrusts dipping in the same direction as bedding are called "synthetic", those dipping in the opposite direction "antithetic".

Moreover, there is a differentiation between thrusts whose dip is greater or smaller than that of bedding (in their footwall). Thus a division of thrusts into four types is possible (Fig. 11).

All four types are found in the Ruhr Carboniferous; frequently different segments of one thrust belong to different types. In some cases different parts of one thrust even can dip in opposite directions (Fig. 12). In the past these thrusts have been interpretated as "folded thrusts", i.e. these thrusts should have existed already before the folding and then should have been folded passively.

With respect to these thrust it is not appropiate to distinguish thrusts in regard to the direction of their dip, but to the direction of their "vergence", i.e. the direction of displacement of the hangingwall. According to this definition the Sutan-thrust is directed northward, while the Erin-thrust (Fig. 13) is directed southward. Statistically, about 60 % of all thrusts in the Ruhr Carboniferous are directed northward, and some 40 % are directed southward. This relationship reflects the slight northward directed vergence of folds in the Ruhr Carboniferous

RELATIONS BETWEEN THE DIPS OF THRUSTS AND BEDDING IN THE RUHR CARBONIFEROUS

Evaluating more than 650 thrusts in the Ruhr Carboniferous, about 15% belong to type I (antithetic; dip of thrust < dip of bedding), 25% are of type II (antithetic, dip of thrust > dip of bedding), 51% are of the type III (synthetic, dip of thrust > dip of bedding) and only 9% are of type IV (synthetic, dip of thrust < dip of bedding). Obviously the angle between thrusts and bedding plays an important part.

Drozdzewski (1979) first recognized that there is a clear relation between the dip of bedding and the angle between bedding and thrusts. Focussing on synthetic thrusts only, he found that the angle between bedding and thrusts decreases with increasing bedding-dip. This investigation has been extendend to antithetic thrusts by Wrede (1980) and led to a relationship between bedding-dip and the dip between bedding and thrusts as displayed in Fig. 14.

It is clear that in the antithetic part of the diagram the angle between bedding and thrusts ("Ã") increases with increasing bedding dip. Thus in case of vertical dipping strata "antithetic" thrusts cut nearly perpendicular across bedding while "synthetic" thrusts are largely substituted by bedding-parallel slips. This is illustrated by the example of the "Alstaden-thrust" (Fig. 15). In the steep northern flank of the Levin anticline this antithetic thrust is nearly horizontal, while in the southern flank the thrust is synthetic and the angle between bedding and thrust is relatively small.

Summarizing, angles between bedding and thrusts are not constant but subject to change in relation to the bedding dip.

The angle between bedding and thrust "β" and the dip of a thrust ("δ") are connected by the equation

$$\delta = \beta - \epsilon$$

(δ = dip of thrust, ϵ = bedding-dip, β = angle between bedding and thrust) (cf. section 6, Fig. 18).

Based on this equation, it is possible to outline a direct relation between the dip of bedding and the dip of thrusts considering the respective directions of dip. In the Ruhr Carboniferous dips of bedding as well as dips of thrusts roughly can be divided in two directions: northward and southward dipping. Thus the diagram displaying the relation between bedding-dips and thrusts-dips is divided in four parts, two of them representing synthetic thrusts, two representing antithetic. Fig. 16 displays the distribution of thrusts in the Ruhr Carboniferous in this relation.

The four types of thrusts defined above (section 4) can be determined in this diagram too (Fig. 17). So it can be deduced that not only the dip of thrusts regularly is subject to change during the folding process (i.e. the increase of bedding-dip), but likewise the character of thrusts (type I - IV) may be changed. Especially a thrust crossing an axial-plane ($\epsilon = 0°$) changes its type in a characteristic way, as it has already been demonstrated on the Alstaden thrust.

An important figure is given by the dip of thrust at the $\epsilon = 0°$-Line. This "specific thrust angle" determines the character of thrusts in the antithetic part of the diagram: If this angle is more than 26.5°, a thrust which is synthetic on one limb of a fold will be antithetic on the other (type III - type II - type I; Fig. 13) ("unconformly deformed thrust"). If this angle is below 26.5°, the thrust will be partly synthetic on the opposite limb, but the dip of thrust will be smaller there than bedding-dip (type IV "conformly deformed thrust"). The Sutan thrust (Fig. 12) is an example of this.

A NUMERICAL MODEL FOR THE RELATIONS BETWEEN THRUST AND BEDDING

The distribution of plots in the diagrams (Fig. 14, 17) can roughly be described by the indicated graphs which represent "specific thrust angles" of 20° resp. 30°. These curves can be synthezised mathematically using a simplified tectonic model of simultaneous thrusting and folding (Wrede 1980, 1982). Movements during the folding process mainly took place on bedding planes; lengths of fold limbs, thickness of strata and stratigraphic throws of thrusts are regarded to be more or less independent from folding (Brix et al. 1988). Displacements, however, are strongly dependent on the dip of strata in the case of deformed ("folded") thrusts (Nehm 1930: Figs. 4 & 5). For this reason it is not possible to use the "displacement/distance method" suggested by Williams and Chapman (1983) for the prediction of thrust-tips. In these cases the graph of the d/p-diagram does not form a continuous function trending towards a zero displacement but a wave-like curve with minima in the position of axes of anticlines and maxima in the synclines (Fig. 18).

In the following calculations the factors thickness of strata and throw are kept constant. This provided, dependence between dip of thrust and dip of bedding can be deduced from Fig. 19:

h stratigraphic throw of the thrust (constant)

a bed-length between the thrust and a fold-axis (constant)

ϵ bedding-dip

δ dip of thrust

(In the indicated case of an antithetic thrust both ϵ and δ positive, while in the synthetic case ϵ will be negative).

β angle between bedding and thrust.

In the case of horizontal bedding, β equals δ and is defined by δ (o) = h/a. This angle δ (o) is the "specific thrust angle"γ determing the character of thrusts in different fold-limbs. This angle is strongly material dependent and for the present will be regarded to be constant and characteristic for each thrust.

The relation between bedding dip and the angle β (between thrust and bedding) is determined by the following equations:

$\tan \gamma = h/a$

$a = h * \cot \gamma$

$a = p + q$

$p = \cot (90^{\circ}-\epsilon) * h$

$q = a - \cot (90^{\circ}-\epsilon) * h$

$q = h * (\cot\gamma - \cot (90^{\circ}-\epsilon))$ (1)

$q = \cot \beta * h$ (2)

$$(1) = (2)$$

$$\cot \beta * h = h * (\cot \gamma - \cot (90^\circ - \epsilon))$$

$$\cot \beta = \cot \gamma - \cot (90^\circ - \epsilon)$$

$$\tan \beta = \frac{1}{\cot \gamma - \cot (90^\circ - \epsilon)} \qquad \beta = \arctan \frac{1}{\cot \gamma - \cot (90^\circ - \epsilon)}$$

This relation between the angle β and bedding dip can be computed to a relation between bedding-dips and thrust-dips:

$$180^\circ - \beta + \epsilon + \delta = 180^\circ$$

$$\delta = \beta - \epsilon$$

$$\delta = \arctan \frac{1}{\cot \gamma - \cot (90^\circ - \epsilon)} - \epsilon$$

Thus a direct dependence between bedding-dip ϵ and thrust-dip δ is defined which is only influenced by the "specific thrust angle" γ, which is material dependent. A closer view of the diagrams (Figs. 12 & 14) shows that this specific angle in reality is not a constant but is subject to slight change during the folding. This reflects strain dependent anisotropies of the rock material.

Nevertheless, the indicated graphs in the diagrams ($\gamma = 20^\circ$ resp. 30°) are in good coincidence with the distribution of plots.

CONCLUSIONS

A complex interactivity between folding and thrusting can be observed in the excellently exposed and documented Ruhr Carboniferous.

Thrusts here obviously are elements of the folding process. They die out both upwards and downwards, being substituted there by other tectonics elements (folds, other thrusts). Thus tectonic stockwerks can be defined by the presence or absence of thrusts in certain tectonic levels.

Most of the thrusts in the Ruhr Carboniferous cannot be interpreted in terms of "ramp-flat" geometry, because they do not run into bedding-planes. Fold structures in the hangingwall and footwall of thrusts are very similar: Fold axes in general can be traced across the fault-planes (Fig. 18). This is a very important difference to folds being described from other areas to be fault-bend folds or fault propagation folds (e.g. Suppe 1983, 1985).

Thrusts in the Ruhr Carboniferous are not related to a common basal detachment, which, according to new seismic data, is absent below this area (Franke et al. 1990). According to the paper of Brix et al. (1988) the southward increasing strain is compensated for by tectonic deformation of a relatively thick geological unit without major detachment.

Two mechanisms can be distinguished creating thrusts during the folding process: Some thrusts substitute folds along the strike or - related to stockwerk tectonics - in a vertical direction. Most of the thrusts, however, accomodate volume problems in the centre of locked folds by mass-transport towards the fold limbs. Development of these thrusts starts when orogenic shortening exceeds the value of about 15% (i.e. the average dip of strata exeeds 30°).

Within the active fold belt thrusts-planes develop in close relation to the folding: dips of thrusts, angles between thrust-planes and bedding, and displacements depend from the varying dips of strata. The analysis of these relations and interactivities between folding and thrusting helps us to understand the tectonic mechanisms within a fold and thrust belt like the Ruhr Carboniferous and enables us to predict the characteristics of thrusts in dependence from the fold pattern.

REFERENCES

Bradley, D.C. 1989: Description and analysis of early faults based on geometry of fault-bed intersections. Jour. Struct. Geol., 11, 1011-1019.

Brix, M.R., G. Drozdzewski, R.O. Greiling, R. Wolf, V. Wrede, 1988: The N Variscan margin of the Ruhr coal district (Western Germany): structural style of a buried thrust front? - Geol. Rdsch., 77, 115-126.

Büttner, D., H. Engel, D. Juch, W.-F. Roos, L. Steinberg, A. Thomsen, M. Wolff (1985): Kohlenvorratsberechnung in den Steinkohlenlagerstätten Nordrhein-Westfalens und im Saarland. BMFT-Forsch. Ber. T 85-147, 208 p., Krefeld.

Drozdzewski, G., 1979: Grundmuster der Falten- und Bruchstrukturen im Ruhrkarbon. - Z. dt. geol. Ges., 130, 51-67.

Drozdzewski, G., O. Bornemann, E. Kunz, V. Wrede, 1980: Beiträge zur Tiefentektonik des Ruhrkarbons, 192 p., Krefeld (GLA).

Drozdzewski, G., O. Bornemann, E. Kunz, V. Wrede, 1981: Ergebnisse des Untersuchungsvorhabens "Tiefentektonik des Ruhrkarbons". - Glückauf, 117, 935-942.

Drozdzewski, G., H. Engel, R. Wolf, V. Wrede, 1985: Beiträge zur Tiefentektonik westdeutscher Steinkohlenlagerstätten, 236 p., Krefeld (GLA).

Drozdzewski, G., V. Wrede, 1989: Die Überschiebungen des Ruhrkarbons als Elemente seines Stockwerkbaus, erläutert an Aufschlußbildern aus dem südlichen Ruhrgebiet. Mitt. geol. Ges. Essen, 11, 74-88.

Franke, W., R.K. Borthfeld, M. Brix, G. Drozdzewski, H.J. Dürbaum, P. Giese, W. Janoth, H. Jödicke, Chr. Richert, A. Scherp, A. Schmoll, R. Thomas, M. Thünker, K. Weber, M.G. Wiesner, H.K. Wong (1990): Crustal structure of the Rhenish Massif: results of deep seismic reflection lines DEKKORP 2-North-Q.-Geol. Rdsch., 79, 523-566.

Hewig, R. (1988): Gebirgsbau; in O. Stehn: Erläuterungen zu Bl. 4509 Bochum. - Geol. Kt. Nordrh.-Westf. 1:25.000.

Kunz, E., R. Wolf, V. Wrede, 1988: Ergänzende Beiträge zur Tiefentektonik des Ruhrkarbons, 65 p., Krefeld (GLA).

Nehm, W., 1930: Bewegungsvorgänge bei der Aufrichtung des rheinisch-westfälischen Steinkohlengebirges. Glückauf, 66, 789-797.

Rich, J.L., 1934: Mechanics of low-angle overthrust faulting illustrated by Cumberland thrust block, Virginia, Kentucky and Tennessee. Am. Assoc. Petr. Geol. Bull., 18, 1584-1596.

Scholz, J., 1956: Zur tektonischen Analyse der mitgefalteten Überschiebungen im niederrheinisch-westfälischen Steinkohlengebirge. Z. dt. geol. Ges., 107, 158-201.

Seidel, G., 1957: Entwurf einer genetischen und morphologischen Systematik der großtektonischen Störungen des Ruhrkarbons. Mitt. westf. Berggewerkschaftsk., 12, 111-145.

Stahl, A., 1949: Zum Problem der mitgefalteten Wechsel des Ruhrgebiets. Glückauf, 85, 448-455.

Suppe, J., 1983: Geometry and kinematics of fault-bend folding. Am. Jour. Science, 283, 684-721.

Suppe, J., 1985: Principles of structural geology. 537 p., Englewood Cliffs.

Williams, G., Chapman, T., 1983: strains developed in the hangingwall of thrusts due to their slip/propagation rate: a dislocation model. Jour. Struct. Geol., 5, 563-571.

Wrede, V., 1980: Zusammenhänge zwischen Faltung und Überschiebungstektonik, dargestellt am Beispiel der Bochumer Hauptmulde im östlichen Ruhrgebiet. - 128 p., Thesis, TU Clausthal.

Wrede, V. 1982: Genetische Zusammenhänge zwischen Falten- und Überschiebungstektonik im Ruhrkarbon. Z. dt. geol. Ges., 133, 185-199.

Wrede, V. 1987: Einengung und Bruchtektonik im Ruhrkarbon. - Glückauf Forsch.-H., 48, 116-121.

Wrede, V., 1988: Tiefentektonik der Wittener Hauptmulde im östlichen Ruhrkarbon. - In: Ergänzende Beiträge zur Tiefentektonik des Ruhrkarbons, 35-52, Krefeld (GLA).

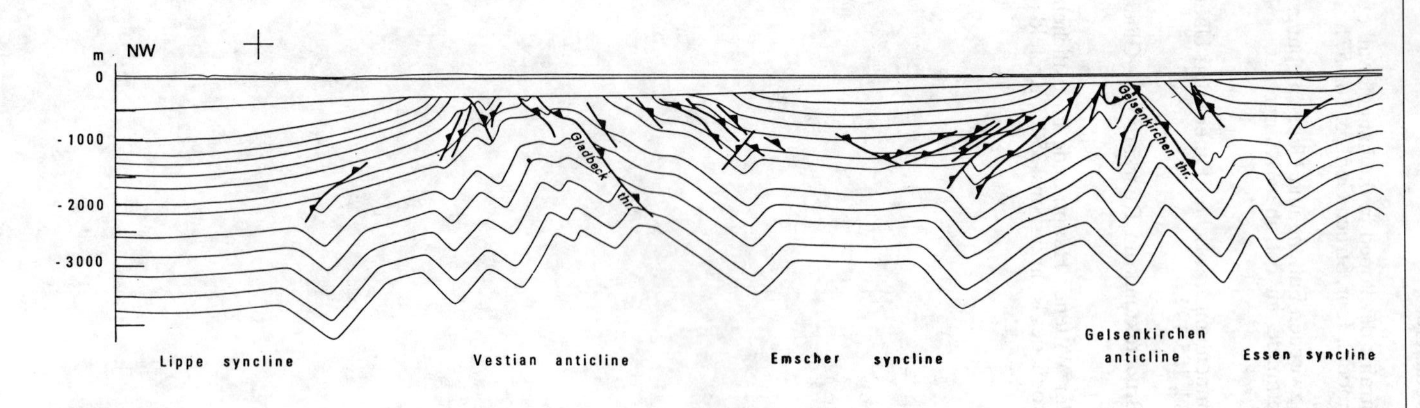

FIGURE 1
Cross-section of the Ruhr Carboniferous. Disharmonic folding and concentration of thrusts in the second tectonic stockwerk are clearly to be seen. (Sections in this paper are not exaggerated.)

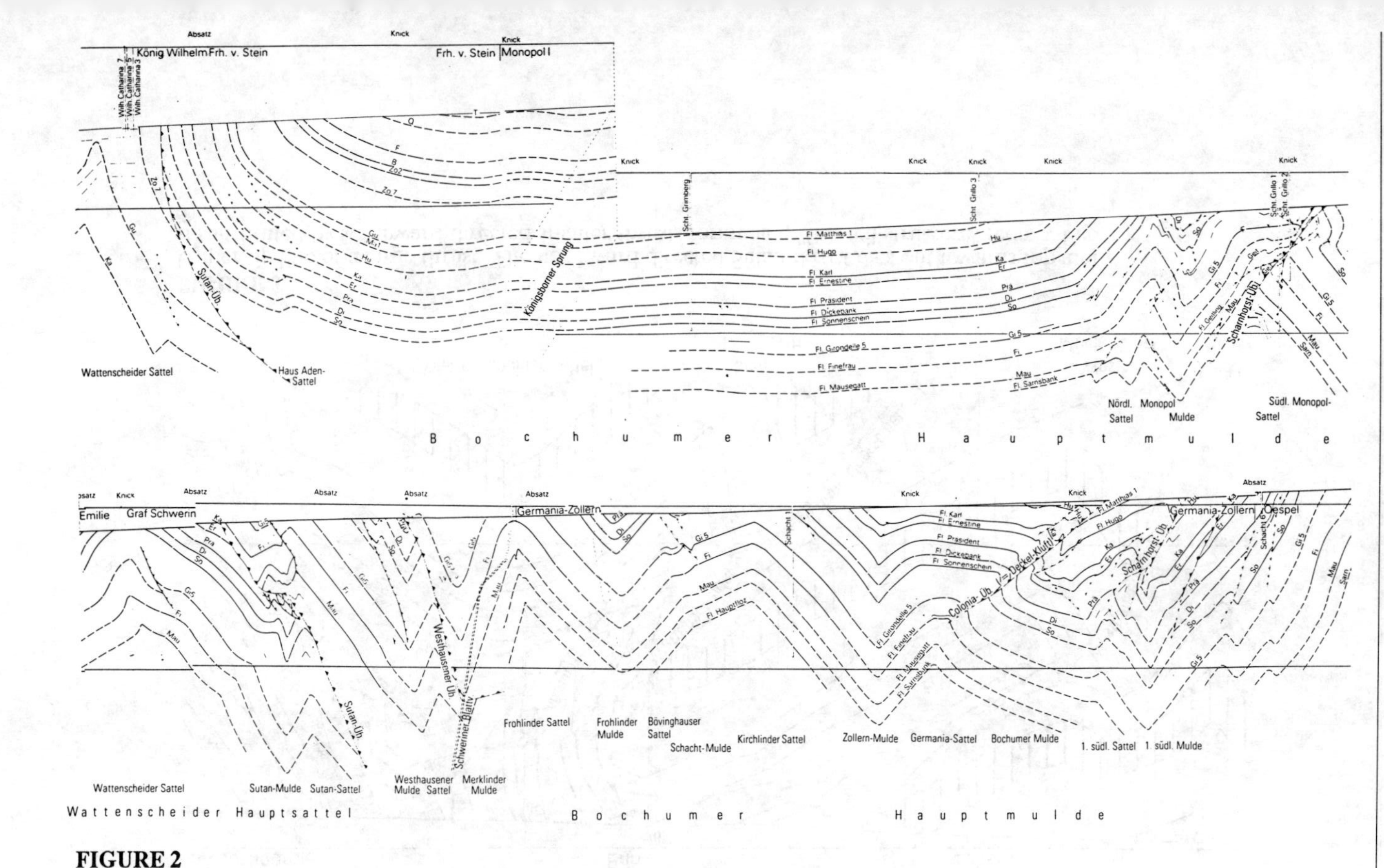

FIGURE 2
Dependency of stockwerk arrangement on axis undulation. Both sections display the Bochum Syncline, the upper one is positioned within the Hamm axis-depression, the lower one within the Dortmund culmination.

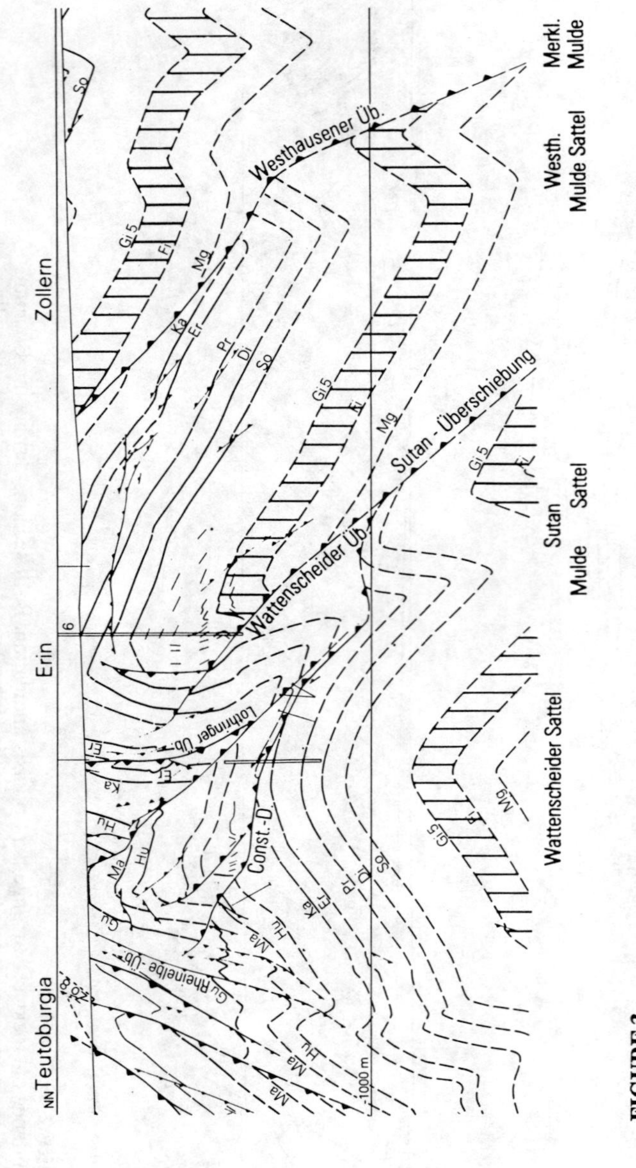

FIGURE 3
Upper tip line of a large thrust. The northward directed Sutan-thrust dies out towards top and is substituted by southward directed smaller thrusts formung a "fish teil-structure" (comp. Fig. 6).

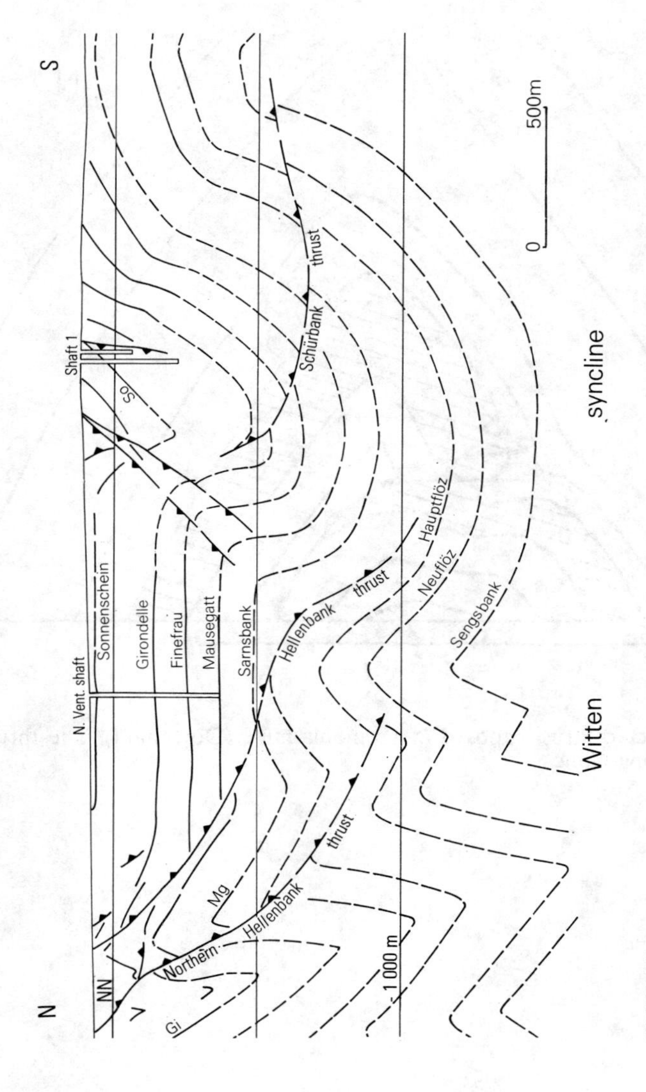

FIGURE 4
Section across the Witten syncline showing thrust tips in the basal part of the second tectonic stockwerk.

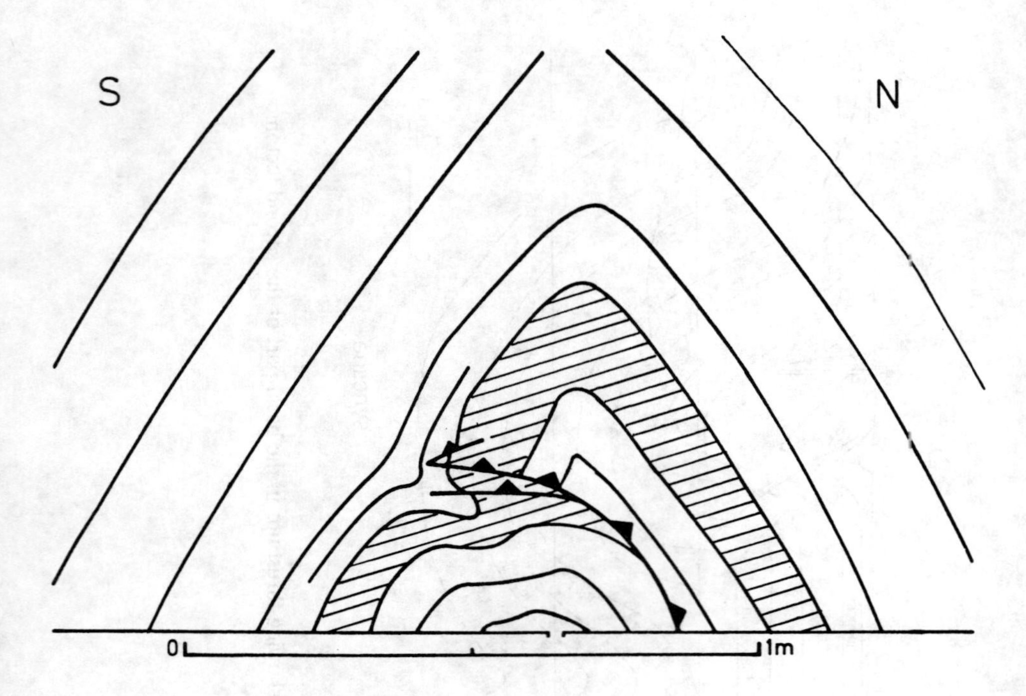

FIGURE 5
Tip of a southward directed thrust exposed at Tremonia-mine (Dortmund). The thrust does not extend into the bedding-planes.

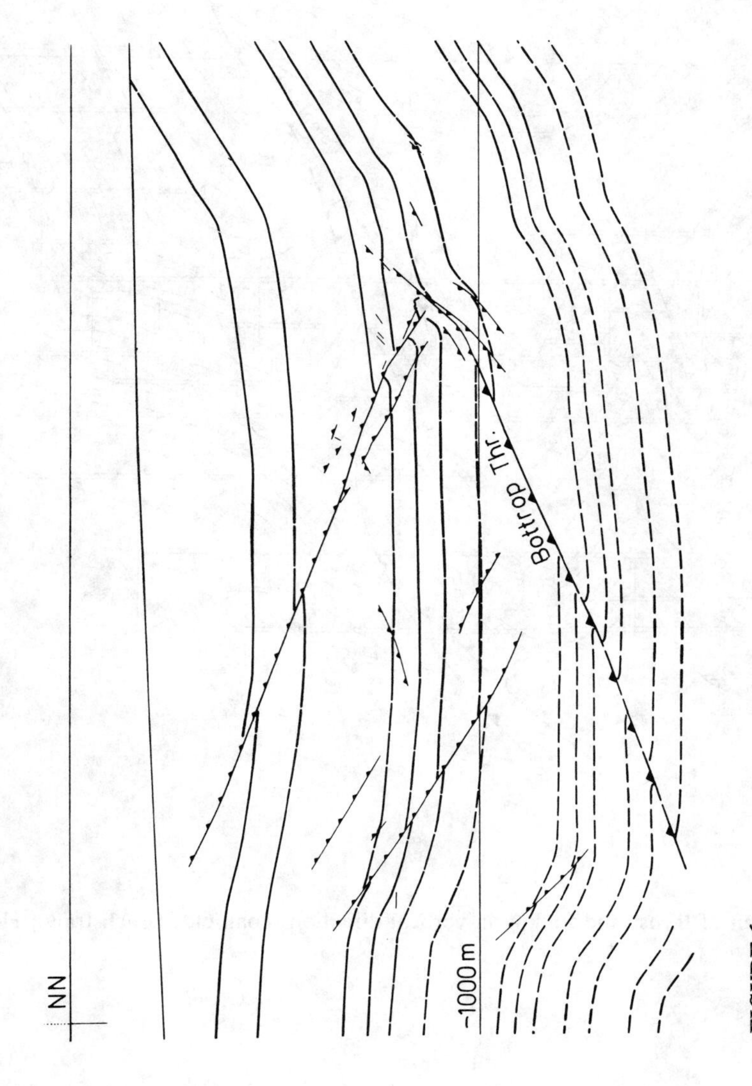

FIGURE 6
System of opposite directed thrusts forming a "fish tail-structure".

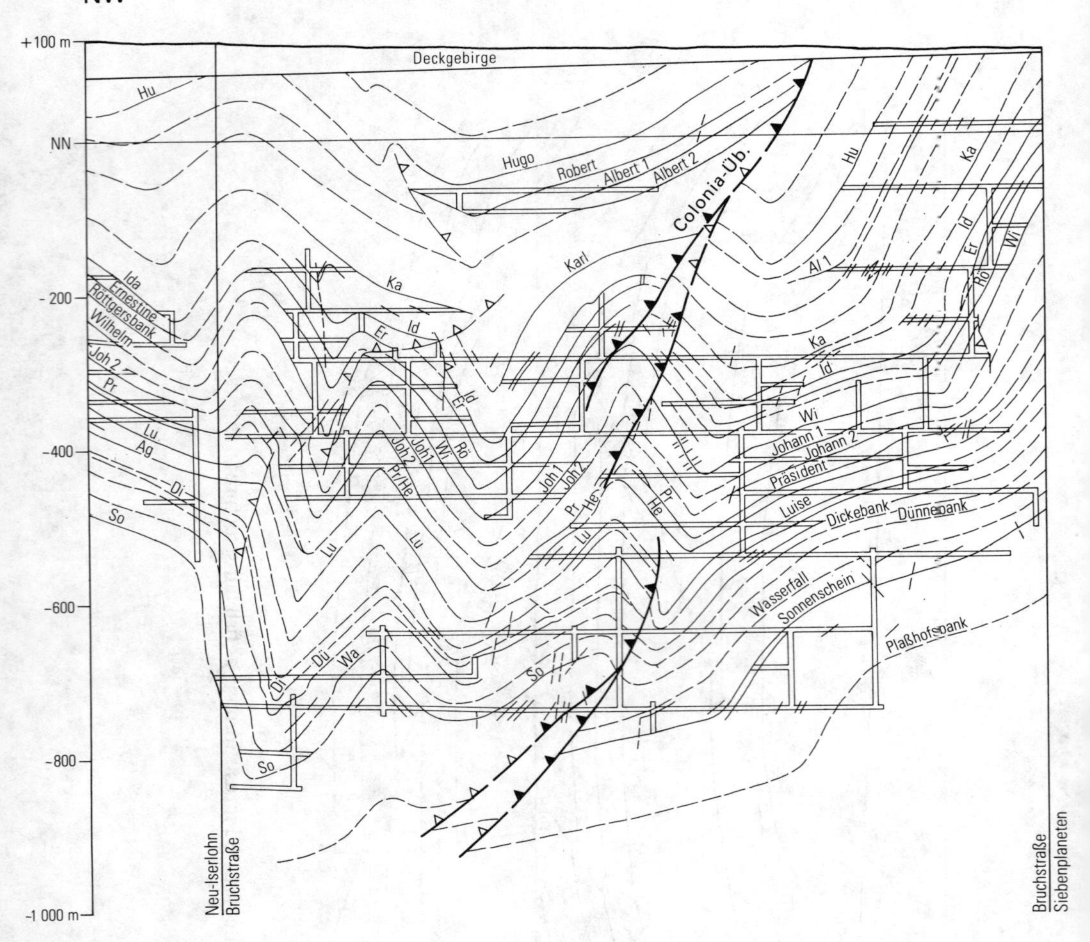

FIGURE 7
Mutual substitution of thrust and folding in vertical direction (coalfield "Bruchstraße; Hewig 1988).

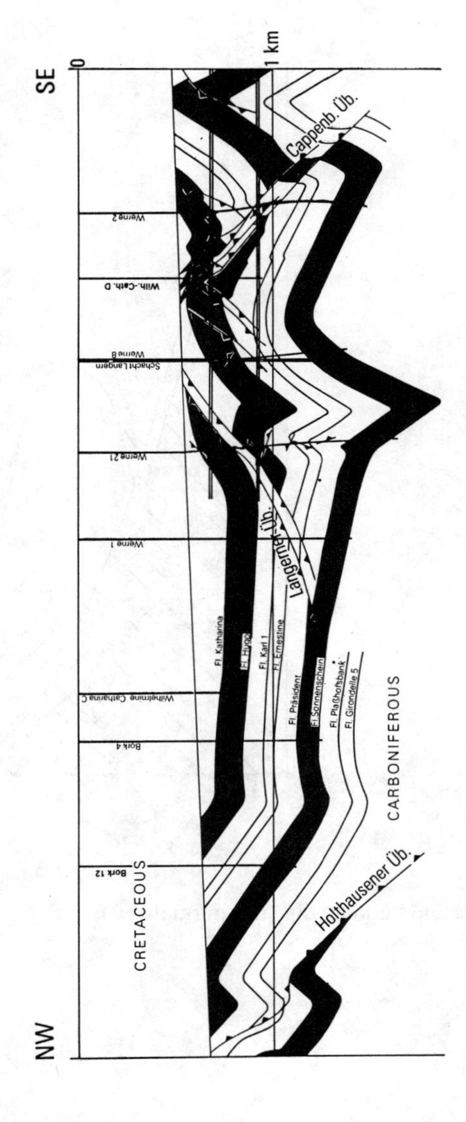

FIGURE 8
Langern-thrust, an example of a fault thrusting the hinge of a boxfold.

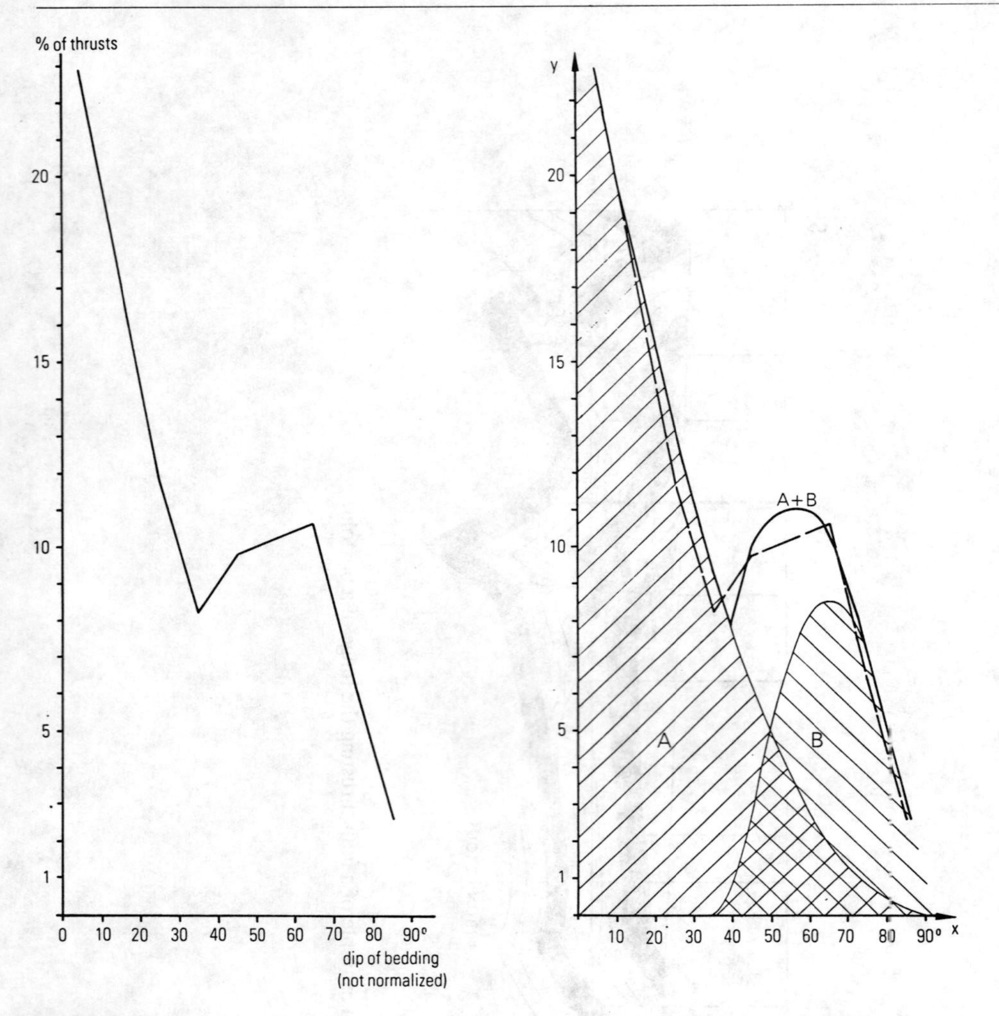

FIGURE 9
Relation between number of thrusts and bedding dip (not normalized).

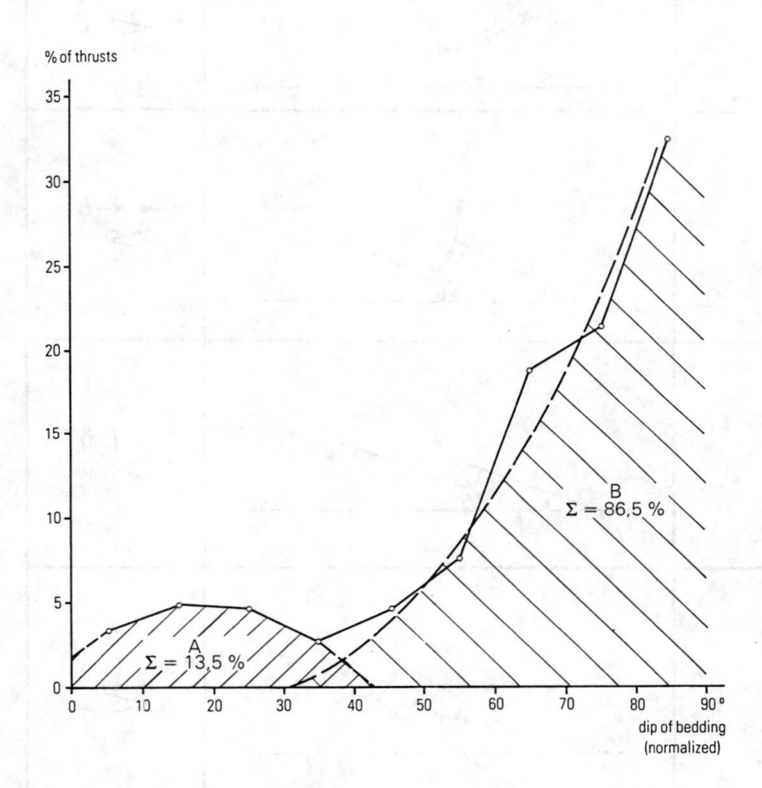

FIGURE 10
Relation between number of thrusts and bedding dip (normalized)

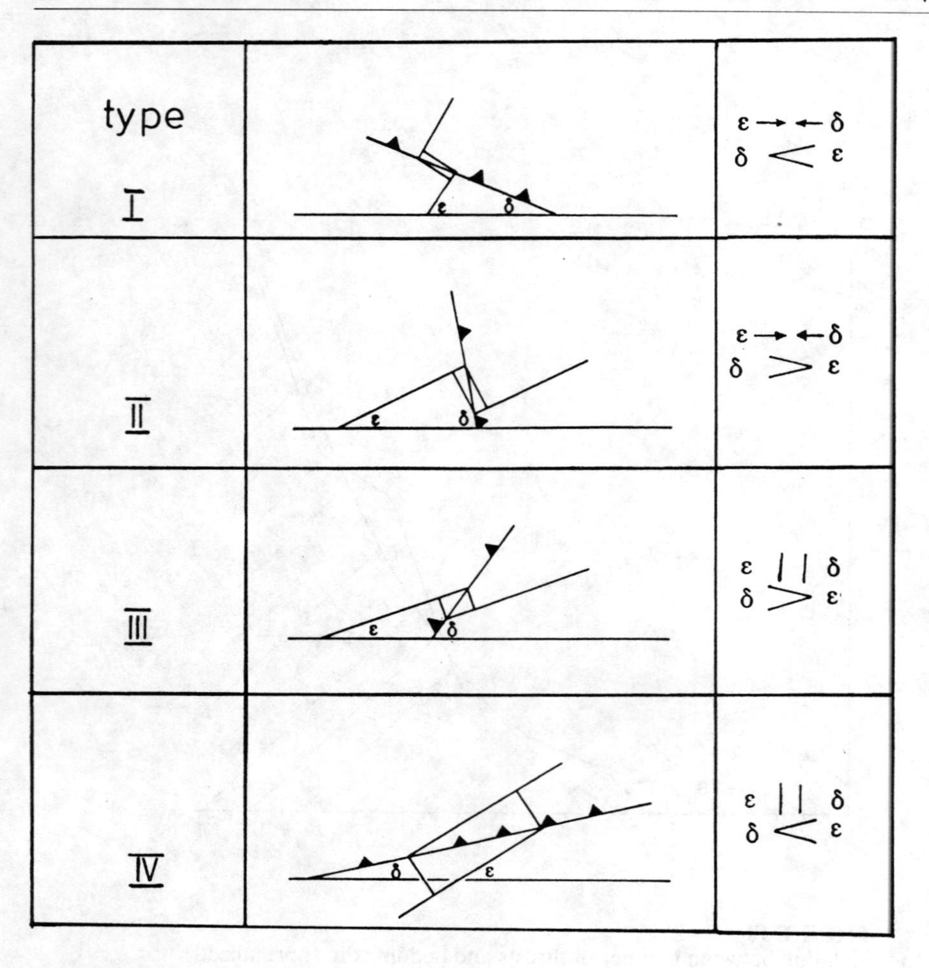

FIGURE 11
Classification of thrusts which effect orogenic shortening in respect to relative direction and amount of thrust and bedding dips.

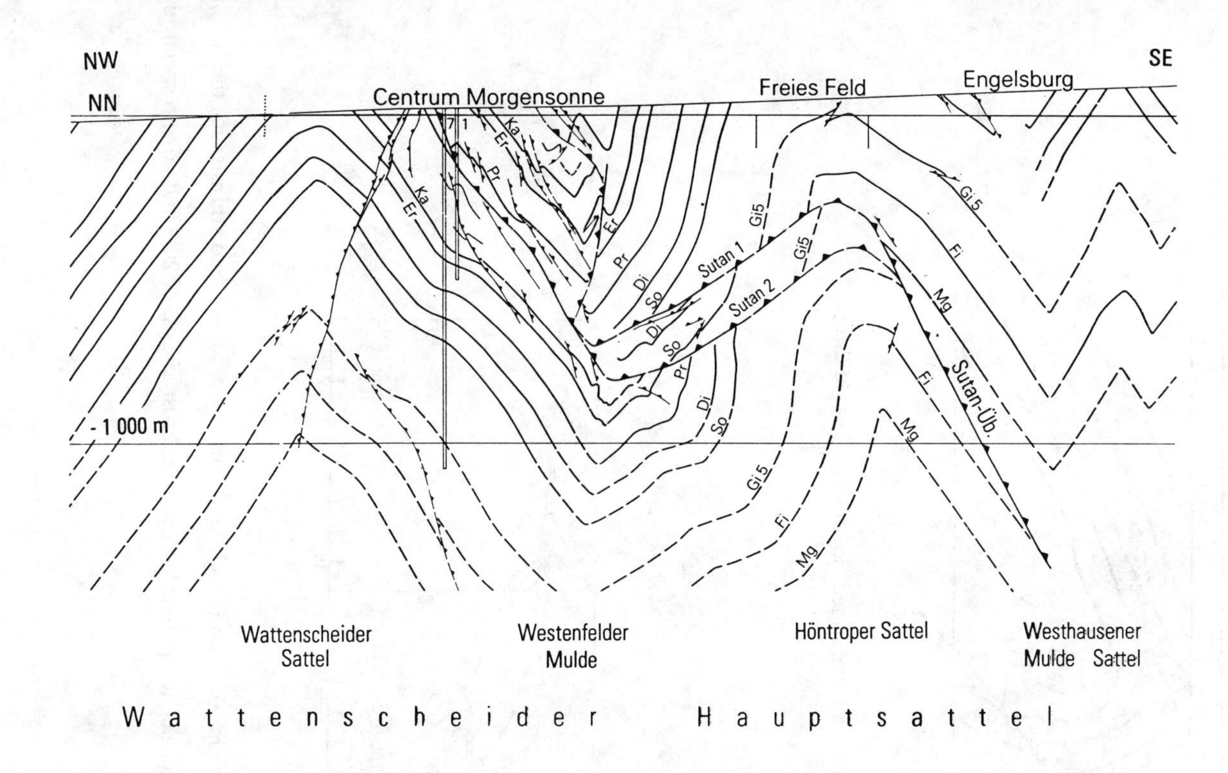

FIGURE 12
Sutan-thrust, an example of a "conformly deformed" thrust.

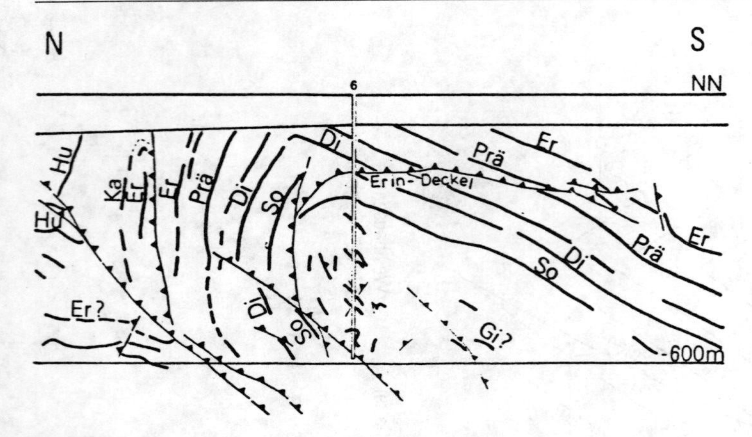

FIGURE 13
"Erin-Deckel", a "conformly deformed" southward directed thrust.

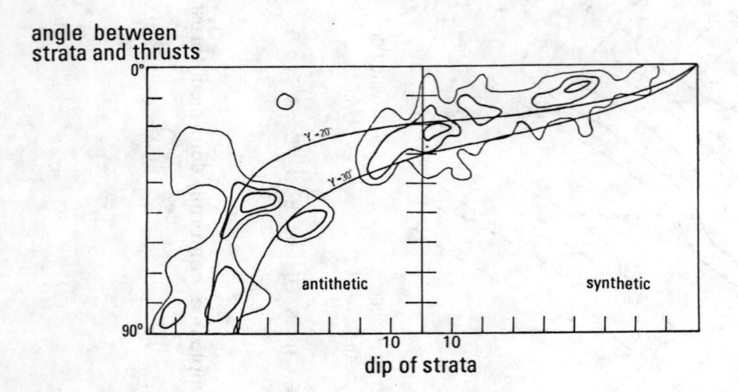

FIGURE 14
Relation between dip of strata and angle between strata and thrusts in the Ruhr Carboniferous.
Indicated lines are calculated graphs for "specific thrust angles" of 20½ and 30½, respectively (see text).

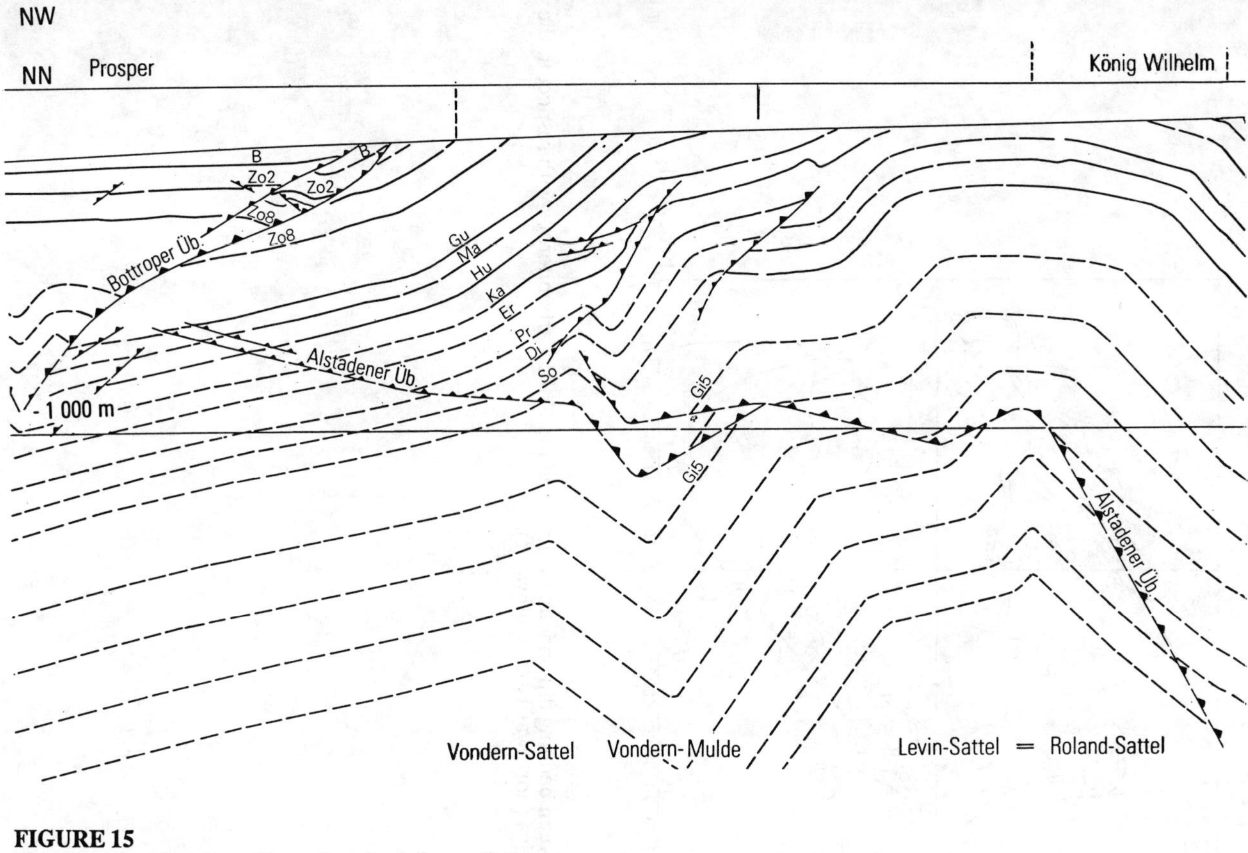

FIGURE 15
The Alstaden thrust, an "inconformly deformed" thrust.

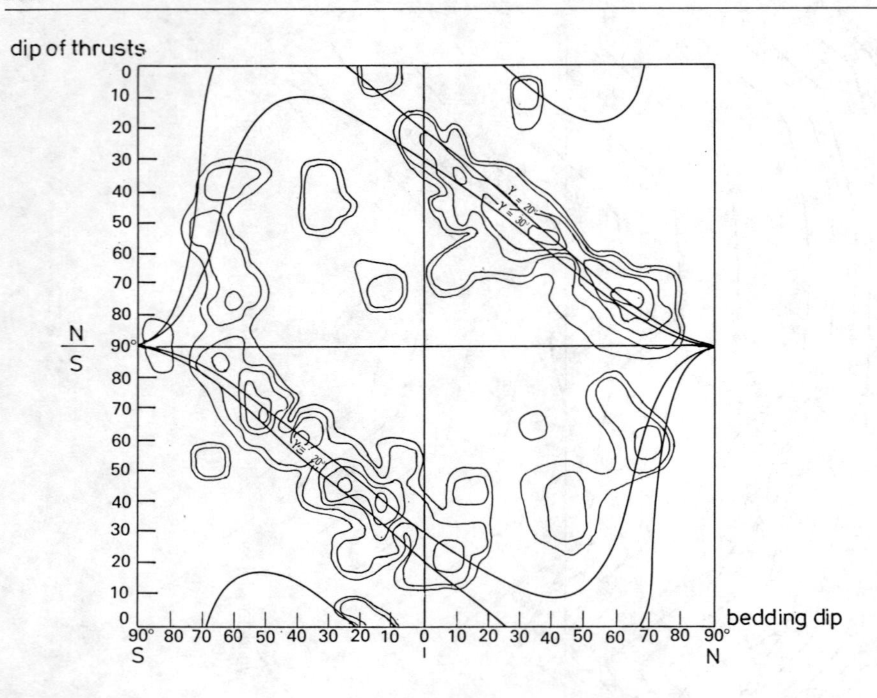

FIGURE 16
Relation between bedding dip and dip of thrusts in the Ruhr Carboniferous with respect to the direction of dips (northward and southward).

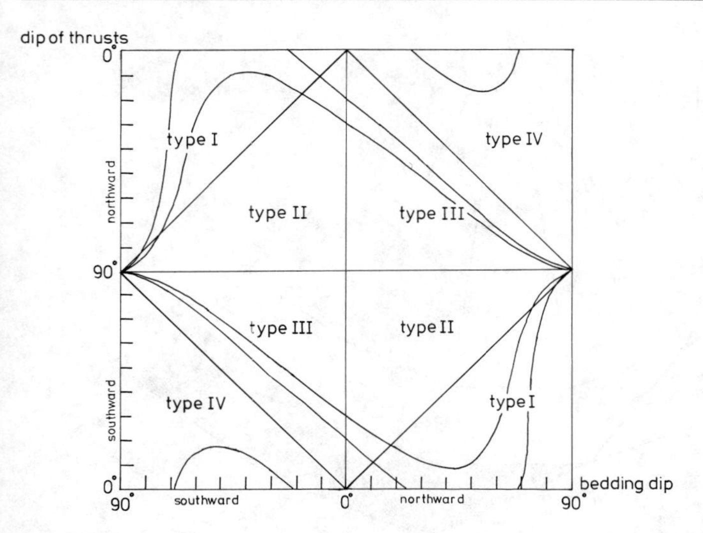

FIGURE 17
Distribution of the four types of thrusts in the diagram of Fig. 16, indicating the changes in character of thrusts during folding.

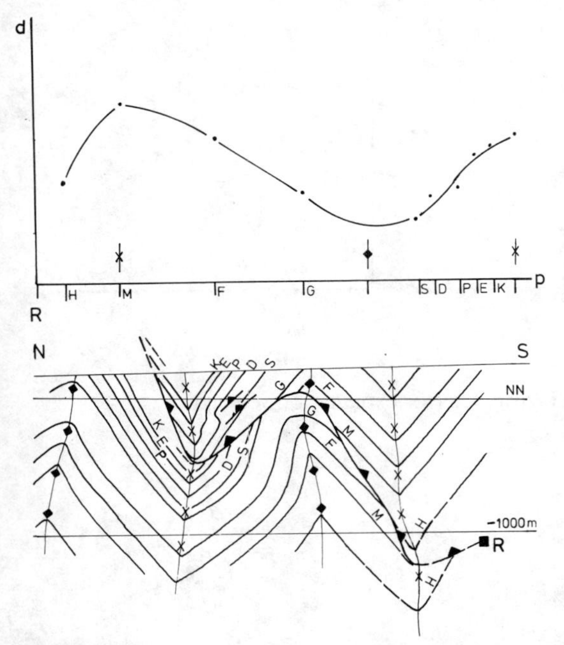

FIGURE 18
Displacement/distance plot (according to Williams & Chapman 1983) for the Sutan-thrust: displacements (d) vary in dependence from folding.
(p) indicates distance of single seams from an arbitrary reference point R.

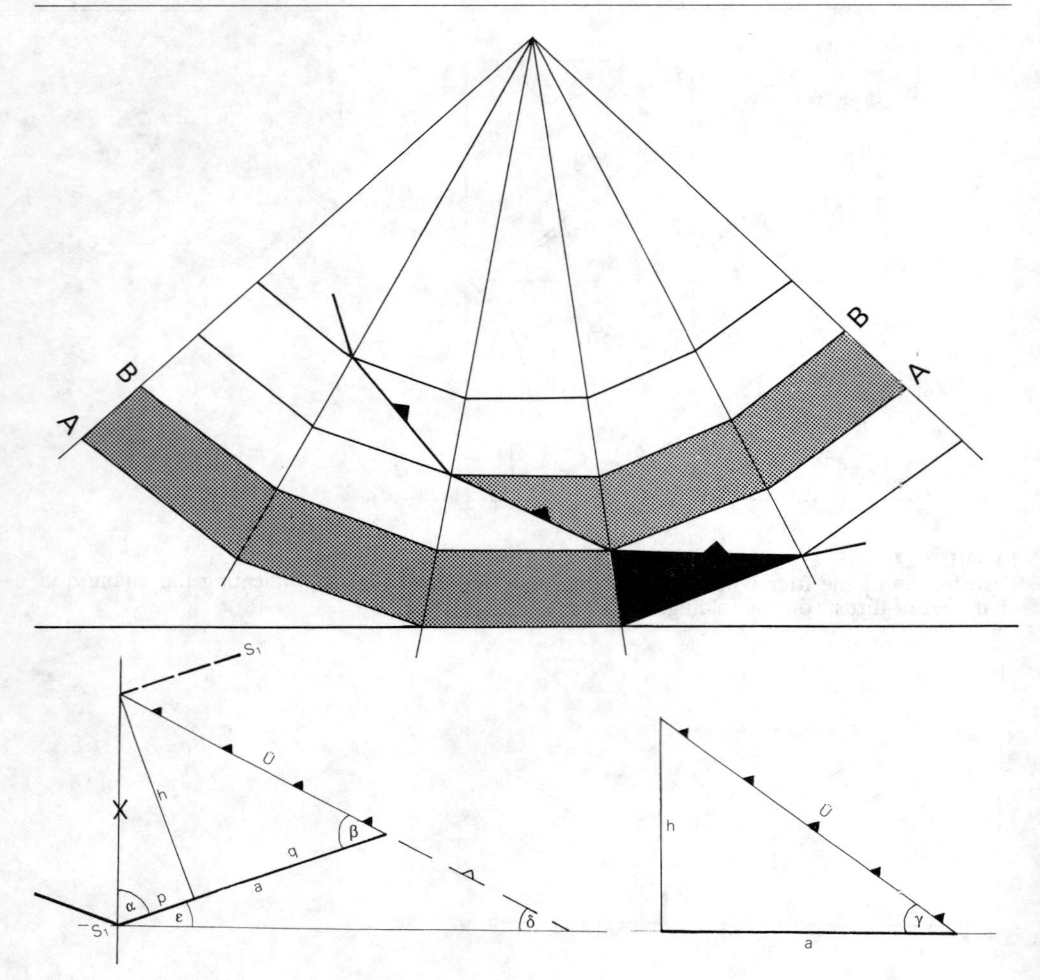

FIGURE 19
Geometric relations between bedding and thrust dips (for explanation see text).

On the Structure of the Variscan Front in the Eifel-Ardennes-Area

V.Wrede[1] / G.Drozdzewski[1] / J.Dvorak[2]
1 Geologisches Landesamt Nordrhein-Westfalen, De Greiff-Str.195, D-4150 Krefeld, Germany
2 Ustredni ustav Geologicky, Leitnerova 22, CS 60200 Brno, CSFR

ABSTRACT

The tectonic structure of the Ardennes is markedly different to that exposed in the Ruhr Carboniferous east of the river Rhine. Here the Variscan Orogen dies out gradually towards the north, folds being the main tectonic element. According to recent investigations the Ruhr Carboniferous can be described as an autochthonous foreland. West of the Rhine the structure of the Variscan front is dominated by thrusts, and in the Aachen area they reach stratigraphic throws of up to 1.5 - 4 km (Aachen Thrust, Venn Thrust).

Further west, in the Belgian Ardennes and northern France, the hangingwall of the Midi Thrust System has been described as the "Dinant Nappe", for which a transport of up to hundreds of kilometers has been discussed. This nappistic model is supported by different seismic results indicating a decollement at about 4 - 6 km depth beneath the Ardennes. But in fact, this interpretation creates a number of problems and contradictions.

As a result of intensive exploration in the Lower Rhine Embayment it has been concluded that the structures exposed in the Aachen area can be linked to tectonic structures in the Ruhr District. Thus an eastern termination of the "Dinant-Nappe" does not exist.
A comparison of the orogenic shortening in the areas on both sides of the Lower Rhine Embayment shows that despite the differences in structural style the total shortening of strata is similar in both regions. Thrusts of the Eifel-Ardennes region are closely related to folds, and their throws diminish laterally as well as towards depth. Several deep boreholes located in the hangingwall of the Midi Thrust System give only dubious hints to an extension of the thrust-plane towards the south.

INTRODUCTION

The tectonic structure of the Variscan front in the Ardennes is very different to that exposed in the Ruhr Carboniferous east of the river Rhine. Here folds are the main tectonic element, varying in wavelength from decameters to tens of kilometers (Fig. 1). Nearly all possible shapes of folds are present: transitions from concentric to chevron or box-folds being common. Connected with the folding a large number of thrusts exist, ranging in scale from meters or decameters to 40 km along the strike. The maximum throw of these thrusts is about 1 km with a maximum displacement of less than 2.5 km. These thrusts are bound to stockwerk-tectonics and are distributed in various stratigraphic levels, dying out both upwards and at depth. Such structures are also frequently antithetic in relation to the dip of bedding. Numerous observations prove a contemporary development of thrusts in relation to the folding process (Drozdzewski et al. 1980; 1985; Kunz, Wolf, Wrede 1988). No regional detachment can be ovserved in the Ruhr Carboniferous or in the northern part of the Sauerland south of it (Brix et al. 1988). Orogenic shortening in the Ruhr Carboniferous gradually decreases from SE, where it reaches an amount of up to 60 %, to NW, where the amount of shortening is less than 5 % (Wrede 1987 a).

The Variscan front of the Ardennes is dominated by thrusts. The Dinant Synclinorium has been thrust over the Synclinorium of Namur in the north by the Midi-Thrust-System, which towards the west can be traced to the coast of the British Channel in France (Fig. 15). To the east this thrust system is connected to the Aachen thrust at the northern edge of the Eifel Mountains. Due to seismic exploration these thrusts have been extended southwards (Meissner, et al. 1981; Meissner et al. 1983; Durst 1985; Betz et al. 1988; Cazes et al. 1988): A strong seismic reflector was found in the Aachen-Venn area at a depth of about 4 km (fig. 2). Although this reflector extends farther towards north, it has been connected and identified with the Aachen thrust and reaches some 30 km to the south (Durst 1985: fig. 2; Betz et al. 1988: Fig. 4). In northern France a similar reflector can be traced for some 100 km southward. Based on these geophysical results the hangingwall of the Midi Thrust System has been described as the "Dinant-Nappe" (Bless et al. 1980 a; 1982; Cazes et al. 1985; Raoult & Meilliez 1986). According to this tectonic model a displacement of the allochtonous unit of up to some 100 km is dicussed, supposing the Dinant nappe to be derived from the northern flank of the Armorican-Mid-German High (Bless et al. 1980 a; Cazes et al. 1988, Dreesen et al. 1989). So the tectonic structure of the Variscan front in the western part of the Rhenish Massif seems to be substantially different from that in the eastern part. The areas are divided by the neotectonic structure of the Lower Rhine Embay where the Palaeozoic is covered by Tertiary and Quaternary sediments. In this area the Aachen and Erkelenz coal-fields and the structure of the Krefeld High are situated, thus holding a key-position for the understanding of the nature of the Variscan front in Central Europe.

Moreover, the sedimentary facies development of the "autochthonous" and "allochthonous" units indicates that these units were part of one sedimentary basin. e. g. the Upper Devonian of the "allochthonous" Dinant-Synclinorium is in direct continuation with the "autochthonous" Namur basin. Likewise, in the Aachen area the sedimentary development of Inde- and Wurm-Synclinoria, both separated by the Aachen Thrust, is similar to that found in comparable structures of the Ruhr Carboniferous where large thrusts are absent.

So the differences in the structural style between the areas east and west of the Lower Rhine Embayment cannot be explained by nappe-scale transports. These differences were in fact controlled by the Caledonian Brabant Massif. This consolidated structure blocked up the northward directed movement of the Variscan Front in the western part of Central Europe, thus forming a thrust dominated tectonic situation.

TECTONICS OF THE AACHEN AREA

Amongst others, two important thrusts exist in the Aachen area: the Venn thrust and the Aachen thrust, which is regarded as the extension of the Midi thrust. The stratigraphic throw of the Aachen thrust is about 1400 m (Herbst 1962) whereas that of the Venn thrust reaches an amount of up to 4 km (Thome 1955; Schmidt 1956). The Aachen thrust is exposed in the center of the Aachen Anticline separating the Wurm- and Inde-Synclines. The Venn thrust is situated in the northern flank of the Venn Anticline south of the Inde Syncline (Wrede & Zeller 1988). As a result of recent investigation within the Aachen and Erkelenz coalfields (Wrede 1985 a) and drilling within the Lower Rhine Embayment the folds of the Aachen area can be connected to fold-structures exposed in the eastern part of the Rhenish Massif (Wrede 1985 b, in press; Wrede & Hilden 1988): The Aachen Anticline extends into the Velbert Anticlinorium, where no large thrusts are known. The Bochum Synclinorium in the Ruhr Carboniferous is the eastern continuation of the Wurm Syncline, while the Inde Syncline is in the position of the Herzkamp Syncline of the Ruhrarea. The Venn anticlinorium, finally, can be correlated with the Remscheid-Altena Anticlinorium, both exposing Lower Palaeozoic strata (fig. 3,4).
Farther west, the Wurm Syncline is the equivalent of the Namur basin, whereas the Inde Syncline can be regarded as the eastern extension of the Syncline of Verviers, which is part of the allochthonous Dinant nappe (Graulich & Dejonghe 1986).

Thus along strike the main tectonic elements can be traced in from the eastern part of the Rhenish Massif to the Aachen area and beyond to the Ardennes without any break. Gradually, however, the orogenic shortening by folds is substituted by thrusts. The "allochthonous" Inde Syncline and Venn Anticline and the "autochthonous" Wurm-Syncline can be traced to parallel arranged, autochthonous structures in the east. So, no space is left for the eastward extension of hypothetic fold structures that might be hidden below the Aachen thrust or the Dinant nappe (M. Teichmüller & R. Teichmüller 1979: Fig. 2). Thus neither an eastern limitation of the Dinant nappe exists (Paproth & Struve 1982) nor an eastward extension of the Midi-Aachen thrust system in the Bergian or Sauerland Mountains, as suggested e. g. by Matte & Hirn (1988: Fig. 1).

The total orogenic shortening of the Ruhr and Aachen areas is similar. In the Aachen area, however, the gradient of orogenic shortening is higher in the southern parts and dies out quickly towards north, whereas in the Ruhr Carboniferous this gradient towards north diminishes slowly with a rate of 1 to 2 % per km (Fig. 5).

Another fact hardly compatible with a nappe model of the Aachen region is the observation that the large thrusts diminish in throw both along strike and downwards. This is clearly to be

seen at the Venn thrust, which totally dies out within the hinge culmination of the Venn anticlinorium, where lower tectonic stockwerks are exposed (Knapp 1980). In the Geul-Valley at the Belgian-Dutch border the Aachen thrust likewise is exposed in a low tectonic level. Here it displays a much smaller throw than in the area of the Aachen hinge depression and is split into several branches (Verhoogen 1935). This has been confirmed recently by the results of an ore-exploration programme which has been carried out in the Belgian part of this area.
This observation contradicts an interpretation of these thrust as "off-splays" of a basal décollement summarizing the throws of individual thrusts towards south. In this case one would expect an increasing throw of the faults towards depth.

So it is very likely that the Aachen and Venn thrusts also die out downwards in the northern flank of the Venn-Anticlinorium and do not extend to the south below this structure. Otherwise the Aachen and Venn anticlinoria would be rootless folds and the thrusts would cut arbitrarily the stratigraphic succession (Teichmüller & Teichmüller 1979, Walter 1982) leading to unrestorable cross-sections (Fig. 6).

Xenoliths of the Eifel Volcanic District represent material from the total crust. Voll (1983) deduced a complete profile of the crust from these xenoliths which gives no hint of a break in the sequence. He clearly rejects a southern continuation of the Aachen thrust below the Wehr depression, as assumed by Woerner et. al. (1982).

BOREHOLES

Three boreholes in Belgium and Northern France have exposed the Midi thrust in the subsurface: EPINOY 1, JEUMONT 1 and WEPION. In the EPINOY and WEPION boreholes the thrust-system is bound to the overturned limb of an inclined syncline. In these cases Lower Devonian and Ordovician strata are thrust over Lower Carboniferous and Upper Devonan strata (Fig. 7). In the Jeumont borehole Lower Devonian is thrust over Westphalian and Namurian strata in normal position (Raoult 1988). All these boreholes are situated less than 10 km southward of the outcrop of the Midi Thrust System. Two more boreholes in Belgium, GRAND HALLEUX and HAVELANGE, are situated 35 and 20 km respectively south of this outcrop.

In about 3000 m depth the GRAND HALLEUX borehole exposed a flat lying, nearly isoclinal syncline within Revinian strata instead of discrete thrust-planes (Vangustaine 1970; Graulich 1980). In this area, and in this tectonic stockwerk thrust tectonics of the upper levels are obviously substituted by folds (Breddin 1971) (Fig. 8).

Within the 5600 m deep section of the HAVELANGE borehole no larger thrust has been observed. Here likewise a strongly inclined fold-limb has been exposed. Some faults in the deeper part of the section have been interpretated to be an equivalent of the Midi thrust (Graulich, Leclerq, Hance 1989) (fig. 9). However, strata of upper Siegenian age overlie Gedinnian strata along these fault planes, thus indicating a stratigraphic gap.

With a thrust, this would only be possible if the fault were younger than folding. This fold/thrust relation would be very uncommon within the Rhenohercynian.

The inverted bedding in the lowest part of the section can more easily be explained by normal and or strike-slip faulting than by thrust-sheets. So here again the thrust wich are exposed in the upper tectonic stockwerks are substituted by folds. (Fig. 10).

The EPINOY borehole and the borehole ST. GHISLAIN, which were positioned in the autochthonous unit north of the Midi-outcrop, proved the existence of relatively thick layers of anhydrite within Viséan strate (Rouchy et al. 1986). In the EPINOY borehole this evaporitic layer is exposed within an overturned fold-limb below the Midi-thrust. It is very conspicious that this evaporitic layer has not been used as shear plane for thrust movements.

PALAEOGEOGRAPHY

The northern edge of the Ardennes displays a remarkable break in thickness of Lower Devonian strata: South of the Variscan front more than 2000 m of Lower Devonian are known, while north of it the Lower Devonian and parts of the Middle Devonian are missing. At that time a flexure-like structure, possibly combined with synsedimentary faults must have existed in this area, separating a stable platform in the north from a subsiding basin in the south.
The strata of the Upper Devonian have a key-position for the sedimentary development of the Ardennes. Tsien (1974) investigated the Middle Devonian and Frasnian (Fig. 11), while Thorez & Dreesen (1967), Dreesen & Thorez (1980), Thorez et al. (1988) have made deteiled biostratigraphic and petrographic investigations of the Fammenian Condroz-Fracies. Within the allochthonous Dinant-Synclinorium a facial development can be ascertained from alluvial and deltaic deposits in the NE passing into lagoonal, sabhka- and barrier deposits, sub-lagoonal tidal-flats to a fully marine environment in the south. The terrestrial facies in the north-eastern part of the Dinant Synclinorium is closely related to the facial development of the autochthonous Vesdre-Massif and the Namur-Synclinorium. So no gap exits between the sediments of the allochthonous units south of the Namur-Synclinorium, which is hidden below the nappe, would have to be regarded to be dry land with no deposition during the Famennian!

Similar relations in the facial development of the autochthonous and allochthonous units are found during the Eifelian, Givetian and Frasnian (Kasig & Neumann-Mahlkau 1969). The palaeogeography of the Dinantian is determined by local basins and shoals, which may be a result of synsedimentary fault-block tectonics (Bless et al. 1980 b; Swennen et al. 1982). Alternative models have been developed by these authors for the pattern of these basins and shoals. None of them constrains a horizontal displacement of more than some tens of kilometers at the Midi Thrust System. The Dinantian limestones of the northern Ardennes hardly can be derived from the Mid-German-High. There only a thin layer of black shales represents this formation in the wells Saar 1 and Gironville 101 (Paproth 1976; Kneuper 1976) and in the Eisen Barite deposit (Müller & Stoppel 1981). Bouckaert & Herbst (1960) clearly pointed out that the Namurian of the Aachen region is related closely to that of other areas around the Brabant Massif. Conclusively they state that - concerning the Aachen thrust - "no larger nappistic transport can be presumed".

A very interesting comparison of the sedimentary development of the Westphalian both in the autochthonous and allochthonous units has been recently made possible in the Aachen area. The borehole Frenzer Staffel 1, which was drilled in the allochthonous Inde-Syncline in 1985, for the first time exposed the uppermost part of the Westphalian A2 in this structure (Wrede & Zeller, 1991). So it is possible to compare the thickness of the Westphalian A2 of the Inde district with that exposed in the autochthonous Wurm district. The thickness of the Westphalian A2 is 85 m higher in the Inde Syncline than in the Wurm Syncline. This difference in thickness in relation to a distance of about 10 km between Inde-hyphen and Wurm-Synclines leads to a southward progressive gradient of thickness of about 9 m per km. Significantly this gradient is nearly equal to that developed in the southern Ruhr Carboniferous (Strack & Freudenberg 1984) (Fig. 12). Within the Westphalian A1 the gradient of thickness is 5 - 6 m per km progressively towards north in the Aachen area. Precisely this reversal in direction and nearly identical gradient are also recorded in the Ruhr Carboniferous. This coincidence does not seem to be accidental. On the contrary, other results from recent coal-exploration in the Aachen and Erkelenz coalfields likewise confirm that the Wurm area, the Inde area, the Erkelenz and Ruhr Carboniferous are parts of one coherent sedimentary basin (Arndt et al. 1988, Strack 1988). The comparability of sedimentary development on the southern flank of this basin in different regions leads to the consequence that the post-sedimentary tectonic shortening cannot be substantially different. Thus a similar orogenic shortening of both Ruhr and Aachen areas seems to be likely again. The sedimentary development as well as the reconstruction of the fold and thrust structures result in a shortening of about 60 %. So the original distance separation the Wurm- and Inde-Syncline, which is about 10 km today, during time of sedimentation can be extimated about 25 km.

CONCLUSIONS

The tectonic style of the Variscan front is different in the eastern and western parts of the Rhenish Massif. The thrust-dominated tectonics in the northern Ardennes imply a nappe model for this area. This model is supported strongly by geophysical results. It can be demonstrated, however, that there is a gradual transition between this thrust-dominated area in the west and the fold-dominated, autochthonous Variscan foreland, exposed in the Ruhr Carboniferous. This refutes the model of a nappe-scale displacement for the Diant nappe of 100 km or more. Downwards a
substitution of thrusts by folds can be observed. These observations indicate a close relationship between folding and thrusting and make larger displacements unlikely.

Moreover, a number of palaeogeographic arguments indicate that not larger displacements occurred at the so called Dinant nappe. The sedimentary development of the Ardennes, compared with that of the Rhenish Massif east of the Rhine clearly reflects the influence of the Brabant Massif as a part of the Caledonian consolidated basement: A partly incomplete sequence of post-Caledonian Palaeozoic strata, deposited in a near-coastal environment, rich in carbonates and even eveporites points to a reduced mobility of this basement in the northern Ardennes. It has already been demonstrated by Wrede (1987 b) that in the Aachen area the tectonic style is controlled strongly by the uplifted block of the Brabant Massif. This shallow structure not only caused intensive thrusting instead of folding by blocking the northward directed movement of the Variscan front. It also influenced both the pattern of pre-Variscan and Variscan faults and the strike and distribution of fold axes in the Aachen area and in the subsurface of the Lower Rhine Embayment (Fig. 13, 14). All these features loose intensity towards east, where the Caledonian basement is deeply buried. Considering all the geological facts, there is no constraint to assume a displacement of more than some tens of kilometers for the Midi Thrust System. The differences in the tectonic style between the Ardennes and the eastern part of the Rhenish Massif can be explained by the uplifted Brabant Massif which, spur-like, interferes with the Variscan front.

Recently Bless et al. (1989 a, b) have given up their idea of a large-scale displacement of the Dinant nappes. They now favour a par-authochtonous model for this structural unit. They assume the large thrusts at the northern border of the Ardennes (inter-alia Midi-Aachen thrust, Faille Bordière) to be interverted listric normal faults of early Devonian or older origin. This new model is based mainly on the fact that until Devonian times strata thicknesses in the Dinant basin south of the thrust belt were much higher than in the Namur basin north of it. In the Namurian and Lower Westphalian thickness increases towards the north. However, these authors still expect the existence of a vast detachment (with limited horizontal displacement) separating the Dinant Basin from its basement. In our opinion, the observed decrease of throw of the thrust downwards contradicts the correlation of these thrusts with the seismic reflector at depth. The true nature of this reflector will be a matter of further discussion until it is exposed by a borehole. The results e.g. of the "KTB-Vorbohrung" in the Oberpfalz area have proved that interpretation of deep-seismic reflectors still is an open question (Emmermann 1990). Structural interpretation of the crust must not be based on seismic data alone or the transfer of theoretical models developed elsewhere, but should, above all, keep in account the geological facts of the discussed area. In the case of the Eifel-Ardennes region these facts more likely seem to contradict a nappe or décollement interpretation of the seismic reflector than to support it.

REFERENCES

Arndt, R., F. Lehmann, V. Weber, W. Skala, G. Wallrafen, C. Frieg, Ch. Gelbke, R.W. Heil, E. Kunz, W. Müller, L.E. Reimer, Ä. Strack 1988: Erkundung des Südteiles der Erkelenzer Anthrazitlagerstätte. Forsch. Ber. 03265 18A, 76 p, Bonn (BMFT) (unpublished).

Betz, D., H. Durst, T. Gundlach, 1988: Deep structural seismic reflection investigations across the northeastern Savelot-Venn Massif. Ann. Soc. géol. Belg. 111, 217-228.

Bless, M.J.M., J. Bouckaert, E. Paproth, 1980 a: Environmental aspects of some pre-Permian desopsits in NW Europe. Meded. Rijks Geol. Dienst, 32: 3-13.

Bless, M.J.M., J. Bouckaert, E. Paproth 1980 b: Tektonischer Rahmen und Sedimentationsbedingungen im Prä-Perm einiger Gebiete NW-Europas. Z. dt. geol. Ges., 131, 699-713.

Bless, M..J.M., J. Bouckaert, E. Paproth, 182: Recent exploration in pre-Permian rocks around the Brabant-Massif in Belgium, the netherlands and the federal republic of Germany. 63 p. Maastricht (Mus. Nat. Hist.).

Bless, M.J.M., J. Bouckaert, E. Paproth, 1989 a: Die Midi-Aachener Überschiebung: eine Inversionserscheinung? Nachr. dt. geol. Ges., 41, 13-14.

Bless, M.J.M., J. Bouckaert, E. Paproth, 1989 b: The Dinant nappes: A model of tensional listric faulting inverted into compressional folding and thrusting. - Bull. Soc. Belg. Geolog., 98, 221-230.

Bouckaert, J., G. Herbst, 1960: Zur Gliederung des Namurs im Aachener Gebiet. Fortschr. Geol. Rhld. Westf., 3, 369-384.

Bredding, H., 1971: Tiefentektonik und Deckenbau im Massiv von Stavelot-Venn (Ardennen und Rheinisches Schiefergebirge) Geol. Mitt., 12, 81-120.

Brix, M.R., G. Drozdzewski, R.O. Greiling, R. Wolf, V. Wrede, 1988: The N Variscan margin of the Ruhr coal district (western Germany): structural style of a buried thrust front? Geol. Rdsch., 77, 115-126.

Cazes, M., A. Mascle, J.-F. Raoult, 1988: Interprétation de la vibrosismique réflexion. Programme ecors; profil nord de la France, 53-77.

Cazes, M., G. Torreilles, Ch. Bois, B. Damotte, A. Galdéano, A. Hirn, A. Mascle, P. Matte, P. van Ngoc, J.F. Raoult, 1985: Structure de la croute hercynienne du Nord de la France: premiers résultats du profil ECORS, Bull. Soc. géol. France, 8, 925-941.

Cazes, M., G. Torreilles, B. Damotte, 1988: Implanation du profil et différentes opérations sur le terrain. - Programme ecors, profil nord de la France, 17-23.

Dreesen, R., J. Thorez, 1980: Sedimentary environments, conodont biofacies and paleoecology of the Belgian Famennian. An approach. Ann. Soc. géol. Belg. 103, 97-110.

Dreesen, R., W. Kasig, E. Paproth V. Wilder, 1985: Recent investigations within the Devonian and Carboniferous North and South of the Stavelot-Venn Massif. N. Jb. Geol. Paläont. Abh., 171, 237-265.

Drozdzewski, G., O. Bornemann., E. Kunz., V. Wrede, 1980: Beiträge zur Tiefentektonik des Ruhrkarbons, 192 p., Krefeld (GLA).

Drozdzewski, G., H. Engel, R. Wolf, V. Wrede, 1985: Beiträge zur Tiefentektonik westdeutscher Steinkohlenlagerstätten, 236 p., Krefeld (GLA).

Durst, H., 1985: Interpretation of a Reflection-Seismic Profile across the North-eastern Stavelot-Venn Massif and its Northern Foreland. N. Jb. Geol. Paläont. Abh. 171, 441-446.

Emmermann, R., 1990: Vorstoß ins Erdinnere: Das kontinentale Tiefbohrprogramm. - Spektr. d. Wissensch., Okt. 1990: 60-70.

Graulich, J.-M., 1980: Le Sondage de Grand Halleux. Belg. Geol. Dienst, Prof. Pap. 1980/86, 175, 78 p.

Graulich, J.-M., L. Dejonghe, 1986: Le bien-fonde de la notion de synclinorium de Verviers. Bull. Soc. géol. Bel., 95, 35-43.

Graulich, J.-M., V. Leclercq, L. Hance, 1989: Le sondage d'Havelange. Principales données et aspects techniques. Mém. Expl. Cartes Géol. et. Min. de la Belgique, 26: 65 p.

Herbst, G. 1982: Ein Aufschluß im Oberkarbon an der Aachener Überschiebung. Fortschr. Geol. Rhld. Westf., 3, 115-1158.

Kasig, W., P. Neumann-Mahlkau, 1969: Die Entwicklung des Eifeliums in Old-Red-Fazies zur Riff-Fazies im Givetium und Unteren Frasnium am Nordrand des Hohen Venns (Belgien-Deutschland). Geol. Mitt., 8, 327-388.

Knapp, G. 1980: Erläuterungen zur Geologischen Karte der nördlichen Eifel 1: 100.000. 155 p., Krefeld (GLA).

Kneuper, G., 1976: Regionalgeologische Folgerungen aus der Bohrung Saar 1.Geol. Jb., A 27, 499-510.

Kunz, E., R. Wolf, V. Wrede., 1988: Ergänzende Beiträge zur Tiefentektonik des Ruhrkarbons, 65 p., Krefeld (GLA).

Matte, Ph., A. Hirn, 1988: Généralités sur la chaine varisque d'europe, coupe complète de la chaine sons l'oust de la France. Programme ecors, profil nord de la France, 197-222.

Meissner, R., H. Bartelsen, H. Murawski, 1981: Thin-skinned tectonics in the northern Rhenish Massif, Germany, Nature, 290, 399-401.

Meissner, R., M. Springer, H. Murawski, H. Bartelsen, E.R. Flüh, H. Dürschner, 1983: Combined Seismic Reflection - Refraction Investigations in the Rhenish Massif and their Relations to Recent Tectonic Movements. In: K. Fuchs (ed.): Plateau uplift: 276-287, Berlin, Heidelberg (Springer).

Müller, G., D. Stoppel, 1981: Zur Stratigraphie und Tektonik im Bereich der Schwerspatgrube "Korb" bei Eisen (N-Saarland). Z. dt. geol. Ges., 132, 325-352.

Paproth, E., 1976: Erläuterungen der biostratigraphischen Bestimmungen aus der Tiefbohrung Saar 1 und Versuch einer paläogeographischen Deutung. Geol. Jb., A 27, 393-398.

Paproth, E., W. Struve, 1982: Bemerkungen zur Entwicklung des Givetium am Niederrhein. Paläogeographischer Rahmen der Bohrung Schwazbachtal 1. Seckenberg. leth., 63, 359-376.

Raoult, S.F., 1988: Le front varisque du nord de la France: Interprétation des principales coupes d'après les profiles sismiques, la géologie de surface et les sondages. Programme ecors, profil nord de la France, 171-196.

Raoult, S.F., F. Meilliez, 1986: Commentaires sur une coupe structurale de l'Ardenne selon le méridien de Dinant. Ann. Soc. Géol. Nord, CV, 97-109.

Rouchy, J.M., C. Pierre, F. Groessens, C. Monty, A. Laumondais, B. Moine. 1986: Les evaporites pre-permiennes du segment varisque Franco-Belge: Aspects palaeogeographiques et struturaux. Bull. Soc. Belg. de Geol., 05: 139-149.

Schmidt, W., 1956: Neue Ergebnisse der Revisions-Kartierung des Hohen Venns. Beih. geol. Jahrb., 21, 146 p. Hannover.

Strack, Ä., 1988: Stratigraphie in den Explatationsräumen des Steinkohlenbergbaus. Mitt. Westf. Berggewerkschaftsk., 62, 210 p., Bochum.

Strack, Ä., U. Freudenberg, 1984: Schichtenmächtigkeiten und Kohleninhalte im Westfal des Niederrheinisch-Westfälischen Steinkohlenreviers. Fortschr. Geol. Rhld. Westf., 32, 243-256.

Swennen, R., S. van Orsmael, L. Jacobs, K. op de Beek, J. Bouckaert, W. Viaene, 1982: Dinantian Sedimentation around the Brabant Massif, sedimentology and geochemistry. Publ. Naturhist. Gen. Limburg, 32, 16-23.

Teichmüller, M., R. Teichmüller 1979: Ein Inkohlungsprofil entland der linksrheinischen Geotraverse von Schleiden nach Aachen und die Inkohlung in der Nord-Süd-Zone der Eifel. Fortschr. Geol. Rhld. Westf., 27, 323-355.

Thome, K.N., 1955: Die tektonische Prägung des Venn-Sattels und seiner Umgebung. Geol. Rdsch., 44, 266-305.

Thorez, J., R. Dreesen, 1986: A model of a regressive depositional system around the Old Red Continent as exemplified by a field trip in the Upper Famennian "Psammites du Condroz" in Belgium. - Ann. Soc. géol. Belg., 109, 285-323.

Thorez, J., E. Goemaere, R. Dreesen, 1988: Tide - and wave - influenced depositional environments in the psammites du Condroz (Upper Famennian) in Belgium. In: de Boer et al. (ed.): Tide-influenced Sedimentary Environments and Facies, 389-415.

Tsien, H.H., 1974: Paleoecology of Middle Devonian and Frasnian in Belgium. Int. Symp. on Belg. Micropal. Limits, Namur 1974, 12, 53 p.

Vangustaine, M., 1970: L'appartenance an Revinien inférieur et moyen des roches noires de la profonde du sondage de Grand-Halleux et leur disposition en un pli couche. Ann. Soc. Géol., Belg. 93, 591-600.

Verhoogen, J., 1935: Le prolonge ment oriental des failles du massif de la Vesdre et du massif de Herve. Ann. Soc. géol. Belg. 58 (B): 111-118.

Voll, G., 1983: Crustal Xenoliths and their evidence for crustal structure underneath the Eifel Colvanic District. - In: Fuchs, K. et al. (ed.): Plateau Uplift, 336-342.

Walter, R., 1982: Europe-1: proposal for a deep test-hole through the Stavelot-Venn Anticline, Federal Republic of Germany. Publ. natuurhis. Gen. Limb., 32, 59-62.

Wörner, G., H.U. Schmincke, W. Schreyer, 1982: Crustal xenolithes from the Quarternary Whr volcano. (East Eifel). N. Jb. Miner. Abh., 144, 29-55.

Wrede, V., 1985 a: Tiefentektonik des Aachen - Erkelenzer Steinkohlengebietes. Beitr. z. Tiefentekt. westdt. Steinkohlenlagerst., 9-103.

Wrede, V., 1985 b: Die Fortsetzung der Aachener Überschiebung nach Osten - eine Arbeitshypothese. Fortschr. Geol. Rhld. Westf., 33, 297-306.

Wrede, V., 1987 a: Einengung und Bruchtektonik im Ruhr-Karbon. Glückauf Forsch.-H., 48, 116-121.

Wrede, V., 1987 b: Der Einfluß des Brabanter Massivs auf die Tektonik des Aachen-Erkjelenzer Steinkohlengebietes. N. Jb. Geol. Paläont., Mh. 1987 177-192.

Wrede, V. (in press): Die Tektonik des prä-permischen Untergrundes von Krefelder und Venloer Scholle. Fortschr. Geol. Rhld. Westf., 37.

Wrede, V., H.-D. Hilden, 1988: Geologische Entwicklung. Geologie am Niederrhein, 7-14.

Wrede, V., M. Zeller, 1988: Geologie der Aachener Steinkohlenlagerstätte. 77 p., Krefeld (GLA).

Wrede, V., M. Zeller, (1991): Die stratigraphische Einstufung der Bohrung Frenzer Staffel 1 (1985). Geol. Jb., A 116, 73-86.

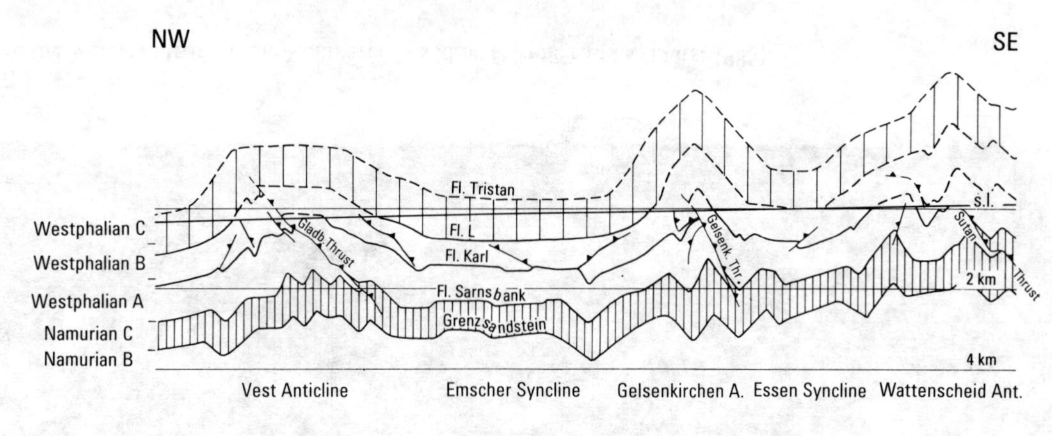

FIGURE 1
Typical section of the Variscan front in the Ruhr Carboniferous: fold dominated tectonics without detatchment.

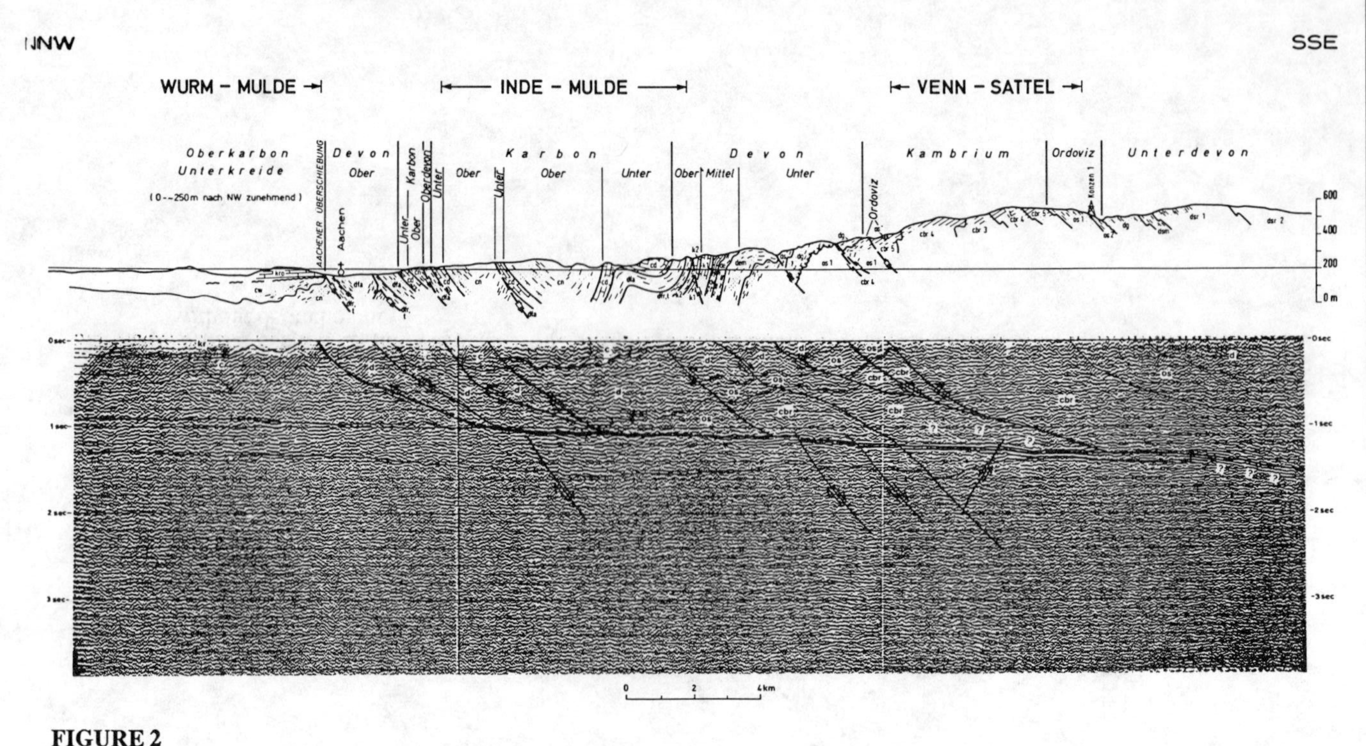

FIGURE 2
Section of the Variscan front in the Venn area based on seismic results (Durst 1985).

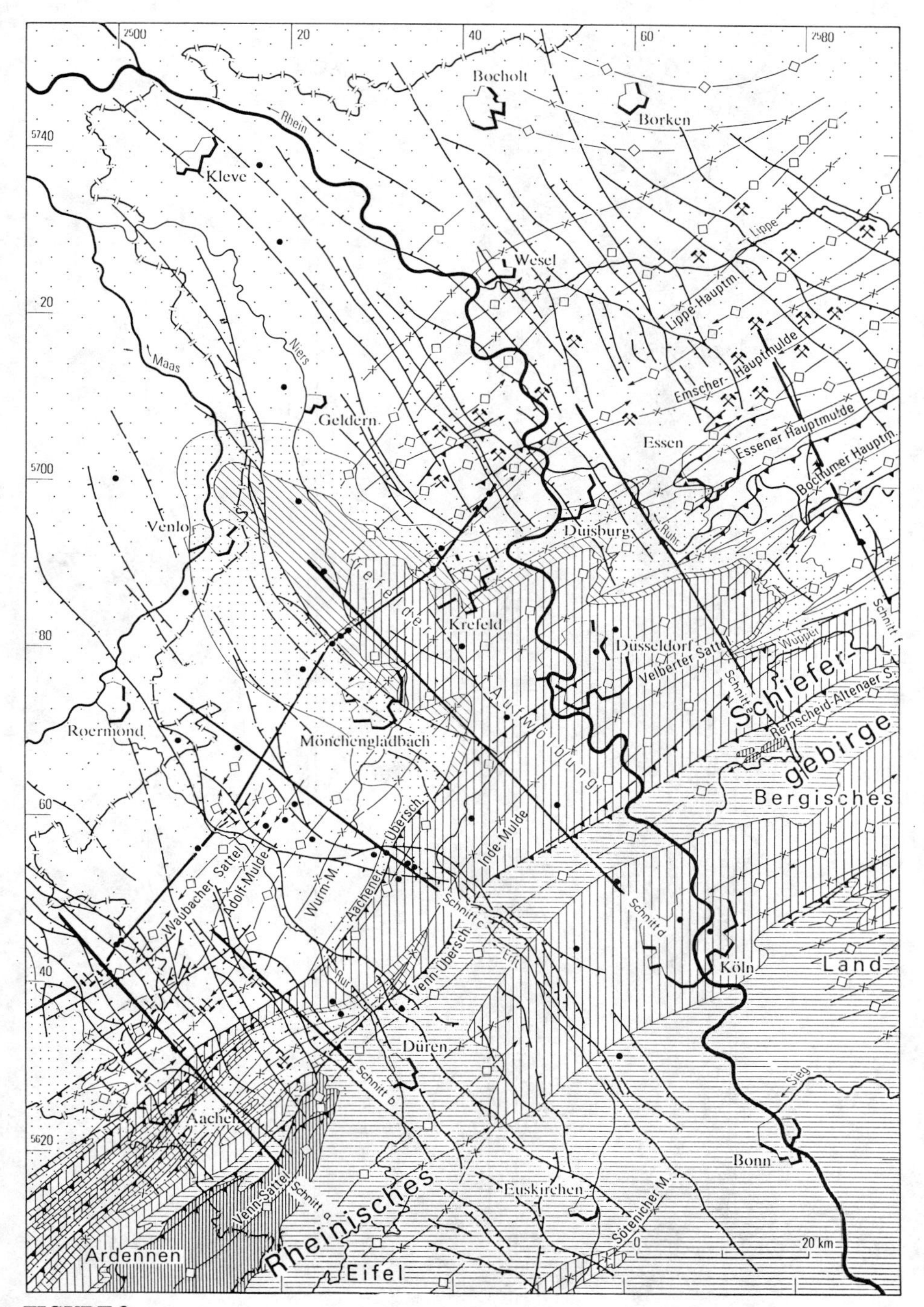

FIGURE 3
Structural map of the palaeozoic subsurface of the Lower Rhine Embayment (Wrede & Hilden 1988).

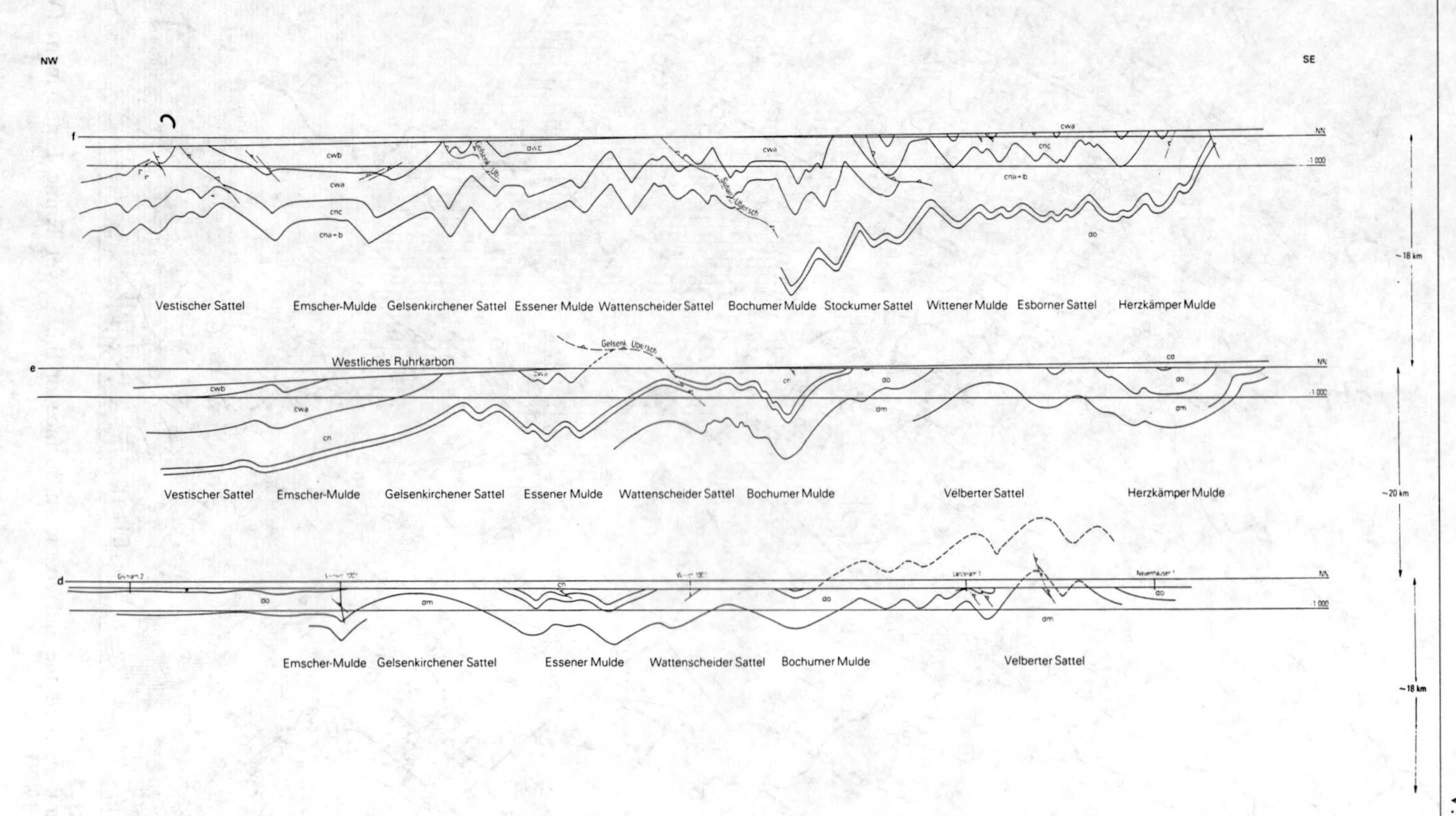
NW
SE
f
e
d
NN
-1000
~18 km
~20 km
Vestischer Sattel
Emscher-Mulde
Gelsenkirchener Sattel
Essener Mulde
Wattenscheider Sattel
Bochumer Mulde
Stockumer Sattel
Wittener Mulde
Esborner Sattel
Herzkämper Mulde
Westliches Ruhrkarbon
Velberter Sattel
cwb
cwa
cnc
cna-b
do
dm
cn
cd
Sutan-Überschiebung
Gelsenk. Überschiebung
Gelsenk. Üb.

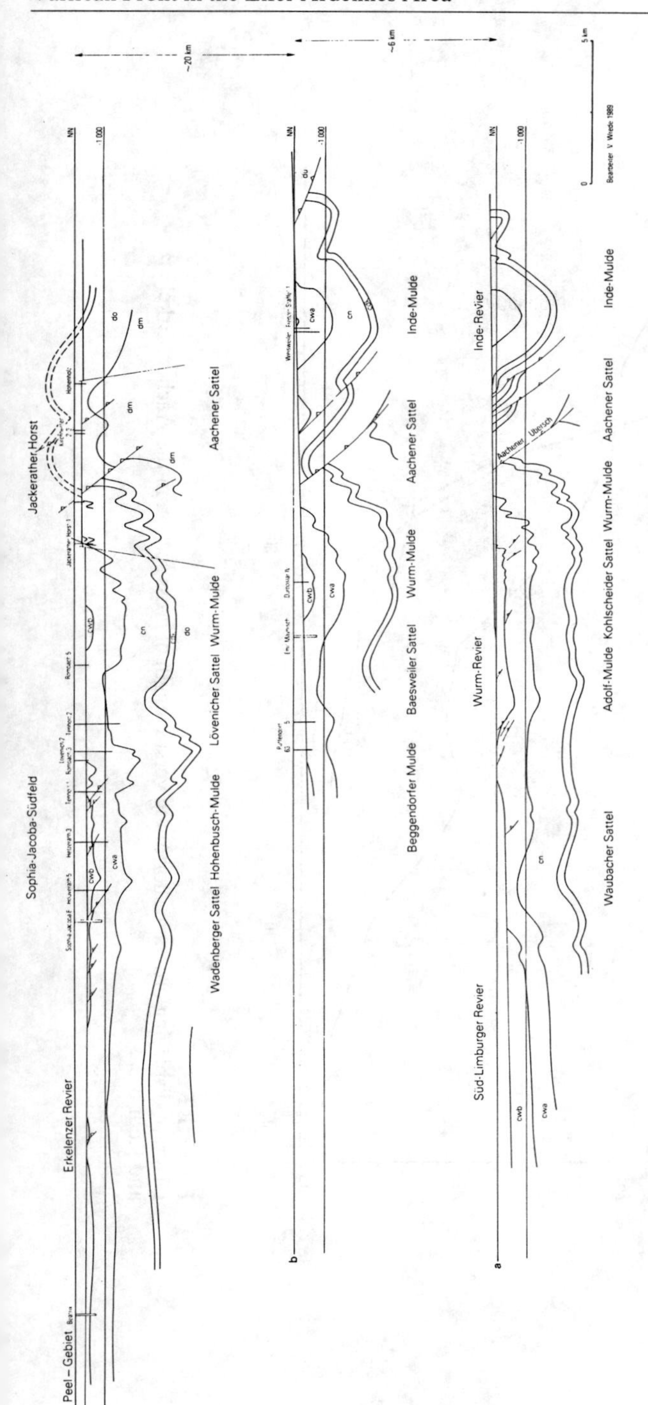

FIGURE 4
Sections of the Variscan front in the Eifel, the Lower Rhine Embayment, the Bergian Mountains, and the Ruhr Carboniferous (Position of sections indicated in fig. 3).

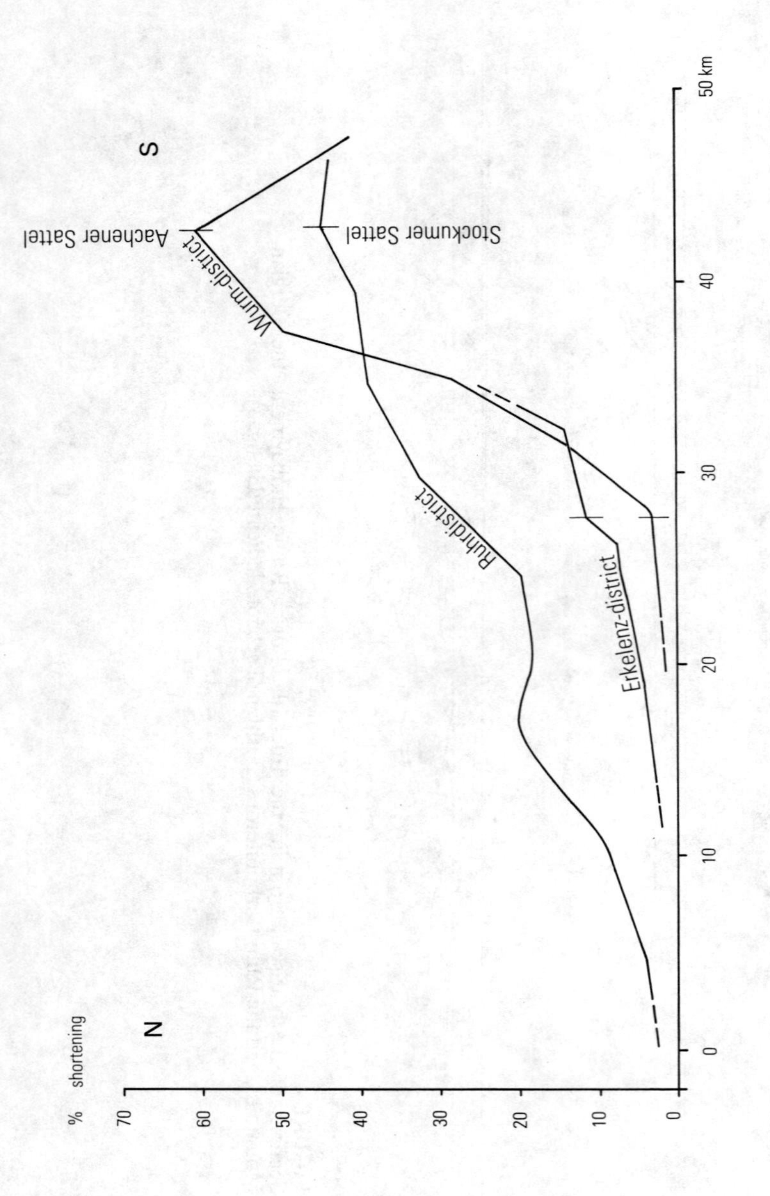

FIGURE 5
Diagram displaying orogenic shortening versus length of cross section in the Aachen, Erkelenz and Ruhr areas.

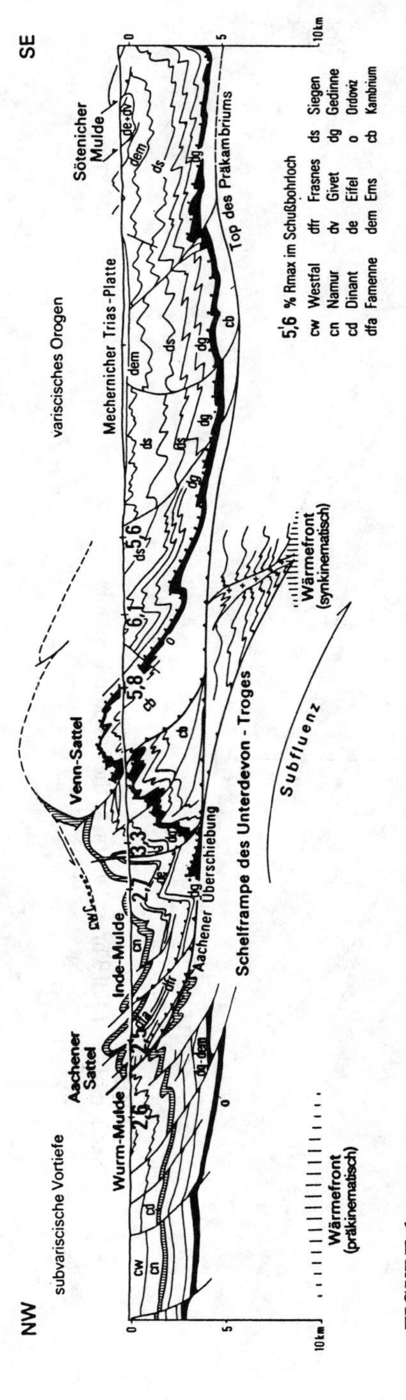

FIGURE 6
Sections of the Aachen-Venn area according to a nappistic model (Teichmüller & Teichmüller 1979).

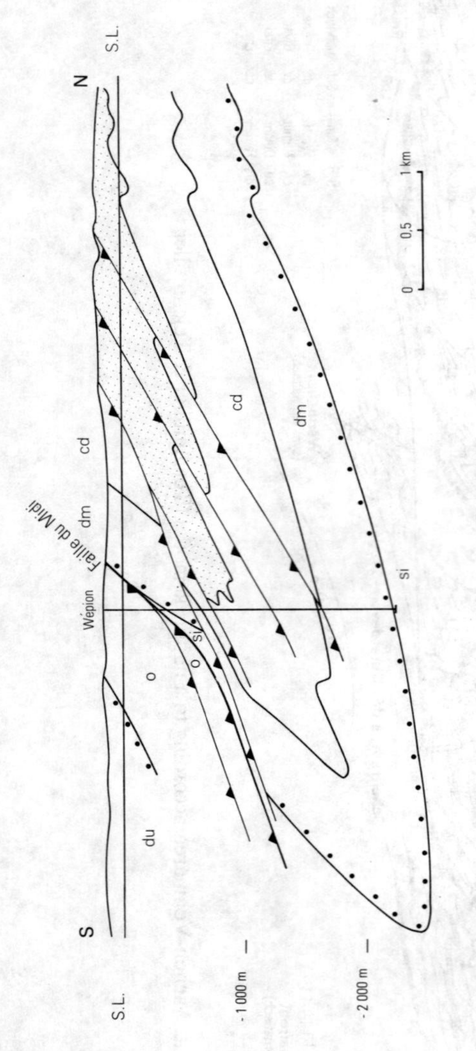

FIGURE 7
Section of the Midi Thrust System at the Wépion-borehole

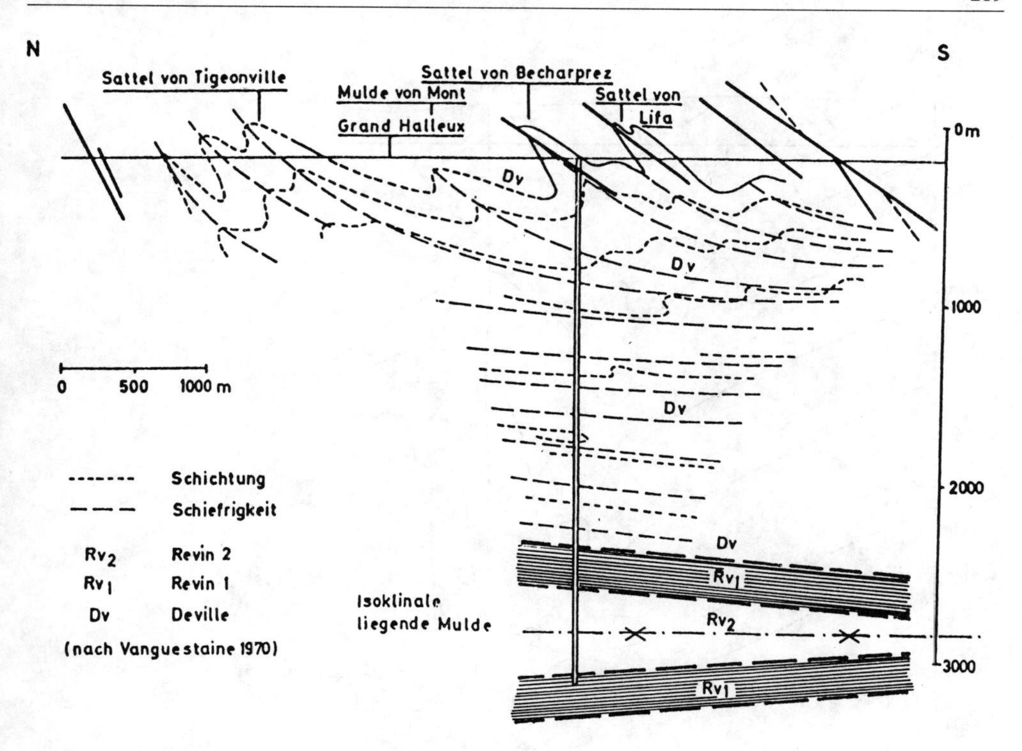

FIGURE 8
Section of the borehole Grund Halleux (Breddin 1971)

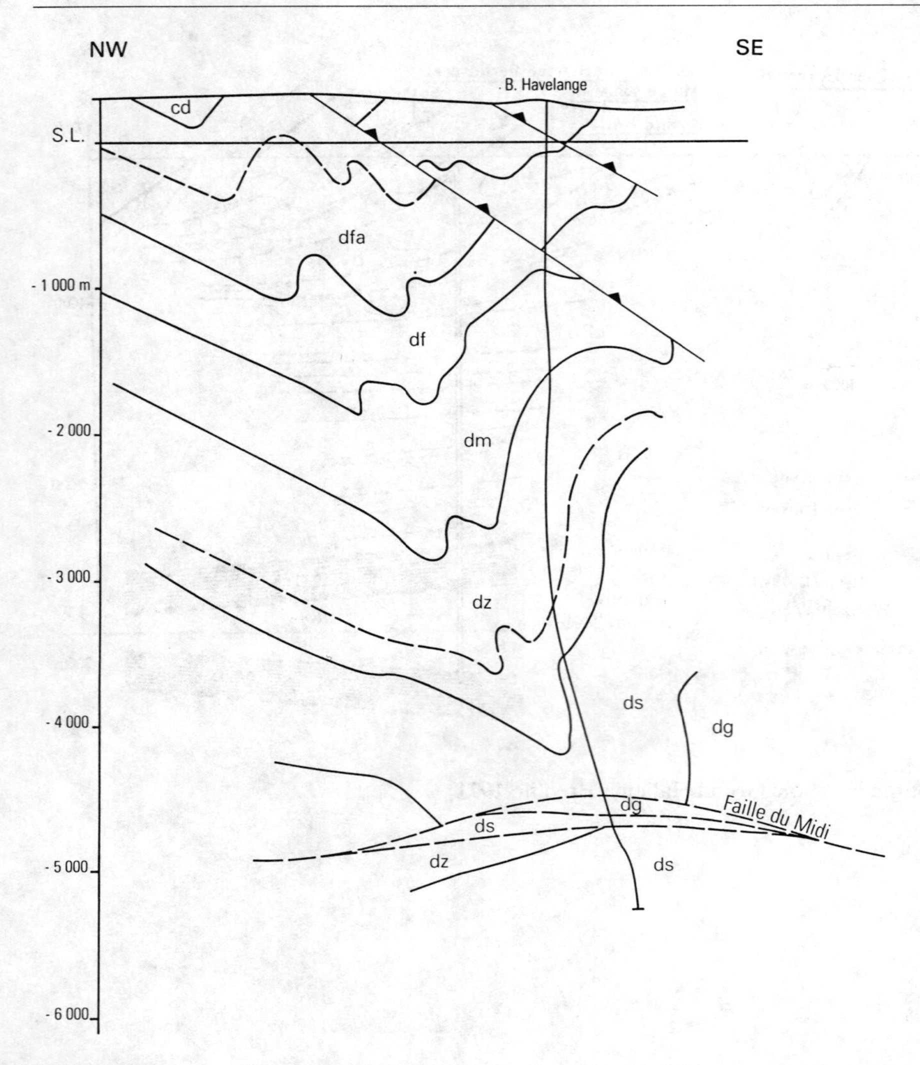

FIGURE 9
Section of the Havelange borehole. Faults in the deeper part of section are interpreted as thrusts (Graulich, Leclerq, Hance 1989).

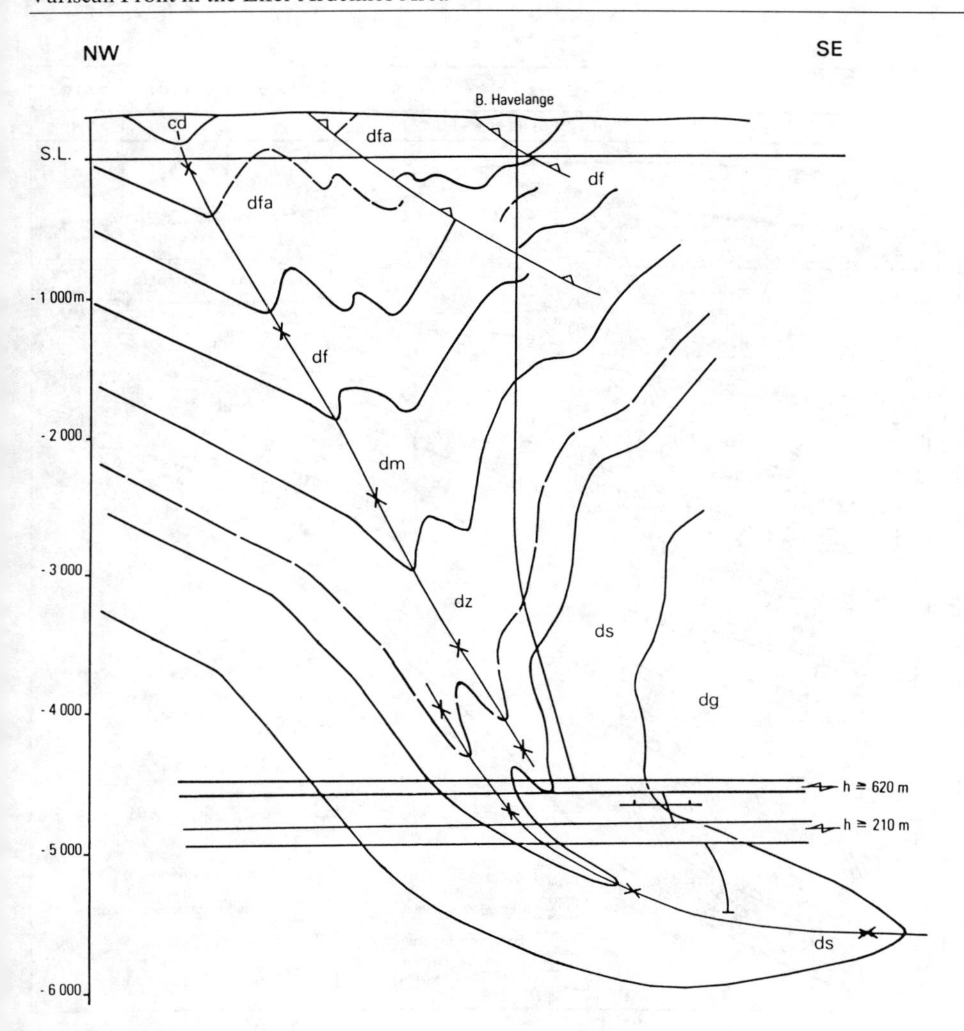

FIGURE 10
Alternative interpretation of the Havelange section. Faults are interpreted to be normal or strike-slip faults. The indicated throws are minimal assumptions. Fold strcture has been restored.

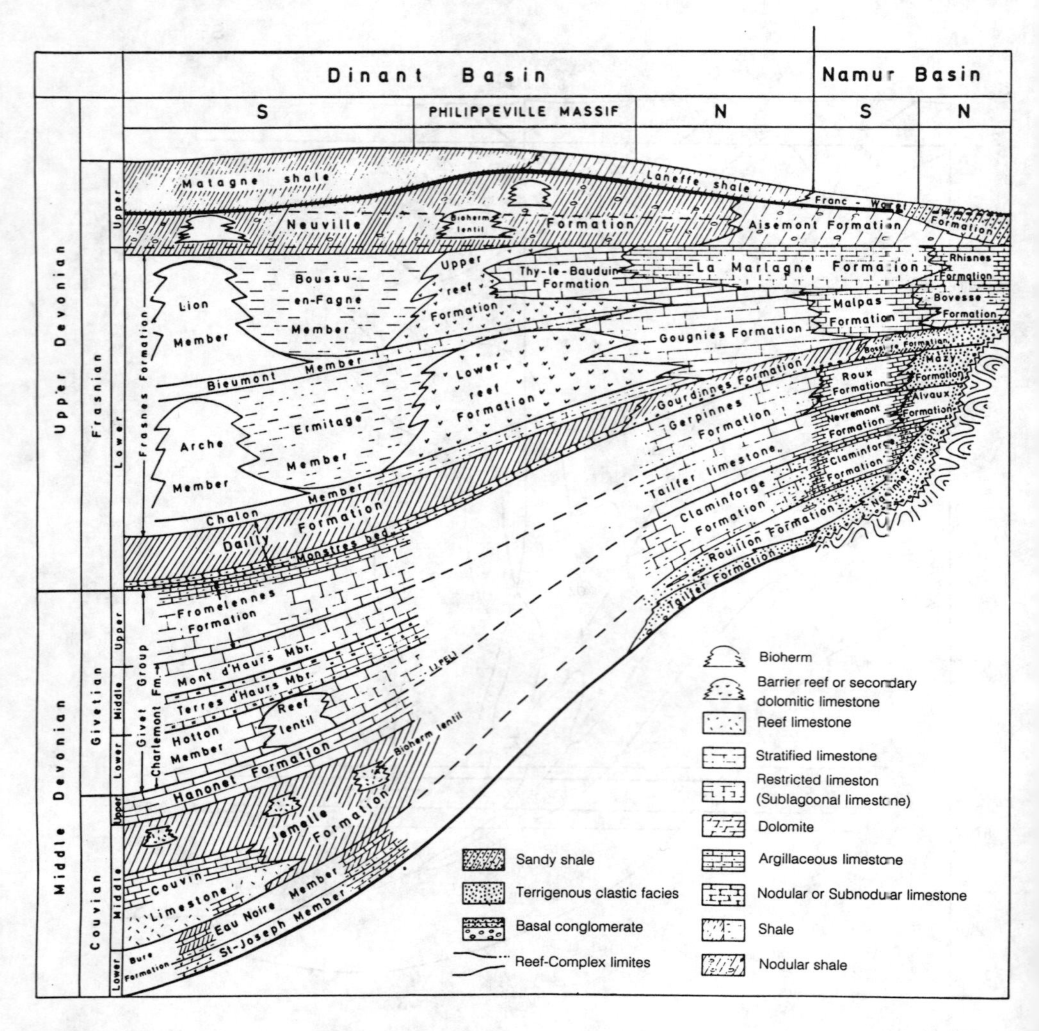

FIGURE 11
Facies of the Middle and Upper Devonian in the Dinant and Namur Basins (Tsien 1974).

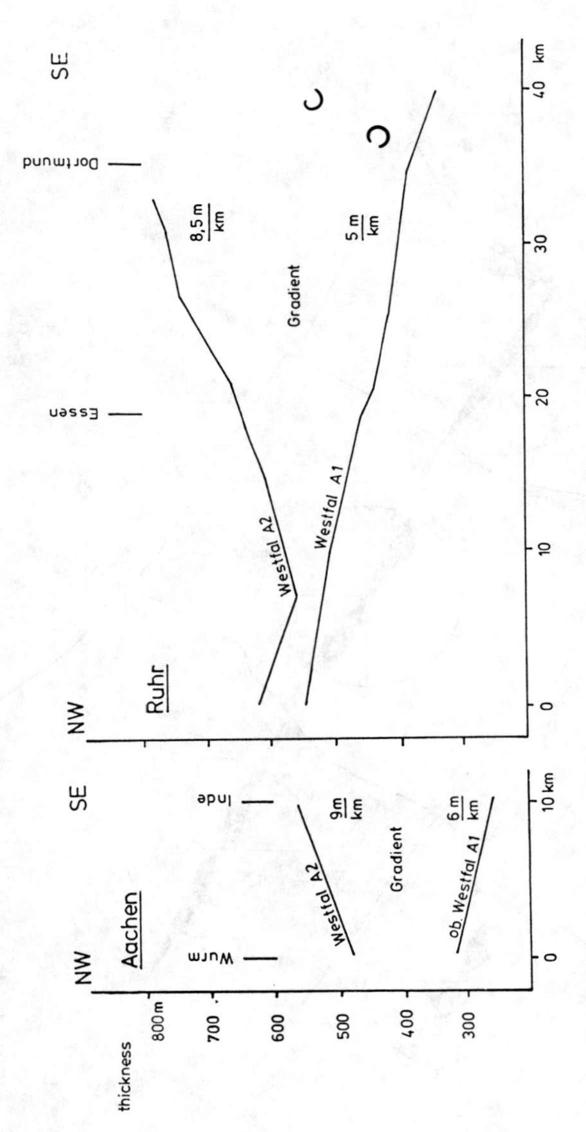

FIGURE 12
Development of sediment-thickness for the Westphalian A 1 and A 2 in the Aachen area and the Ruhr Carboniferous.

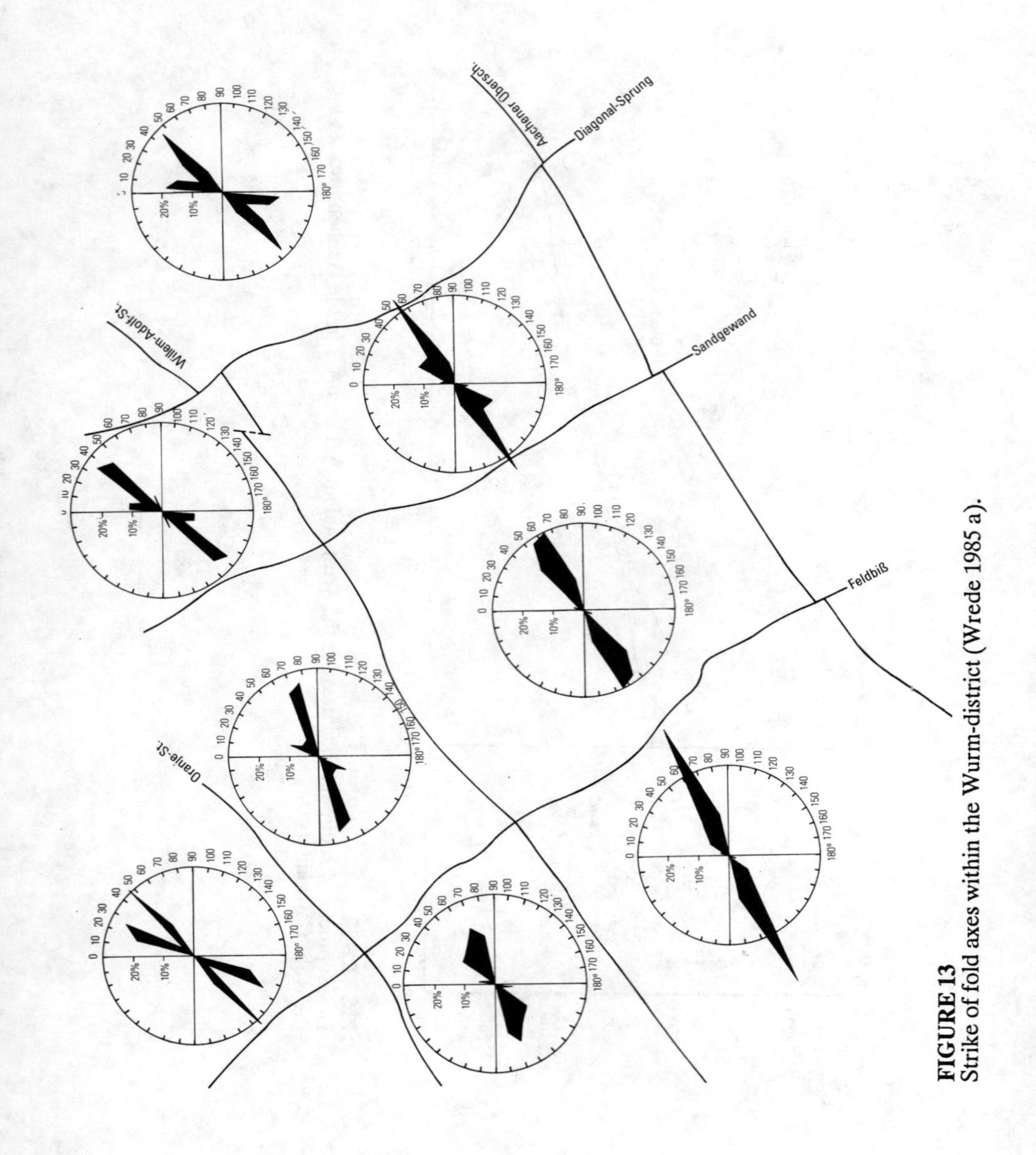

FIGURE 13
Strike of fold axes within the Wurm-district (Wrede 1985 a).

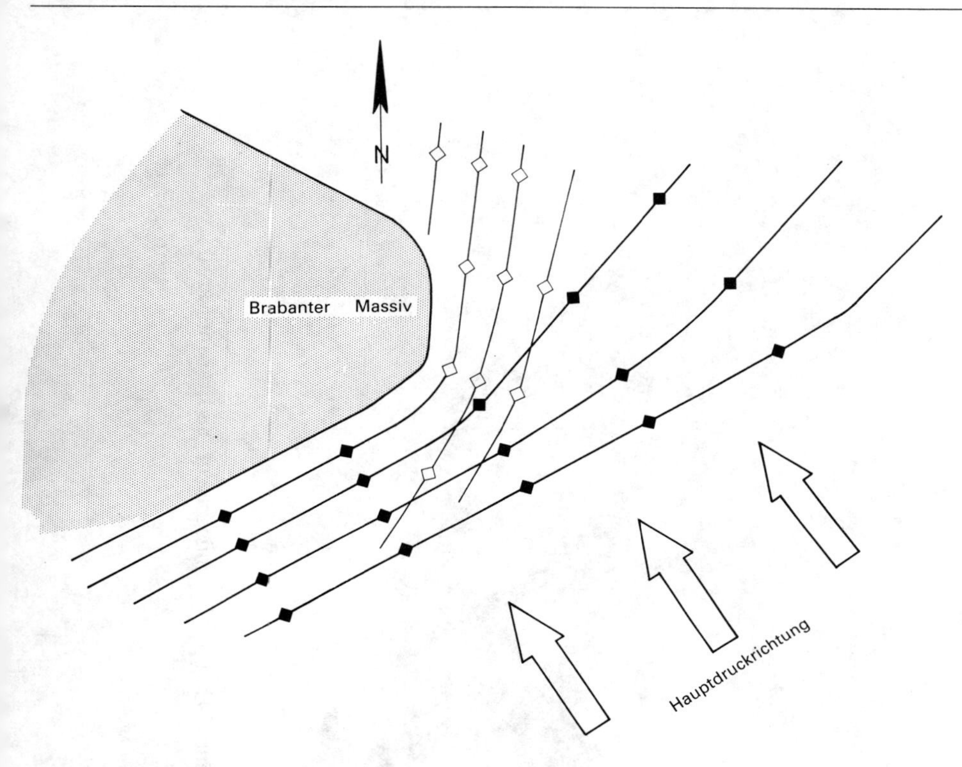

FIGURE 14
The influence of the Brabant massif to strike directions of fold axes in the Liége-Aachen-Lower Rhine-area (s. fig. 13, 15) (schematic; Wrede 1987 b).

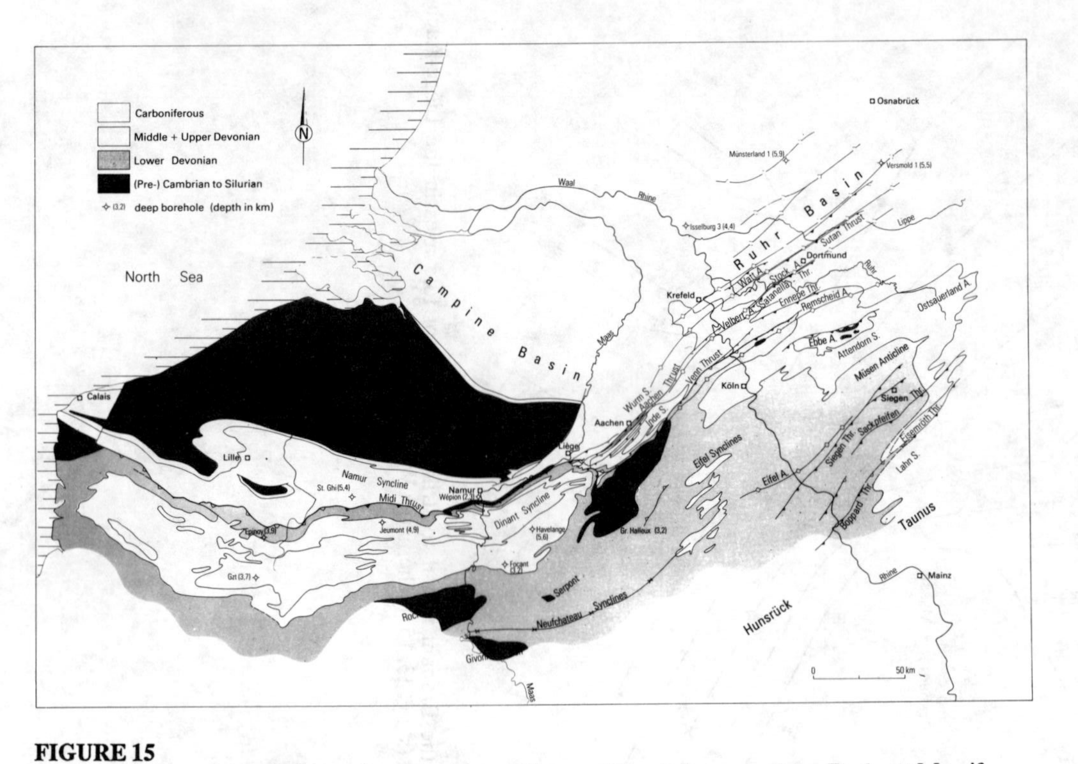

FIGURE 15
Structural map of the Variscan front in Central Europe. The influence of the Brabant Massif on the Variscan structure is clearly seen.

Displacement Gradients of Thrusts, Normal Faults and Folds from the Ruhr and the South-Wales Coalfields

P. Gillespie
Department of Geology, University of Wales, P.O. Box 914, Cardiff, UK
Present address: Department of Earth Sciences, University of Liverpool,
P.O. Box 147, Liverpool L69 3BX, UK

ABSTRACT

Subsurface data from the Ruhr Coalfield and the South Wales Coalfield provide an excellent opportunity to describe the displacement variations of folds and faults. This paper details techniques for the description of these variations. The techniques should be of use to mining and structural geologists in fault prediction.

The displacement variations on an individual fault surface may be contoured to produce a displacement contour diagram. Both thrusts and normal faults show systematic displacement variations with maximum displacement values in the centre of the fault, decreasing to an approximately elliptical tip-line loop. Normal fault ellipses have axial ratios of about 2:1 with the long axis horizontal. Thrusts have similar orientations but have axial ratios of up to 18:1. Similarly, the amplitude variations of a fold may be contoured in the fold axial plane to produce an amplitude contour diagram.

By plotting the maximum displacement against the width (horizontal dimension) of different faults it is possible to show that displacement is approximately proportional to width for both faults and folds over several orders of magnitude. Horizontal displacement gradients for faults and folds are generally between 2 and 20 %. The vertical displacement gradients of normal faults are about twice these values. However, the high aspect ratios of thrusts mean that their vertical displacements can be very much higher.

INTRODUCTION

At any one place on a fault there is a localised displacement of a given, quantifiable value. Individual faults do not extend indefinitely and therefore they must either;
1) diminish to zero, 2) meet with another structure or, 3) meet the free surface.
Similarly the shortening across an individual fold is a form of localised displacement which must vary around the fold axial plane. Studying displacement variations is one of the most informative ways of describing geological structures.

In this paper displacement variation techniques are developed and applied to data from the Ruhr Coalfield and the South Wales Coal Basin (Fig. 1). The data from the Ruhr Coalfield are extremely well recorded on mine plans and sections and have been compiled and published in the "Tiefentektonik" volumes (Drozdzewski et al. 1980, 1985). These provide closely spaced sections of more than 1 km in depth at a scale of 1:20,000. They show the effects of faults and folds on named coal seams. The structure of the area is dominated by WSW-ENE trending thrusts and folds cut by orthogonal normal faults. The thrusts and folds developed simultaneously in the Variscan (Wrede, this volume) and do not cut the Permian unconformity. The normal faults have been active from Variscan times until the present day (Wolf in Drozdzewski et al. 1985).

The South Wales Coalfield (Fig. 1) represents the lateral equivalent of the Ruhr Coalfield (Gayer et al., this volume). The data is in the form of records from deep mine and opencast operations. East-west trending Variscan thrusts are cut by orthogonal normal faults, but the timing of fault movement is not so well known as in the Ruhr as there is no overlying sequence. The large upright anticlines so common in the Ruhr are absent in South Wales. The South Wales information includes both field data and recorded subsurface data provided by British Coal.

The paper is divided into two parts, in which different techniques for studying displacement variations are given. The first part describes displacement contour diagrams, which are a technique for describing the displacement variations on individual structures. The second part concerns displacement versus width plots which allow an examination of the scaling parameters of structures.

This new approach to the analysis of geological structures provides a means for understanding regional geology. The implications for the geology of the Ruhr Coalfield and the South Wales Coalfield are discussed.

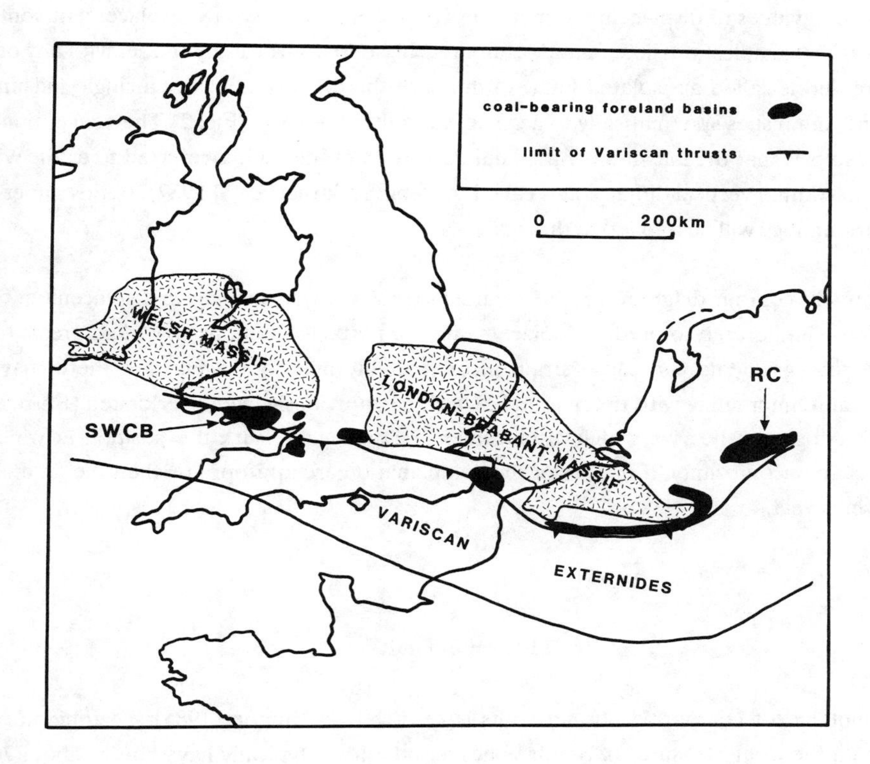

FIGURE 1
Location map showing the South Wales Coal Basin (SWCB) and the Ruhr Coalfield (RC) marked with the present-day extent of proved coal-bearing foreland basins, and the northern limit of Variscan thrusts with displacements greater than 1 km.

DISPLACEMENT CONTOURS

Variations in values of displacement on a fault surface can be shown by displacement contour diagrams. The simplest possible fault does not reach the free surface or interact with any other structure and is called an isolated fault. In this case the displacement has a single maximum value and diminishes systematically to a single, closed tip-line loop (Fig. 2) The convention for normal faults is that the maximum horizontal dimension of the fault is referred to as the width and the maximum vertical dimension is called the length (Barnett et. al 1987). In this paper the same terminology will be applied to thrusts.

Displacement contour diagrams are the clearest way of showing changes in displacement on a fault. They can be used to predict displacement on a particular fault in unworked areas. High displacement gradients can cause strain incompatibility and so suggest that there may be another fault, interacting with the first fault, which has previously been overlooked (Barnett et al 1987). Alternatively, a very high displacement gradient may mark the position at which a fault cuts an unconformity. If neither of these explanation are appropriate, then the data itself may be incorrect.

1 Normal Faults

The technique for contouring normal faults is described in Rippon (1985). Tectonic normal faults from the Coal Measures of South Wales and the Ruhr typically have dips of about 70½. Under these conditions the throw (vertical displacement) has a similar value to the true displacement and it is therefore most convenient to contour throw values. The values are projected onto the vertical plane striking parallel to the fault to produce a strike projection plane.

An example of a strike projection of throw for a normal fault is given in Fig. 3(a). This fault occurs in the Nant Helen opencast site in the South Wales Coalfield where throw values were recorded on extraction plans of six successive seams represented by the subhorizontal rows of data points on the diagram. The fault has a width of 380 m and the contours show a systematic decrease of throw from a single maximum value of about 8.5 m to zero. What is recorded is probably the lower half of a roughly elliptical fault plane.

The aspect ratio of a fault is defined as the ratio of width to length (Fig. 2). The normal fault from Nant Helen has an estimated aspect ratio of 2.6. This is comparable with values for

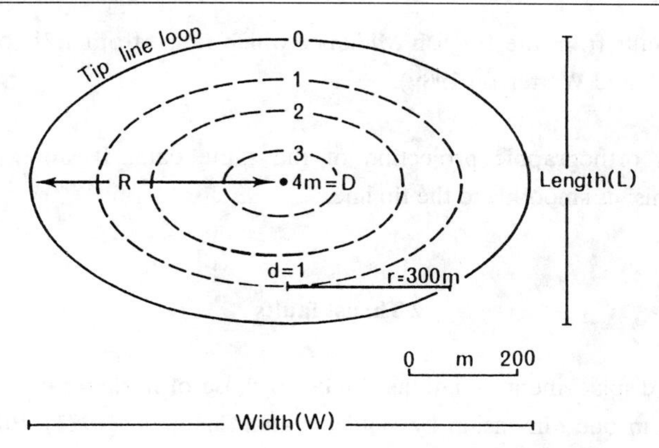

FIGURE 2
Displacement contours on an ideal fault plane (after Barnett et al. 1987).

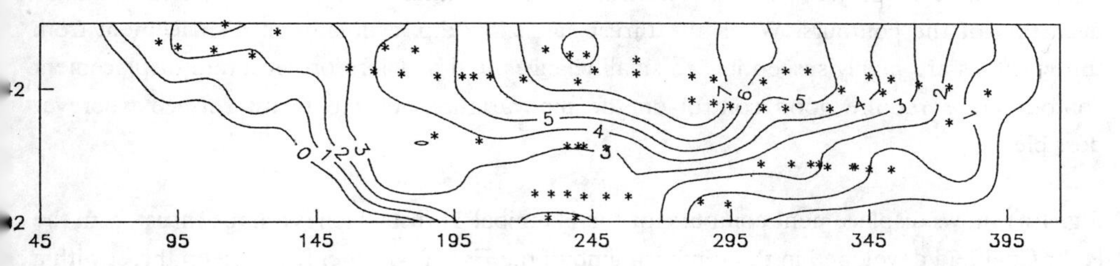

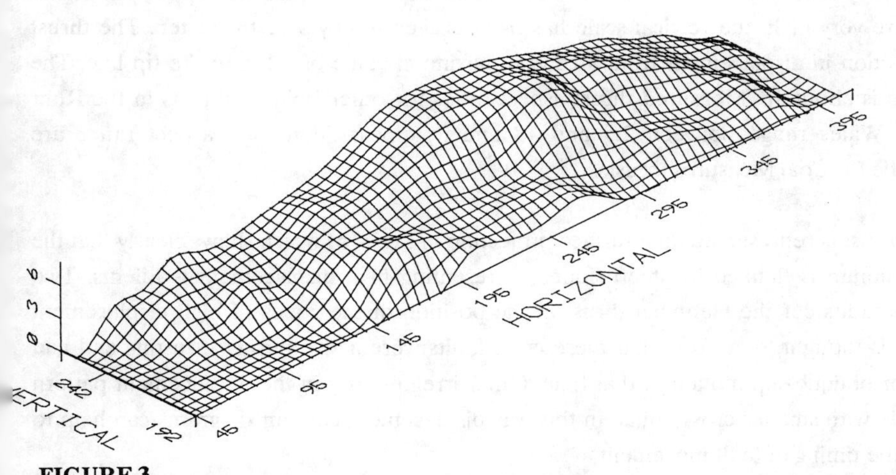

FIGURE 3
(a) Strike projection of throw on a normal fault from Nant Helen opencast site, South Wales. All figures are in meters; (b) orthographic projection of the same data.

coalfield normal faults from the English Midlands which range from 1.25 to 3.0 with a mean value of 2.25 (Walsh and Watterson 1989).

Figure 3(b) is an orthographic projection of the same data. It shows clearly how the displacement diminishes smoothly to the tip line.

2 Thrust faults

Thrust faults have displacement variations similar to those of normal faults. These variations have been studied in one dimension by Williams and Chapman (1983), Pfiffner (1985) and Ellis and Dunlap (1988). With adequate data, thrusts can be contoured for displacement variations in the same way as normal faults. However, when the thrust surface dips at less than 45½ it is best to project onto the horizontal plane rather than the vertical, to minimise distortion of the contours. When the thrust has a low dip, estimates of displacement from throw values are highly susceptible to small changes in dip. Therefore accurate displacement contour maps require good control on the dip variations of the thrust surface wherever possible.

Fig. 4(a) shows displacement contours for the Hannibal Thrust which is a large thrust from the Ruhr Coalfield developed in the southern limb of the Essen syncline. It is a steep thrust with a dip of greater than 45½ for most of its length and therefore a strike projection of displacement was used, the same projection as in the normal fault example. As the vertical displacement gradients are very high, the vertical scale has been exggerated by a factor of ten. The thrust shows reduction in displacement from a central maximum value of 1 km to the tip line. The aspect ratio is about 18. Measured thrust aspect ratios for other isolated thrusts in the Ruhr and South Wales range between 2.5 and 18 and it appears that high aspect ratios are characteristic for Coal Measures thrusts.

The same thrust is represented on a surface projection in Fig. 4(b) which shows clearly that the central maximum is flanked by steep slopes representing high displacement gradients. Two major cross faults cut the Hannibal thrust at the position of the zones of high displacement gradient. It is thought, therefore, that these cross faults were active during thrusting and had strike-slip or oblique-slip motion at that time. Other irregularities in the displacement pattern also coincide with smaller cross faults. In this way displacement contour diagrams can help to determine the timing of fault movement.

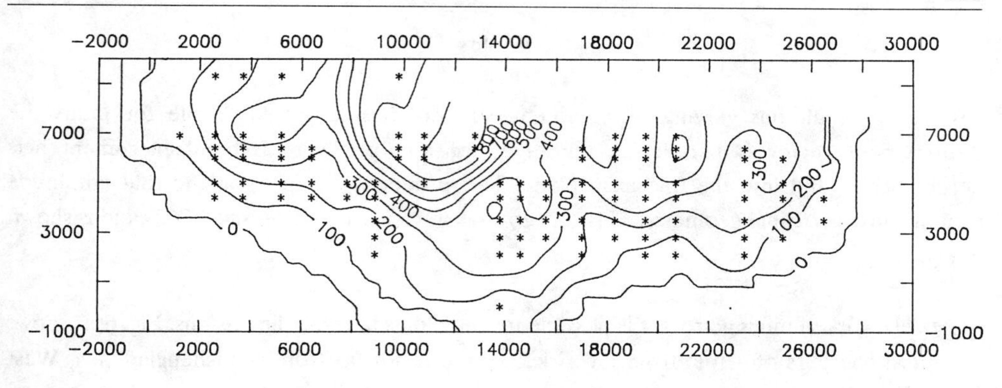

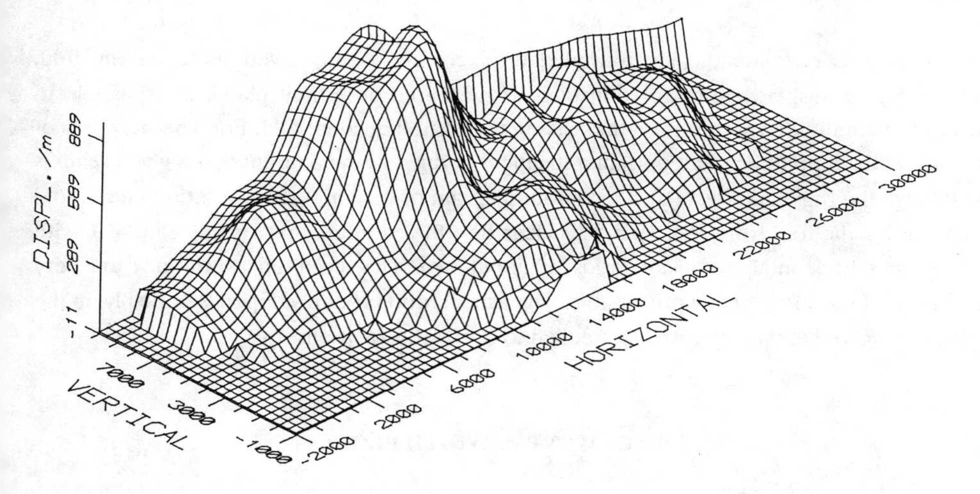

FIGURE 4
(a) Strike projection of displacement for the Hannibal Thrust from the Ruhr coal measures using data from Drozdzewski et. al. (1980). Vertical exaggeration x10. All figures in metres; (b) orthographic projection of same data with the same vertical exaggeration.

3 Folds

Just as any fault has varying displacement, so also must any fold. While for faults the displacement vector is parallel to the fault, for an upright fold the displacement acts perpendicular and into the fold axial plane. This produces variations in the fold amplitude resulting in a periclinal geometry. This concept was introduced by Ramsay (1962) and is shown in Fig. 5.

In heavily worked mining areas it is possible to study folds in three dimensions. Fig. 6(a) shows structural contours of a fold from the Erkelenz mine, not far from Mönchengladbach, West Germany. It is an open, upright pericline proved in five coal seams.

It is not easy to estimate the layer-parallel displacement at any given place on the fold. However, just as displacement variations can be measured on a fault plane, it is possible to measure the amplitude variations on the axial plane of an individual fold. For The purposes of this analysis, the amplitude of an upright anticline is defined as the amount that a given seam is raised above the regional. Fig. 6(b) is a contour diagram of the amplitude variation on the fold axial plane for the fold from Erkelenz. It is clear that the contours are roughly elliptical, with the long axes horizontal and axial ratios of about two. The amplitude variations are very systematic and therefore highly predictable. The aspect ratio probably depends critically on the lithologies involved and further work is needed to clarify this relationship.

DISPLACEMANT vs. WIDTH PLOTS

Faults and folds occur over a vast range of scales and it is reasonable to assume that the large structures have grown through a series of stages represented by the smaller structures. By plotting the maximum displacement against the width (horizontal dimension) for a given kind of structure on a log/log plot, we can investigate the way in which the structure changes as it grows.

Displacement vs. width plots have previously been presented for normal faults (Ruzhich, 1977, Walsh and Watterson, 1988), thrusts (Elliott, 1976) and strike-slip faults (Ranalli, 1977).

In most circumstances the individual structures are not known completely in 3D and there is only information about one or more horizontal chords along the fault surface (or axial surface) which do not pass through the centre of the structure. This will cause anomalously low estimates of displacement/width, but should not seriously affect the spread of points on the plot (see Walsh and Watterson 1988 for discussion).

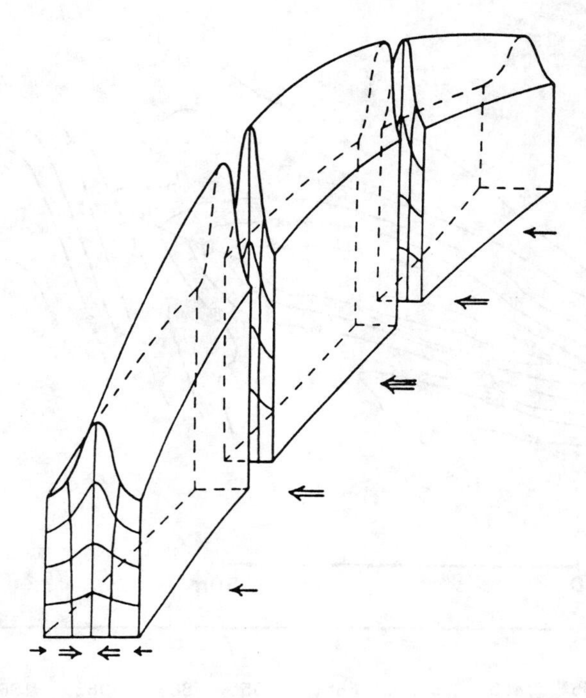

FIGURE 5
The effect of fold amplitude variations on 3D geometry of the structure (after Ramsay, 1962). Arrows represent different layer-parallel displacement vectors.

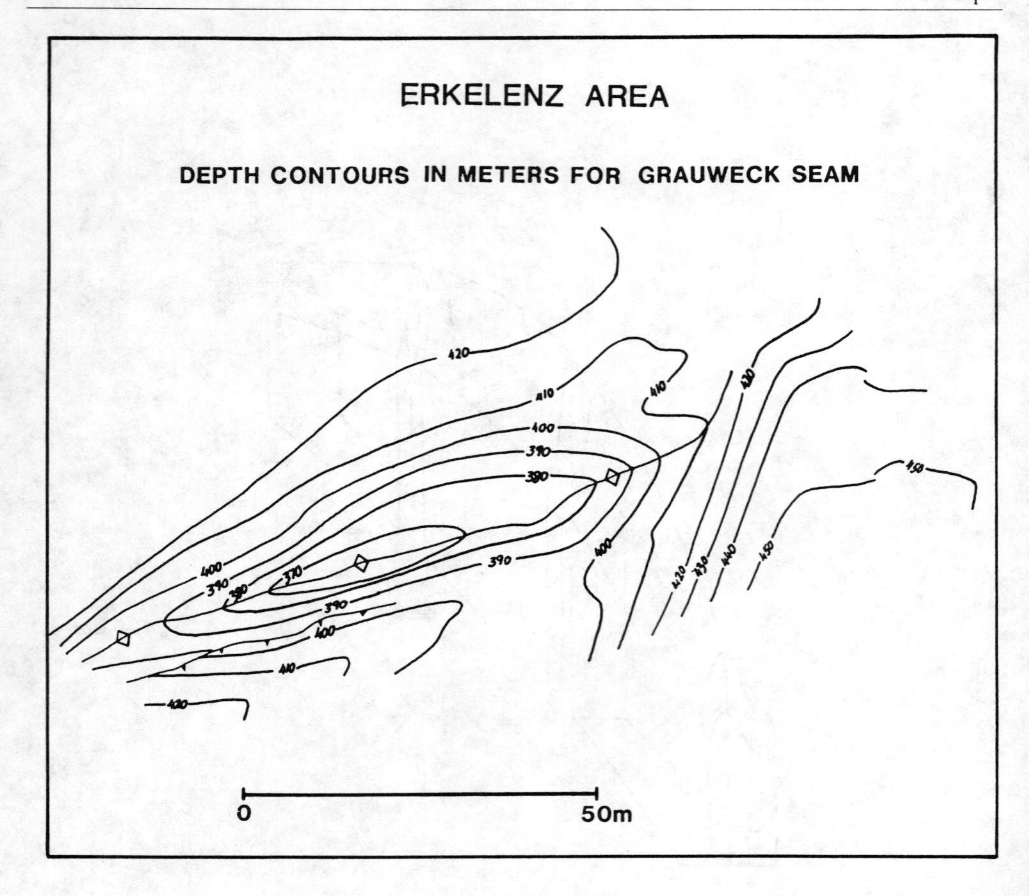

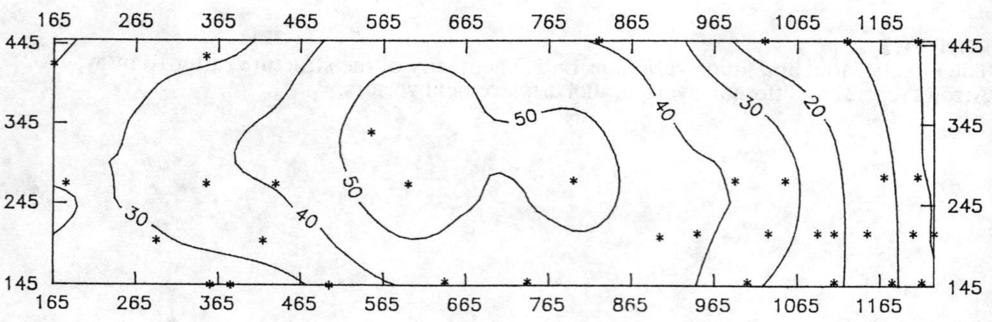

FIGURE 6
(a) Depth contours in meters for Grauweck Seam in the Erkelenz Mine showing an upright open anticline; (b) contours of fold amplitude on the axial plane of the fold in (a).

1 Normal Faults

A plot of displacement vs. width for normal faults is given in Fig. 7(a). All of the faults chosen are completely isolated: they do not interact with adjoining faults. The 82 points shown come from a variety of sources. The "Ruhr" data (n=12) come from the maps and sections of Wolf (in Drozdzewski et. al. 1985). The "South Wales" data (n=10) come from opencast records, opencast field data and Woodland and Evans (1964). The "Devon" faults (n=2) are field data from the Bude Formation at Bude. The "Baykal Rift" data (n=58) is given in Ruzhich (1977) and includes data from the Baykal and Transbaykal rift zones of the USSR, which is a seismically active region. The dividing lines on the graph are lines of equal displacement/width values. For displacements of between 3 m and 1 km, the data lie on a trend which is parallel with the dividing lines. Hence, to a first approximation, D = a.W where D is displacement, W is width and "a" is a constant with a value of about 0.01. The Ruhr data and the South Wales data plot along the same trend. However, for displacements smaller than 3 m there is a spread of points between "a" values of 0.01 and 1.0. This may be a fundamental property of small faults, but equally it may occur because the Baykal Rift data comes from poorly consolidated sediment: details of the sampled faults are not given in Ruzhich (1977). The faults with displacements of greater than 1 km plot with increasingly high displacement/ width ratios, indicating that there may be a power law relationship of the form $D = a.W^k$ where k is between 1 and 2. Ranalli (1977) found a similar power law relationship for strike-slip faults with displacements of between 1 and 100 km for which he established k = 1.17.

The displacement to width data given for English coalfield normal faults by Walsh and Watterson (1988) lies in a range of displacements between 10 cm and 10 m. The D/W ratios lie between 0.01 and 0.001 and so they are an order of magnitude lower than the range in Fig. 7(a). There is no obvious explanation for this apparent difference in the English Coalfield faults and those presented here.

2 Thrusts

Fig. 7(b) shows a displacement vs. width diagram for thrusts. The total number of data point is 36. The "Ruhr" data (n=9) come from section and maps compiled from all available mine records in Drozdzewski et. al. (1980, 1985). The "South Wales" data are the Nant Helen Nr. 1 Thrust taken from opencast and deep mine data and the Llanharan Thrusts described in Woodland and Evans (1964). The "Rockies" data is taken from the foothills of the Canadian Rockies and is given in Elliott (1976). Published maps and sections from the Southern Anthracite Field, Pennsylvannia were used for the "Appalachian" data (n=2). The Southern Anthracite Field is a Westphalian coal basin deformed in the Variscan which is comparable to the South Wales Coal Basin and the Ruhr Coalfield. The smallest thrusts represented are from the Flats Stone Band in the foreshore of Kimmeridge Bay. These may have formed by diagenetic expansion of the dolostone unit and not by tectonic strain (Bellamy 1977, Leddra et al. 1987), but in the absence of any other reliable data for small thrusts, these have been included.

The trend of the points is very similar to that for the normal faults and the D/W ratios of both graphs fall between 0.01 and 0.1. While to a first approximation the displacement is directly proportional to the width, the trend of the data points is a little steeper than the dividing lines. The gradient of the data point gives the relationship.

$$D = aW^{1.4} \qquad (3)$$

Note that the South Wales Coal Basin and the Ruhr Coalfield data again lie along the same trend.

3 Folds

In Fig. 7(c) displacement is plotted against width for folds. The displacement is estimated by taking a line length balance across the fold from between the levels at which the bedding reaches the regional. The displacement recorded is the maximum valus along a chord in the axial plane. This technique rests on the assumtion that folding has occurred by a mechanism that does not involve change in line length, such as flexural slip. While it is likely that there has been layer parallel strain in some of the folds plotted in Fig. 7(c), the errors will be much less than an order of magnitude and should not greatly affect the positions of data on a log/log plot. The width is taken as total extent of the fold axis.

The data comes from the Ruhr Coalfield (n=6) taken from Drozdzewski et al. (1980 + 1985) with other data from the South Wales opencast records, St. Quen's Bay, Jersey (Helm, 1983) and Bude, Devon.

The plot shows that, to a first approximation, the displacement is directly proportional to the width, with a constant of proportionality of about 0.03. While this is only a tentative conclusion, if correct it puts an important constraint on fold geometry as it implies that folds are scale invariant in this respect.

The sampled areas are characterised by low to intermediate strains. It is well established that in areas of high shear strain, sheath folds, with much higher D/W, are developed (see Skjernaa, 1989 for a review) and so the relationship in Fig. 7(c) is not applicable.

DISPLACEMENT GRADIENTS

An average value for the horizontal displacement gradient of normal faults can be estimated by the ratio 2D/W (Barnett et.al. 1987). As normal faults typically have aspect ratios of 2, their vertical displacement gradients are given by 4D/W. Using the displacement vs. width plot (fig. 7a), we can therefore estimate typical horizontal and vertical displacement gradients. For normal faults with displacements greater than 1 m, the horizontal displacement gradients range from 1 % to 20 %. The vertical displacement gradients are therefore 2 % to 40 %. As there is a systematic increase in D/W values, more precise estimates of displacement gradients can be made for faults of a given size.

From Fig. 7(b), the horizontal displacement gradients of thrusts range from 2 % to 20 %. Because of the high axial ratios of thrusts, vertical displacement gradients are much higher than those of normal faults, and lie between 20 % to 100 % (Childs et. al. in prep.) These high displacement gradients may reflect the difficulty of propagating a fault at low angles to the bedding anisotropy.

These figures may be used to estimate the displacement of a fault in unworked areas. However, there is almost an order of magnitude scatter in Fig. 7(a) and so the estimate will be poor. Therefore, displacement contour diagrams should be made where possible, as these provide a much more reliable method of displacement prediction for individual faults.

The displacement gradients of folds is also given by 2D/W and from Fig. 7 (c) this value is about 4 %, which is within the range of horizontal displacement gradients for normal faults and thrusts.

DISCUSSION

This work has brought to light some features of regional significance for the geology of the Ruhr Coalfield and the South Wales Coal Basin.

The displacement contour technique helps to determine the relative timing of movement on individual structures. Displacement gradient anomalies on the Hannibal Thrust coincide with

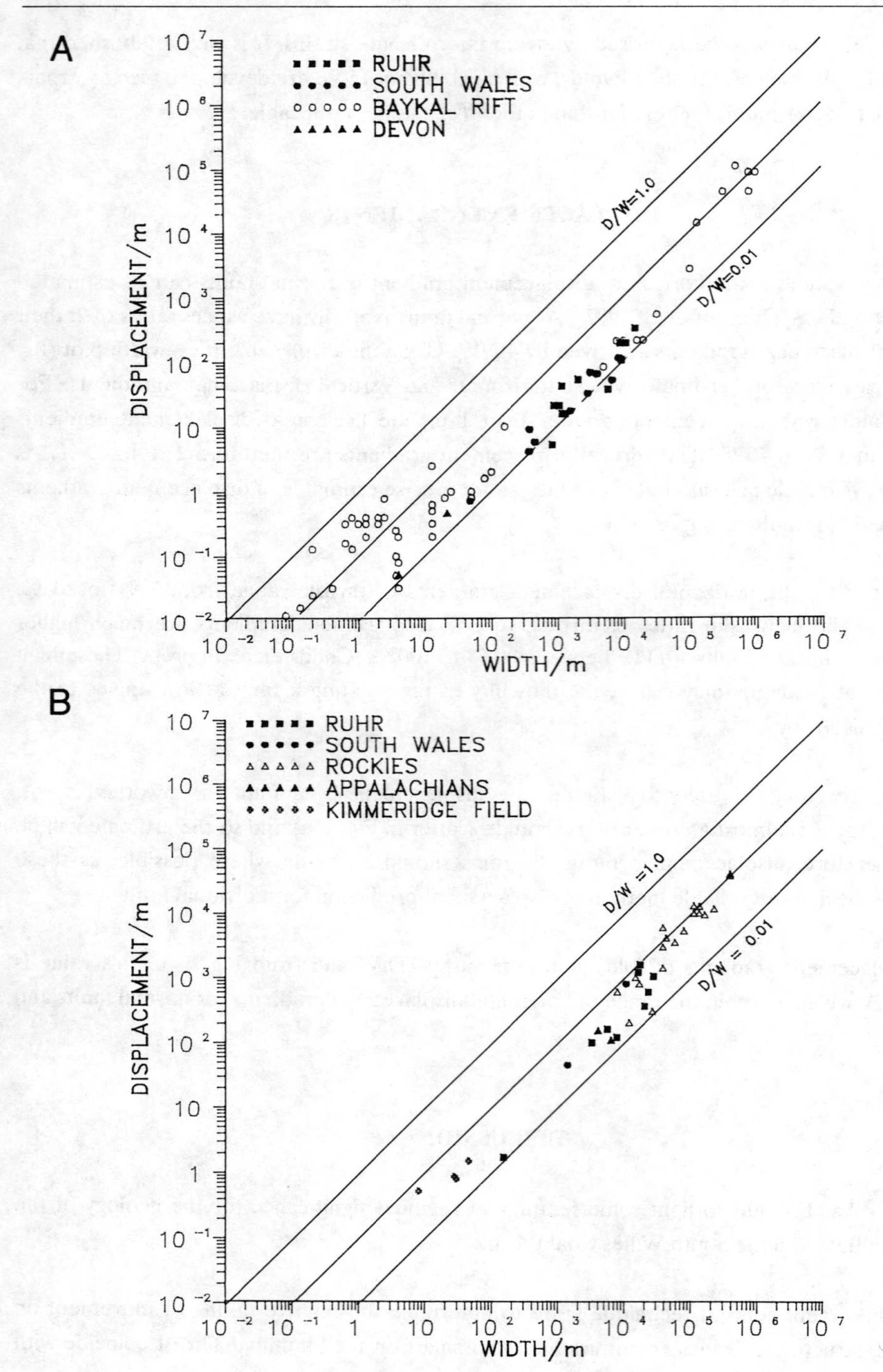
A
RUHR
SOUTH WALES
BAYKAL RIFT
DEVON
D/W=1.0
D/W=0.01
DISPLACEMENT/m
WIDTH/m
B
RUHR
SOUTH WALES
ROCKIES
APPALACHIANS
KIMMERIDGE FIELD
D/W =1.0
D/W = 0.01
DISPLACEMENT/m
WIDTH/m

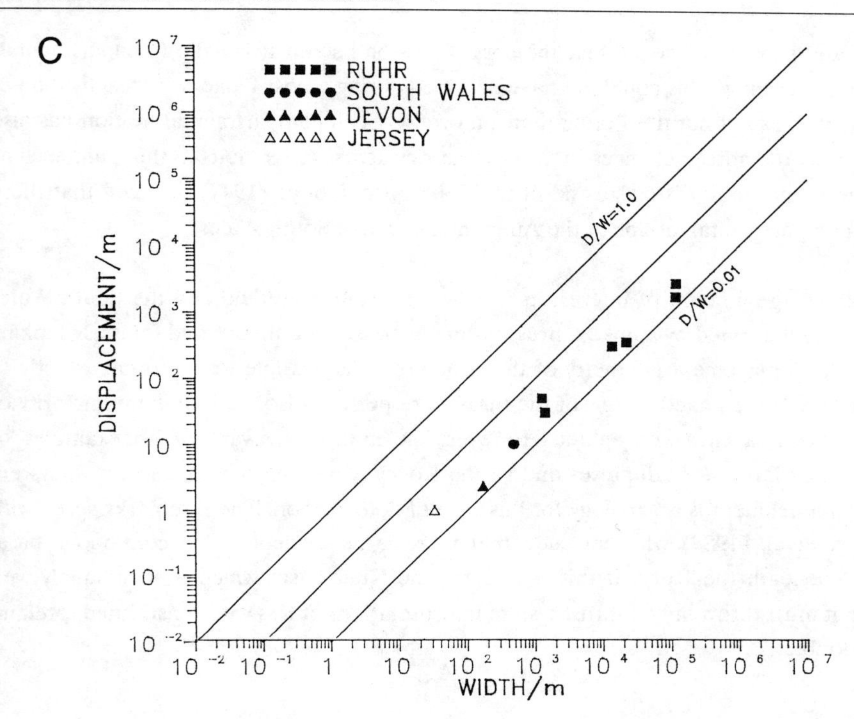

FIGURE 7
Logarithmic plot of displacement versus width for (a) normal faults, dividing lines labelled for constant displacement/width; (b) thrust faults; (c) folds.

cross faults and provide evidence that the cross faults had some strike-slip (compartmental) motion during thrusting. This could not have occurred during a post-Palaeozoic reactivation of the thrust as it does not cut the Permian unconformity. Such compartmental motion has also been inferred by the abrupt changes in fold geometries across cross faults in the Ruhr such as the Wieschermühle Fault (Drozdzewski et al. 1980). Also Trotter (1947) deduced that there was similar compartmental motion in the Ammanford area of South Wales.

There is still a large amount of interest in whether the Ruhr Coalfield and the South Wales Coal Basin were deformed by a linked thrust system or by isolated thrusts and folds. By looking at the detailed displacement geometry of the structures it is possible to determine exactly to what extent thrusts are linked. A few of the thrusts are perfectly isolated while the majority of the thrusts have branches and splays. However, the author is aware of no examples of imbricate fans or large scale duplexes and so the Rocky Mountain type linked thrust system with a basal detatchment is no analogy for this style of deformation. These remarks agree with those of Brix et al. (1988) who conclude that there is no evidence of a continuous basal detatchment beneath the Ruhr. If this is true for the Ruhr basin which is so uniquely well known, then it must throw linked thrust system interpretations of less well constrained foreland basins into doubt.

CONCLUSIONS

Two techniques have been presented for the analysis of geological structures. Fault plane contouring is a simple exercise which gives a very clear impression of the effects of a fault. It is suggested that it should become standard practice to contour faults in areas of mining interest. Furthermore displacement contouring is also useful in interpreting the regional geology of the area.

Displacement to width plots put constraints on the possible displacement gradients and therefore, along with contour maps, help in predicting the extent of partially worked faults.

Aspect ratios of normal faults range from 1.25 to 3.0 whereas those of thrusts are in the range of 2.5 to 18. Horizontal displacement gradients for normal faults and thrusts are the same, but the vertical gradients on thrusts are generally very much higher.

Fold axial planes can be contoured for amplitude and it can be shown that folds show similar horizontal displacement gradients to faults.

Acknowledgements

The author would like to thank Reinhard Greiling, Volker Wrede, Rodney Gayer for their help and encouragement. John Walsh provided many helpful criticisms of the manuscript. Special thanks to Steve Townsend for his contribution to Fig. 2. This work was carried out under a NERC studentship with the cooperation of British Coal.

REFERENCES

Barnett, J.A.M., Mortimer, J., Rippon, J.H., Walsh, J.J., Watterson, J. 1987. Displacement geometry in the volume containing a single normal fault. Bull. Ass. Petrol. Geol. 71, 925-937.

Bellamy, J. 1977. Subsurface expansion megapolygons in upper Jurassic dolostone (Kimmeridge, U.K.): J. Sed. Pet. 47, 973-978.

Brix, M.R., Drozdzewski, R., Greiling, R., Wolf, R., Wrede,V. 1988. The north Variscan margin of the Ruhr coal district (Western Germany): structural style of a buried thrust front? Geologische Rundschau 77, 115-126.

Childs, C., Gillespie, P.A. Walsh, J.J. & Watterson, J. (in prep.) The displacement geometry of thrust faults.

Drozdzewski, G., Bornemann, O., Kunz, E., Wrede,V. 1980. Beiträge zur Tiefentektonik des Ruhrkarbons. Geologisches Landesamt Nordrhein-Westphalen, Krefeld.

Drozdzewski, G., Engel, H., Wolf, R., Wrede, V., 1985. Beiträge zur Tiefentektonik westdeutscher Steinkohlenlagerstätten. Geo-logisches Landesamt Nordrhein-Westfalen, Krefeld.

Elliott, D. 1976. The energy balance and deformation mechanism of thrust sheets. Phil. Trans. R. Soc. Lond. A. 283, 289-312.

Ellis, M.A. & Dunlap, W.J. 1988. Displacement variation along thrust faults: implications for the development of large faults. J. Struct. Geol. 10(2), 183-192.

Gayer, R.A. & Greiling, R.O. 1990. Comparative evolution of coal bearing foreland basins along the Variscan northern margin in Europe (this volume).
Helm, D.G. 1983. The structure and evolution of the Jersey Shale Formation, St. Quen's Bay. Jersey, Channel Island. Proc. Geol. Ass. 94(3), 201-216.

Leddra, M.J., Yassir, N.A., Jones, C. and Jones, M.E. 1987. Anomalous compressional structures formed during the diagenesis of a dolostone at Kimmeridge Bay, Dorset: Proc. Geol. Ass. 98(2), 145-155.

Pfiffner, O.A. 1985. Displacements along thrust faults. Eclogae Geol. Helv.. 78(2), 313-333.

Ramsay, J.G. 1962. The geometry and mechanics of formation of "similar" type folds. J. Geol. 70, 309-327.

Ranalli, G. 1977. Correlation between length and offset in strike-slip faults. Tectonophysics, 37, T1-T7.

Rippon, J.H. 1985. Contoured patterns of throw and hade of normal faults in the Coal Measures (Westphalian) of north-east Derbyshire. Proc. Yorks. Geol. Soc. 45(3), 147-161.

Ruzhich, V.V. 1977. Relations between fault parameters and a practical application of them. In Mekhanizmy Structur Vostochnoisibiri Novisbirsk Nauka (in Russian).

Skjernaa, L. 1989. Tubular folds and sheath folds: definitions and conceptual models for their development, with examples from the Grapesvare area, northern Sweden. J. Struct. Geol. 11(6), 689-703.

Trotter, F.M. 1947. The structure of the Coal Measures in the Pontardawe-Ammanford area, South Wales. J. Geol. Soc. Lond. 103, 89-133.

Walsh, J.J. & Watterson, J. 1988. Analysis of the relationship be-tween displacements and dimensions of faults. J. Struct. Geol., 10(3), 239-247.

Williams, G. & Chapmann, T. 1983. Strains in the hangingwalls of thrusts due to their slip/propagation rate: dislocation model. J. Struct. Geol. 5(6), 563-571.

Wood, G.H.Jr., Trexler, J.P. and Kehn, T.M. 1969. Geology of the westcentral part of the Southern Anthracite Field and adjoining areas, Pennsylvania. Geol. Survey Prof. pap. 602.

Woodland, A.W. and Evans, W.B. 1964. The geology of the South Wales Coalfield. Part IV. The country around Pontypridd and Maesteg. Memoir Geological Survey U.K. (London, H.M.S.O.).

Variscan Thrust Deformation in the South Wales Coalfield - a Case Study from Ffos - Las Opencast Coal Site

K.Frodsham[1] / R.A.Gayer[1] / E.James[2] / R.Pryce[2]
1 Department of Geology, University of Wales, PO Box 914 Cardiff CF1 3YE, UK
2 British Coal Corporation, Opencast Executive, South Wales Region, Farm Road, Aberaman, Aberdare, Mid-Glamorgan CF44 6LX, UK

ABSTRACT

Detailed structural analysis of 2979 borehole logs and field observations in the developing Ffos Las opencast coal site in the northwest of the South Wales Coalfield has revealed a complex interaction of fold and thrust structures close to the intersection of two high strain zones: the E-W trending Trimsaran Disturbance; and the ENE-WSW trending Llannon Disturbance. The borehole analysis indicates that these disturbance zones are low angle thrusts separated by northward verging folds and thrusts.

Investigations of temporary N-S cross-strike sections within the site has allowed a model of the deformation to be determined. An initial phase of bed-parallel deformation developed low angle thrusts and associated imbricate rashings bands and cleavage duplexes with northwards vergence. These structures developed simultaneously along several of the coal seams within the Middle Coal Measures sequence, causing marked thickening and thinning of the coal seams. As the tip folds of the lower thrusts propagated upwards, higher thrusts were folded causing them to lock-up. Further development of the higher thrusts continued in a break-back sequence with the thrusts cutting folded bedding at a high angle. This style of deformation produces a characteristic but complex thrust geometry in which thrusts cut both up and down stratigraphy in the movement direction. The vertical thickening of the deformed sequence was accommodated upwards by underthrusting of more competent strata higher in tne sequence, producing a passive roof duplex. It is argued that the model for deformation at Ffos Las is typical for coal-bearing sequences deformed in orogenic forelands and applies to the Variscan deformation in much of the South Wales Coalfield. The belts of high strain - disturbance zones - could represent either reactivation of syn-depositional extensional faults, or major thrusts imbricating from a basal decollement beneath the coalfield, or buttressing of Variscan stress against basement structures.

INTRODUCTION

The South Wales Coalfield is an elongate E-W basin area of approximately 2200 km^2, lying at the northern margin of the Variscan orogenic belt (Fig. 1). Recent basin analysis has suggested that the coalfield formed as a foreland basin (e.g. Kelling 1988, Jones 1989, Gayer and Jones 1989) and that much of the deformation in the coalfield is related to the foreland progagation of the Variscan fold and thrust belt of SW England (e.g. Coward and Smallwood 1984, Gayer and Jones 1989). The basin is filled with 3-4 km of Upper Carboniferous (Namurian - Westphalian D) sediments, resting unconformably on Dinantian platform limestones.

The basin fill consists predominantly of marine Namurian shales and sandstones, passing up into mainly argillaceous coal-bearing Lower and Middle Coal Measures of Westphalian A-C age, in which marine influence progressively diminishes. The overlying Upper Coal Measures Pennant Sandstone, of Late Westphalian C-Early Stephanian age, consists of thick units of medium to coarse grained fluvial sandstones with only minor argillaceous interbeds and with no marine incursions (Fig. 2). Thus the coalfield can be regarded as being divided into three major rheological units; soft argillaceous and coal bearing strata sandwiched between the two strong units of the Dinantian Limestones below and the Pennant Sandstone above. It has been suggested (Woodland and Evans 1964, Archer 1968) that this arrangement has had a market affect on the deformation style within the coalfield.

The major structure of the coalfield consists of an E-W syncline, strongly asymmetric towards the north (Fig. 1). Within the productive Lower and Middle Coal Measures of the coalfield northward verging thrusts and folds occur within the gently south dipping limb of the syncline, whilst south verging folds and thrust are developed in the more steeply northerly dipping limb. Jones (1989) has suggested that the asymmetry and the southward verging structures of the southern limb of the coalfield syncline are the result of a major passive roof duplex below the Pennant Sandstone, forming a mountain front structure similar to that described from the Canadian Rockies (e.g. Price 1986). In addition three major and several minor ENE-WSW trending structures cross the syncline. These have been termed disturbance zones because they variously involve low angle thrusts, folds and steep strike-slip faults. They are developed with greater complexity and frequency in the northwestern part of the coalfield. Although their trend is oblique to the normal E-W to WNW-ESE trending Variscan folds and thrusts, there is evidence that the two sets formed simultaneously (e.g. Strahan 1907, Owen 1953), however, some strike-slip movement along ENE-WSW trending structures might be as late as Tertiary (Trotter 1947, Archer 1968). It has been generally argued that the ENE-WSW structures are the result of northward directed Variscan compression against ENE-WSW trending Caledonian structures in the basement (e.g. Trotter 1947). Other structures of note in the coalfield are Lag Faults, low angle normal faults that lie parallel and apparently related to thrusts (Trotter 1947, Archer 1968), and cross faults, high angle faults trending between NW-SE and N-S with downthrows to either the east or west. Although these are important structures in the coalfield, this work contributes little additional information to their understanding.

Ffos Las is a large, currently operational (1990), opencast coal site (OCCS) (c. 2 km by 2 km and 150 deep) (Fig. 3 + 4) lying in the northern limb of the coalfield syncline, close to its western margin and approximately 7,5 km southeast of the Carreg Cennen Disturbance, one of the principal ENE-WSW structures. The particular significance of the site is that it lies at the intersection of two mayor structures: the ENE-WSW trending Llannon Disturbance, and the E-W trending Trimsaran Disturbance. It thus provides an opportunity to understand the relationship between typical Variscan structures and those trending obliquely. Neither of these structures is well defined by previous workers (see e.g. discussion in Archer 1968); exposure in this part of the coalfield is poor and control from mine workings is largely inadequate to describe the structures unambiguously. In the Pennant Sandstone outcrop, the Llannon Disturbance is a northward verging thrust and associated fold (Strahan et al. 1907, Archer 1968) with a net downthrow of c. 60 m to the north. However, in the more easily deformable Middle Coal Measures the structure is more complex and was temporarily exposed in the Carway OCCS (1953-1957) where several strongly asymmetric folds and thrusts were developed. Unfortunately no estimate of the total displacement was possible. The Trimsaran Disturbance is even more poorly constrained, probably because it produced a major zone of deformation that effectively terminated mining operations when encountered and thus was never fully observed in deep mine workings. The levels of specific coal seams in workings to the north and south of the disturbance suggest a total northward downthrow of c. 350 m (Fig. 5), (Archer 1968), and it was suggested by Strahan et al. (1909) that the disturbance was not a single fault "but a belt of plication and overthrusting".

Woodland and Evans (1964) and Archer (1968) argued that the highly variable structures developed within the incompetent Lower and Middle Coal Measures are largely accommodation structures, involving ductile flow, and bearing little relationship to the structures formed in the overlying competent Pennant Sandstone or to the regional structural pattern of the coalfield. Our work in these largely incompetent units at Ffos Las indicates that, although the structures are complex, they are not random or chaotic but rather appear to show a systematic polydeformational sequence. However, in many respects these thrust related structures do not follow the standard "rules of thrusting" (e.g. Boyer and Elliott 1982, Butler 1982), but show features characteristic of this coal-bearing foreland basin style of deformation. It is the purpose of this paper to describe the main elements of the structure at Ffos Las OCCS and, by analysing the geometrical relationships of these structures, to understand the processes involved in thrusting within the incompetent Middle Coal Measures and the relative timing and sequence of the various structural elements. Finally we discuss the significance of the particular style of deformation at Ffos Las OCCS for the structural development of the coalfield as a whole.

DEFORMATION AND SMALL SCALE STRUCTURE OF FFOS LAS

The rocks exposed at Ffos Las are entirely within the Middle Coal Measures and are dominantly finely-bedded, often organic rich mudstones and siltstones, with coals and less common sandstones. Coal comprises about 5 % of the total sequence and provides the basis for correlation both within the site and with surrounding areas (Fig. 2). The disposition of the seams prior to deformation appears to have been essentially regular with the exception of the sequence beneath the Big seam, the main production seam. Hence, in the north of the site the sequence is identical to that at the adjoining former Carway OCCS (Fig. 3), with the Green, Ddugaled and Hwch seams underlying the Big. In the south of the site, above the Llannon thrusts, the Green seam is absent and the Ddugaled and Hwch seams are much closer to the overlying Big. The Green seam was apparently present in the old trimsaran Colliery to the south, suggesting that the sequence is only locally thinned. This may be due to primary sedimentation or to the presence of a very low angle extensional (lag) fault, which are common structures in the Gwendraeth Valley area. The required low angle of dip for this extensional fault suggest that it may have been active whilst the sediments were still unconsolidated, possibly during sedimentation. Another line of evidence, again indicating possible early tectonic activitiy, comes from exploration by British Coal, suggesting that a significant axis of sedimentation lies within the Ffos Las area, producing thickness changes across the site.

The rock sequence at Ffos Las is pervasively deformed by low angle thrusts and associated folds. The intensity of deformation is not uniform, however, but is strongly influenced by the relative competence of the strata. The deformation is achieved dominantly by brittle failure, with the competence of the strata controlling the eventual intensity of fracturing. Hence, while sandstones and many siltstones often suffer only mild fracturing, the remainder of the sequence may undergo very intense small-scale fracturing and/or micro-faulting to produce a variety of small-scale structures, typical of Coal Measure deformation in South Wales.

Micro-fracturing and faulting of the generally organic rich incompetent rocks causes them to become very weak and friable, often with pervasive slickensiding of the whole rock. Slickensided surfaces appear to form relatively easily in the deformation of these organic rich rocks.

Microfaulting of incompetent rocks produces two distinct classes of small-scale structures typical of deformed coal-bearing strata. These are 'rashings' and cleavage duplexes. Rashings is a long-standing mining term for sigmoidally sheared rocks, where bedding planes are displaced along the sigmoidally shaped fractures oriented obliquely to the regional bedding attitude. Bands of rashings occur on a wide range of scales from centimetres up to several metres thick, but often have variable thickness due either to truncation beneath overlying thrusts or to variation in internal displacements causing doming of the roof rocks, analogous to duplexing. Deformation within rashings bands appears to be dominantly by layer-parallel shortening, with a variable amount of layer-parallel shear, which can produce an overprint of classic shear zone fabrics (Fig. 6a + 6b). This deformation is consistent with the field observation that rashings bands usually occur either along or in the immediate footwall to large-scale thrusts.

Structures similar in appearance to the small-scale cleavage duplexes described by Nickelsen (1986) occur, usually in the immediate roof or floor of coal seams. These are similar in appearance to rashings but the sigmoidal surfaces are very evenly spaced secondary slickensided cleavage surfaces rather than bed-parallel fractures, and these cleavage duplexes rarely exceed 50 cm in thickness (Fig. 7a + 7b). In contrast to Nickelsen's observation, the structures seen at Ffos Las do not occur along fault planes, but merely appear to accomodate deformation due to strong layer-parallel shear above and below relatively rigid coal seams. We see no evidence for a separate component of layer-parallel shortening preceding shearing and we envisage the cleavage forming as a response to the shearing process. These structures are thus small shear zones.

Rashings and cleavage duplexes are probably end-members of a discontinuous series of small-scale structures, where final appearance depends upon the relative magnitude of layer-parallel shortening and layer-parallel shear. Thus sigmoidal cleavage and well developed shear zone fabrics are common within rashings bands, suggesting high components of layer-parallel shear. However it is not possible to determine any component of layer-parallel shortening within the cleavage duplexes, and this may be the explanation for their distinctive appearance.

Another lithological control on the pattern of deformation at Ffos Las is the control on thrust geometry. Hence thrusts usually dip at very low angles (less than 10½) to bedding, where it has not been previously deformed, and frequently follow very long flats along the horizons of coal seams (see below, and Fig. 8). In addition to the usual behaviour of thrusts in following ramp/flat trajectories in well bedded sequences, they seem to be especially attracted to the coal seams. This is thought to be due to the easy slip horizons provided by coal/seat earth contacts, probably the weakest planes in the rock pile, combined with overpressuring of fluids within the coals. These fluids could either be aqueous, associated with a hydrodynamic cycle in the developing orogen, or gases (particularly methane) originating form the maturation of the coals. The molecular structure of anthracite with the development of an incipient [002] graphite plane could also facilitate slip.

LARGE SCALE STRUCTURE OF FFOS LAS

Only a relatively small volume of the proposed excavation was extracted during the period of the study, thus limiting the scope of the field investigation. However, data from 2979 exploration boreholes, representing a total of some 300 km of drilled rock, was used to deduce the gross structure of the site (Fig. 9). It is clear that the detailed structure, as determined by field observations, is more complex than is suggested from the borehole evidence alone, as was made clear by the fact that it was sometimes impossible to correlate between boreholes although some were drilled only 5 m apart in disturbed zones.

A typical cross-section contructed from borehole data reveals undeformed rocks only in the extreme southeast of the site where a continuous profile from the Big seam up to the Cefn Coed Marine Band occurs. The major structure is best illustrated by tracing the Big seam from south to north. The undeformed section lies in the hangingwall to a major zone of low angle thrusting, the Trimsaran Disturbance. To the north of this zone the Big seam passes around a series of northerly verging folds, which lie in the hangingwall of a second zone of major low angle thrusts, the Llannon Disturbance, which probably dips at a shallower angle than the overlying Trimsaran Disturbance. In addition to these major zones of deformation, small-scale thrust deformation is reflected in the irregularity of seam thickness. Evidence for the nature of the Trimsaran Disturbance is limited to boreholes. The sequence exposed to date has revealed only folds and thrusts that are thought to be associated with the Llannon Disturbance.

GEOMETRIC CHARACTERISTICS OF DEFORMATION

Stereographic projections of all planar and linear data obtained in the field were used to analyse the basic orientational characteristics of the deformation. Poles to bedding, divided into sub-areas are shown in Fig. 10a, and combined across the whole area in Fig. 10b. Fig. 10 a - e shows poles to thrusts, fold axes and poles to cleavage duplexes respectively. Variations in bedding orientation at different stratigraphic horizons are shown in Fig. 10 f.

The plots reveal:

1) There is a dominant trend to the fold axes across the site (plunging 10½ to 091½), which although varying locally and from east to west, shows no systematic trend. Field work shows

that anomalous fold orientaions (e.g. see Fig. 10 d) are usually related to lateral variations in thrust displacement. There is little evidence to suggest a dominant east-northeasterly orientation to the fold axes, as would be expected for the Llannon Disturbance. Variations in orientation of fold axes appear to result from variations in the initial dip of the folded strata, rather than variations in the orientation of the associated thrusting. Thus the steep northward facing limbs of the folds have a constant orientation across the site, and the variation in the fold orientation is caused by changes in the orientation of the more gently dipping southward facing limbs. The origin of the 10½ E plunge remains unclear.

2) The orientation (strike and dip) of the thrust planes (Fig. 10 c) shows a very similar pattern to that of the bedding planes, suggesting that the thrusts do not cut laterally across stratigraphy to any great extent and that they are generally at a low angle to bedding. It also suggests that the thrusts are folded together with the bedding, implying an early pre-folding development of the thrusts. These deductions are partly borne out by field observation, but the similarity in orientation masks a far more complex relationship between bedding and thrust geometry (see below).

3) Evidence for thrust transport direction is best provided by the orientation of the steep limbs of the thrust-propagation folds and the shear direction within cleavage duplexes and rashings bands (Fig. 10 e). Slickenside striations are very irregular, with multiple development on thrust surfaces and divergent lineations on adjoining shear surfaces, thus making them very difficult to interpret as large scale directional indicators. The folds suggest movement towards 001½ and cleavage duplexes towards 002½, provided that the fold plunge is an early or syn-deformational feature and has not resulted from late-stage tilting.

4) Progressing up stratigraphy from the Green to the Penny Pieces Seam and above, the dips generally become less steep and also more easterly directed. This appears to be a result of a lower intensity of deformation in the higher parts of the sequence (Fig. 10 f).
The stereographic analysis indicates a fairly uniform distribution of fold and thrust orientations, resulting largely from a quite constant movement direction throughout the deformation. This precludes the recognition of both distinct types and differently aged folds and thrusts. The simple geometry, in fact, masks a complex pattern of thrust development and fold interaction, which characterises the Ffos Las site. This complexity is best illustrated by three cross-sections temporarily exposed in the site, and is supported by numerous observations of smaller-scale features exposed as isolated structures during the coal extraction in the developing site.

Cross-Section B (Fig. 11)

This west-facing temporary sidewall (location shown in Fig. 4) revealed a thrust-thickened section between the Big and overlying Penny Pieces Seam. This section is essentially that shown in figure 3A of Salih & Lisle (1988). A number of features typical of the deformation at Ffos Las are apparent. The thrusts are typically at a low angle to bedding, frequently forming long bed-parallel flats and show a characteristic break-back sequence of propagations, best seen in the north of the section. Hence, thrust B truncates the fold structures in its footwall and, likewise, thrust C truncates fold structures seen in thrust slice B. This is not consistent with a simple foreland propagating piggy-back thrust sequence, but rather a foreland directed break-back sequence: A before B before C before D. All these thrusts are folded around a large northerly verging thrust-propagation fold, which has produced classic downwards facing fold and thrust structures in the steep limb. As all thrusts are folded by the structure, it might be expected that the folding was a later phase of deformation. However, evidence from the synclinal region in the higher thrusts around the Penny Pieced Seam (C, D and probably B + D) suggests that the syncline was already forming when the thrusting was initiated. Thus, there is evidence for the simultaneous movement of thrusts on different levels within the site.

Cross-Section A (Fig. 12)

An embayment cut into the sequence in the extreme west of the site (Fig. 4) exposed two comparable across-strike sections, one east and the other west facing, 150 m apart through a lower part of the northerly verging fold structure of the site. In this region the Big Seam was overthrust by a sequence in which, although the stratigraphy was poorly defined, the Hwch Seam was repeated several times. The embayment was almost as deep as it was long and consequently the area reveals important information on both the vertical relationships and along strike variations in the deformation. The sections display a sequence of thrust slices bounded by low angle southerly dipping thrusts. Within these slices the rocks are often strongly deformed in a manner that is not consistent simply with their emplacement along the bounding thrusts. We suggest that these structures were formed by, often pervasive, thrust related deformation prior to emplacement into their current positions above the bounding thrusts. This interpretation would be consistent with the deeper level, low angle thrusts being late stage structures, related to the Llannon Disturbance. As in cross-section B, all the footwall structures appear to be truncated by the low angle thrusts, suggesting a potential break-back sequence for their development. One anomalous feature of the thrust deformation in this region is that it is almost impossible to match the hangingwall cut-offs with footwall cut-offs across the thrusts, probably due to the complexity of the early deformation.

The sections also show a variety of small-scale structures, such as: downward facing folds and thrusts in the steep, north facing major fold limbs; small backthrusts; and duplexing of rashings layers. These small-scale structures are frequently not continuous along strike, such that the detailed structure of the two sections is not comparable (cf. Figs. 12 a + b with 12 c + d). This may be the result of initial highly non-cylindrical structures, in part related to lateral and oblique ramps along the thrusts, or to the later thrusts cutting obliquely through the earlier thrust related deformation. The relatively constant strike of bedding and thrusts revealed in the stereographic analysis (Fig. 10) suggests only gentle non-cylindrism of the structures and thus tends to support the latter alternative.

Cross-Section C (Fig. 13)

A second west facing section has recently (August 1989) been developed with continuous exposure for over 500 m and a maximum height of circa 40 m, in the east of the site (Fig. 4). The section shows a complexly deformed sequence between the Big and the Graigog Rider Seams, which illustrates well the true complexity of thrusting and fold thrust interactions at Ffos Las. The principal thrusts and folds shown in Fig. 13 have been annotated, and their temporal relationships, deduced from cross-cutting geometries, have been indicated in the caption to the figure. A number of important points arise from the section.

1) Thrusts and folds occur on a variety of scales with displacements from microscopic up to in excess of 50 m. Small-scale potentially isolated thrusts occur throughout the sequence, best seen in the Big Rock sandstone.

2) The early-formed thrusts are all at a very low angle to bedding, behaving almost like detachments along coals, with very long flats und short ramps. The later the thrust the more steeply it cuts across bedding. This appears to be solely the result of the prior deformation of the bedding, as thrusts seem to form at a similar angle to datum throughout. This process would explain why many of the early structures in the site appear to show dominantly bed-parallel deformation, whereas the latest structures show relatively little relationship to bedding.

3) The fabric of sheared rocks in most of the rock sequence is northward directed, but the Graigog horizon in this section shows a strong southward directed shearing component (Fig. 6b). This fabric consists of: sigmoidally shaped cleavage dipping up to 36½ NW; the development of sheared 'fish' of more competent siltstone material, showing a southward directed shear; and the crude development of a C-S fabric, again indicating southward shear. The shearing has produced a strange thickening and thinning pattern to the seam. In the north of the section the coal is absent and the rashings layer consists mainly of seat earth. To the

south the coal shows a gross southward directed thickening. This is considered to be direct observational evidence for the phenomenon expected elsewhere in the site, where strong shearing along bed-parallel thrusts has led to thinning and occasional disappearance of coal, which has been detached along a thrust in the seat earth.
The southward directed shearing and potential overthrusting is thought to be a result of wedging of the thickened northward directed thrust sequence below more competent overlying strata, forming a structure similar to a passive roof duplex (Banks and Warburton 1986). This would help to explain why much of the northward verging deformation so characteristic of the sequence below the Soap Seam is absent in the overlying seams. It is apparent that the wedging was an early feature of the deformation, as one northerly verging thrust (T_E in Fig. 13) breaches the southward directed roof thrust (T_A in Fig. 13) and the upper seams, and northerly directed fabrics often overprint the southerly directed ones. Although we have not found direct evidence in the site, it seems likely that as thrusting breaks back to higher structural levels, so the passive roof wedging will also move to higher stratigraphic levels.

4) The thrusts within this section show complex temporal relationships both with one another and with the associated folds (Fig. 13). Amongst these relationship there is evidence for movement of thrusts before, during and after specific episodes of fold formation, and further evidence for break-back propagation of thrusts. The structures also suggest that no folds are continuous up or down sequence due to truncation by later thrusts above and to propagation of the thrust through its tip fold below.

5) The Big Seam is greatly thickened in the region where the larger scale thrusts pass through the coal (T_G in Fig. 13). It was noted in an earliert section that the thicker coal seams, especially the Seam, are very irregular in thickness due to small-scale thrusting, known to the miners as lapping (Fig. 14). In this example the coal appears to be accommodating large displacements of the overlying, and presumably underlying, sequence, by gross thickening along numerous small-scale thrusts. Even within these imbricate/duplex zones the thrusts seem to generate in a break-back fashion. Few regions of thickened coal are as well exposed as this example, but it is often hard to demonstrate any relationship with large-scale thrusts. Many of these thrusts, or coal duplexes, are thought to have developed early in the deformation sequence as isolated structures, originating under the influence of local high stresses. The fact that much of this coal thickening was early in the sequence could again be due to the relative attitude of bedding during deformation rather than any major change in the imposed deformation mechanism.

6) Small-scale thrusts commonly develop as accommodation structures to folds. Again, it is the coal seams which show the greatest displacements, suggesting the occurrence of strong shear strains along coal/seat earth contacts.

THRUST SEQUENCE AND MODEL FOR THE THRUST EVOLUTION AT FFOS LAS

From the evidence presented in the three cross-sections (above), it will be evident that the intensity of thrust deformation precludes the possibility of determining the relative timing of all thrusts within the site. Evidence for the thrust sequence must come mainly from sections like those described, together with evidence of the major structures from borehole investigations.

Early deformation appears to be dominantly bed-parallel, leading to much small-scale overthrusting and bed-parallel shearing (cleavage duplexes) with thrusts lying at a low angle to bedding except where they have propagated through their own tip folds. Initially the northward directed movement of the more incompetent strata was counteracted by southward directed shearing along higher coal seams that acted as detachments. With continued northward movement of the sequence as a whole the passive roof to the deformation was breached by one or more northerly directed thrusts imparting a northward directed shear to the sequence.

All the evidence collected to date implies that the northward directed thrusts developed in a break-back sequence, producing an obvious vertical thickening of the deformed stratigraphy. Thus, with increasing deformation, thrusts tended to lose their affinity with bedding and cut across, often strongly, folded strata, but maintaining a relatively constant thrust dip. The deformation cannot be a simple break-back sequence, as higher level thrusts are often folded by movements on lower level thrusts. It would be possible to propose a whole series of northward directed pulses of break-back deformation to account for this style, but we consider a model whereby movement occurred simultaneously at different stratigraphic levels. Evidence for this simultaneous movement has been presented in cross-sections B + C. We visualise a model whereby movement at different levels, usually concentrated around coal seams, leads to the folding and consequent 'locking-up' of higher level thrusts, which leads to the generation of later thrusts in the hangingwall in break-back fashion (Fig. 15).

The role of the traditional LLannon and Trimsaran Disturbances is uncertain. Although deformation of the Big Seam is much more pervasive than is apparent from a borehole section (Fig. 9), the two disturbances do appear to represent spatially distinct zones of major low angle thrusting. The relative timing of the disturbances awaits exposure, but borehole sections show folding of thrusts arising close to the Trimsaran Disturbance, suggesting that the structure is relatively early. This is in contrast to a continuous break-back sequence of deformation. We envisage the break-back style of deformation as representing footwall collapse of the Trimsaran Disturbance. Later deformation was probably dominated by the low angle thrusts and related folds of the Llannon Disturbance, which could represent a footwall splay of the

Trimsaran Disturbance. Thus the site may have been affected by two overlapping pulses of deformation. However, due to the similarity of the structures produced during each pulse, the problem of proving distinct episodes of deformation is likely to be difficult.

SIGNIFICANCE OF THE MODEL FOR DEFORMATION IN THE SOUTH WALES COALFIELD

The deformation recorded at Ffos Las, although considerably more intense and complex, has many features in common with the deformation style observed at many other sites in the South Wales Coalfield. The development of cleavage duplexes and imbricate rashings bands along coal seams is common throughout almost the entire coalfield, and these structures are frequently associated with low angle thrusting very similar to the style of the early layer parallel thrusts at Ffos Las. In several areas out of sequence thrusts have been recorded (Jones 1989) that may well have a similar break-back origin to those at Ffos Las. Finally the development of southward directed shear as a passive roof duplex to the northward verging and vertically thickened deformed stack is analoguous to the model suggested by Jones (1989) for the southward verging thrusts zone (Woodland and Evans 1964) along the southern margin of the coalfield.

The model for deformation at Ffos Las, we suggest, may thus be applicable to much of the remainder of the coalfield. Initially deformation migrated northwards into the rheologically weak Lower and Middle Coal Measures strata by the development of a system of independent layer-parallel thrusts and shears. These tended to be located along coal seams where high fluid pressures (including methane gas) facilitated slip. The evidence from Ffos Las suggests that these thrusts moved simultaneously and not in a foreland propagating piggy-back style. As thrusting progressed, so a thickened deformed wedge built up, resulting in underthrusting of higher, more competent units, forming a passive-roof duplex. Where strain was greater, complex thrust/fold interactions developed such as that at Ffos Las and along several other 'compression belts' (Trotter 1947) and 'zones of disturbance' (Archer 1968).

It is not known what controlled the build up of strain along particular belts. It may have been due to reactivation (inversion) of early syn-depositional extensional structures such as that inferred for the Ffos Las site. However, the intensity of subsequent contractional deformation is such that it would be difficult to identify the earlier extensional fault. Where early lag faults have been recognised e.g. the Pentremawr Lag Fault in the Cynheidre area (Archer 1968 and British Coal unpub. data), the fault does not appear to have been reactivated as a thrust, although Trayner et al. (1986) suggested that the Jubilee Slide in the Maesteg area may represent such a reactivation. Trotter (1948) suggested that the Carreg Cennen Disturbance was the outcrop of a major thrust detachment that underlies the whole coalfield. Although the existence of Trotter's major thrust has never been substantiated and remains enigmatic, both the Trimsaran and Llannon Disturbances at Ffos Las clearly represent relatively major thrusts and could be imbricates from a basal detachment. A third alternative is that the high strain zones are the result of buttressing against basement structures. The ENE-WSW trend of several of the disturbances could then be explained as rectivation of a Caledonian basement (e.g. Trotter 1947). The WNW-ESE trend of the Trimsaran structure coincides with a ridge of crystalline basement defined by seismic refraction surveys by Mechie & Brooks (1984). The ridge presumably represents an upthrusted horst of basement. Unfortunately the only supporting evidence for this explanation is the coincidence of trend and location. If basement structures have controlled the position of high strain zones, the low angle thrust of the Trimsaran and Llannon Disturbances would have to represent footwall shortcut faults (e.g. Powell 1987).

CONCLUSIONS

1. A model for thrust deformation at Ffos Las involves the following five stages:

I) Early deformation in the Middle Coal Measures at Ffos Las is represented by low angle thrusts and associated tip and propagation folds, imbricate rashings bands and cleavage duplex development, all with a northwards vergence and thrust transport towards 001½.

II) This early deformation was preferentially developed simultaneously along several coal seams that represent rheologically weak, easy-slip horizons, with high fluid pressures (including methane gas).

III) Marked thickening and thinning of coal seams is directly linked to the early layer-parallel deformation.

IV) Progressive deformation resulted in the tip folds of lower thrusts deforming higher level thrusts, causing them to lock-up. Continuing movement on the higher level thrusts gave rise to break-back thrusting.

V) As the complex interaction of folds and thrusts developed, a thickened wedge of deformed strata built up and underthrust higher level less deformed competent strata in a passive-roof duplex.

2. The model for thrust deformation at Ffos Las differs from the foreland propagating, piggy-back style of thrust development in three important ways:

I) Thrusts are developed at several levels in the stratigraphy simultaneously, thus there is no normal sequence of thrust propagation.

II) Thrusts that are folded by movements along a lower thrust are not carried passively, but give rise to break-back propagation at the higher level.

III) Thrusts cut both up and down stratigraphy in the movement direction.

3. The model for thrust deformation at Ffos Las probably applies to much of the Variscan deformation in the South Wales Coalfield.

4 High strain zones in the coalfield could represent:

I) Reactivation of syn-depositional extensional faults.

II) Imbrication from a basal detachment beneath the coalfield.

III) Footwall shortcut faults derived from buttressing against basement structures.

Acknowledgements

KF gratefully acknowledges financial and logistic assistance from British Coal Opencast Executive. The field work has been partly supported by a NERC research grant (GST/02/350) to RAG under the Special Topic: Basin Dynamics. The authors wish to thank the staff at Ffos Las for encouragement and assistance on site, in particular Roger Evans, Renga Puia and Morgan Richards. We also thank Margaret Millen for drafting many of the diagrams.

REFERENCES

Archer, A.A. 1968. Geology of the South Wales Coalfield: the Upper Carboniferous and later formations of the Gwendreath Valley and adjoining areas. Memoirs of the Geological Survey of Great Britain. 216pp.

Banks, C.J. and Warburton, J. 1986. 'Passive-roof' geometry in the frontal structures of the Kirthar and Sulamain mountain belt, Pakistan. Journal of Structural geology. 8, 229-238.

Barclay, W.J., Taylor, K. and Thomas, L.P. 1988. Geology of the South Wales Coalfield, Part V, the country around Merthyr Tydfil (3rd edition). Memoirs of the British Geological Survey, Sheet 231, England and Wales.

Boyer, S.E. and Elliott, D. 1982. Thrust systems. Bulletin of the American Association of Petroloeum Geologists. 66, 1196-1230.

Brooks, M., Trayner, P.M. and Trimble, T.J. 1988. Mesozoic reactivation of Variscan thrusting in the Bristol Channel area. Journal of The Geological Society, London, 145, 439-444.

Butler, R.W.H. 1982. The terminology of structures in thrust belts. Journal of Structural Geology, 4, 239-245.

Coward, M.P. and Smallwood, S. 1984. An interpretation of the Variscan tectonics of SW Britain. In: Hutton, D.W. and Sanderson, D.J. (eds.) Variscan tectonics of the North Atlantic Region. Geological Society of London Special Publication. 14, 89-102.

Gayer, R.A. and Jones, J. 1989. The Variscan Foreland in South Wales. Proceedings of the Ussher Society. 7, 177-179.

Jones, J. 1990. Sedimentation and tectonics in the eastern part of the South Wales coalfield. Unpublished Ph.D. thesis, University of Wales.

Kelling, G. 1988. Silesian sedimentation and tectonics in the South Wales Basin: a brief review. In: Besly, B. and Kelling, G. (eds.) Sedimentation in a synorogenic basin complex. The Upper Carboniferous of NW Europe. Blackie, Glasgow.

Mechie, J. and Brooks, M. 1984. A seismic study of deep geological structure in the Bristol Channel area, SW Britain. Geophysical Journal of the Royal Astronomical Society, 78, 661-689.

Nickelsen, R.P. 1986. Cleavage duplexes in the Marcellus Shale of the Appalachian foreland. Journal of Structural Geology, 8, 361-372.

Owen, T.R. 1953. The structure of the Neath Disturbance between Bryniau Gleision and Glynneath, South Wales. Quarterly Journal of the Geological Society, London. 109, 333-365.

Powell, C.M. 1987. Inversion Tectonics in S.W. Dyfed. Proceedings of the Geologists' Association. 98, 193-203.

Price, R.A. 1986. The southeastern Canadian Cordillera: thrust faulting, tectonic wedging, and delamination of the lithosphere. Journal of Structural Geology. 8, 239-254.

Salih, M.R. and Lisle, R.J. 1988. Optical fabrics of vitrinite and their relation to tectonic deformation at Ffos Las, South Wales Coalfield. Annales Tectonicae. 2, 98-106.

Strahan, A. 1902. On the origin of the river-system of South Wales and its connection with that of the Severn and Thames. Quarterly Journal of the Geological Society London. 58, 207-225.

Strahan, A., Cantrill, T.C., Dixon, E.E.L. and Thomas, H.H. 1907. The Geology of the South Wales Coalfield. Pt.VII. The Country around Ammanford. Memoirs of the Geological Survey of Great Britain.

Strahan, A., Cantrill, T.C., Dixon, E.E.L. and Thomas, H.H. 1909. The Geology of the South Wales Coalfield. Pt.X. The Country around Carmarthen. Memoirs of the Geological Survey of Great Britain.

Trotter, F.M. 1947. The structure of the Coal Measures in the Pontordawe-Ammanford area, South Wales. Quarterly Journal of the Geological Society, London. 103, 89-133.

Trotter, F.M. 1948. The devolatilization of coal seams in South Wales. Quarterly Journal of the Geological Society London. 104, 387-437.

Woodland, A.W. and Evans, W.B. 1964. The Geology of the South Wales Coalfield. Part IV. the country äround Pontypridd and Measteg. Memoirs of the Geological Survey of Great Britain. 391pp.

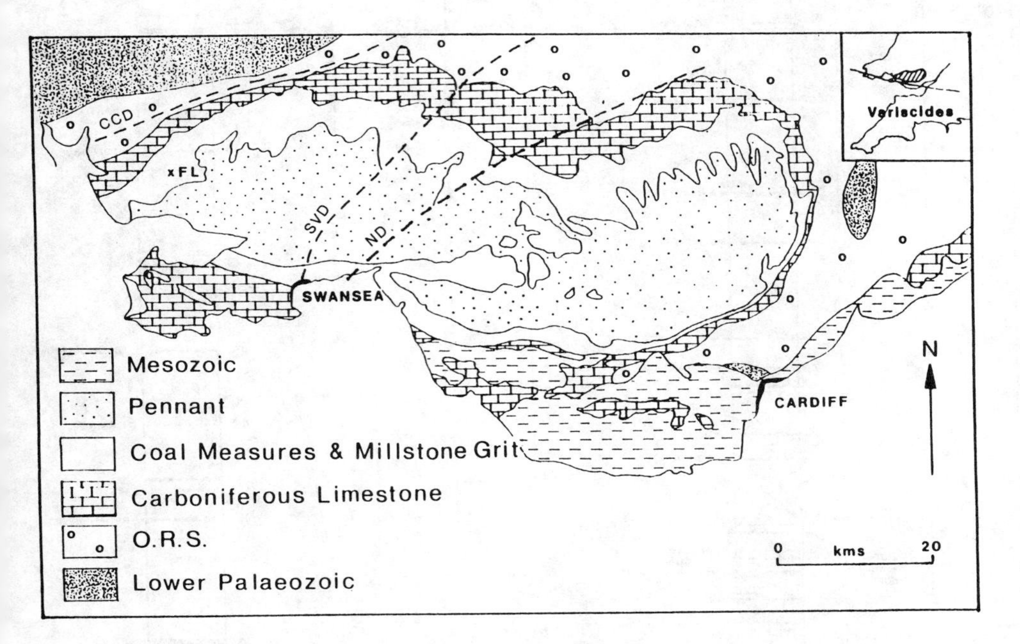

FIGURE 1
Geological map of the South Wales coalfield; FL - Ffos Las, ND - Neath Disturbance, SVD - Swansea Valley Disturbance, CCD - Carreg Cennen Disturbance. Inset shows the relationship of the South Wales coalfield to the SW. England Variscides.

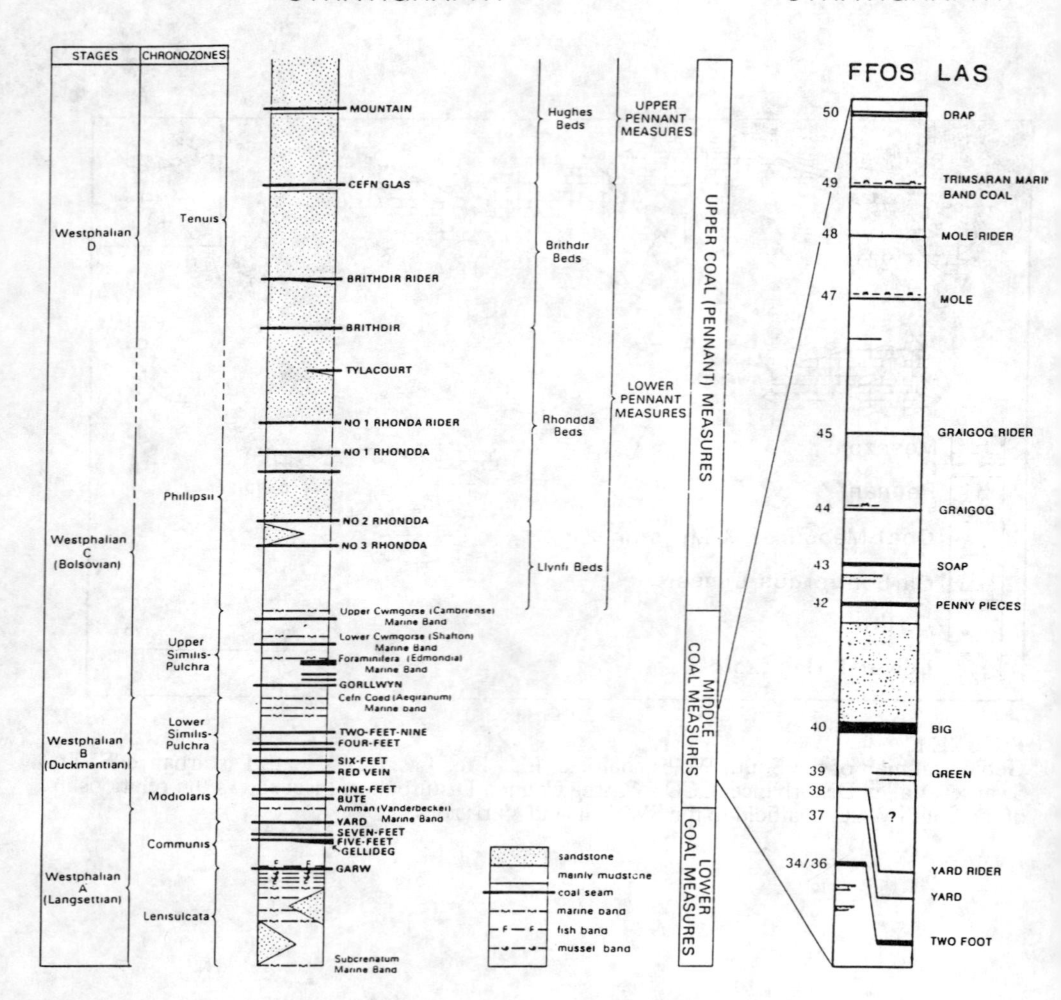

FIGURE 2
Generalised stratigraphy of the Westphalian sequence in the South Wales coalfield (after Barclay et al. 1988), showing the detailed Middle Coal Measures stratigraphy of the Ffos Las OCCS; numbers on the left of the Ffos Las column refer to coal cyclothem sequence of Thewlis (unpublished report British Coal Opencast Executive).

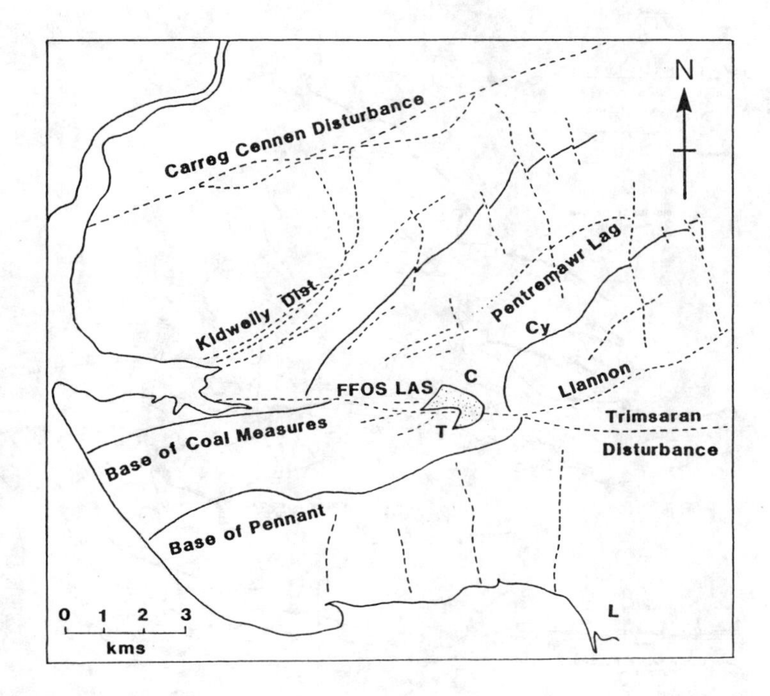

FIGURE 3
Geological map of the northwestern part of the South Wales Coalfield, showing the relationship of Ffos Las Opencast Coal Site to the Trimsaran and Llannon Disturbances.

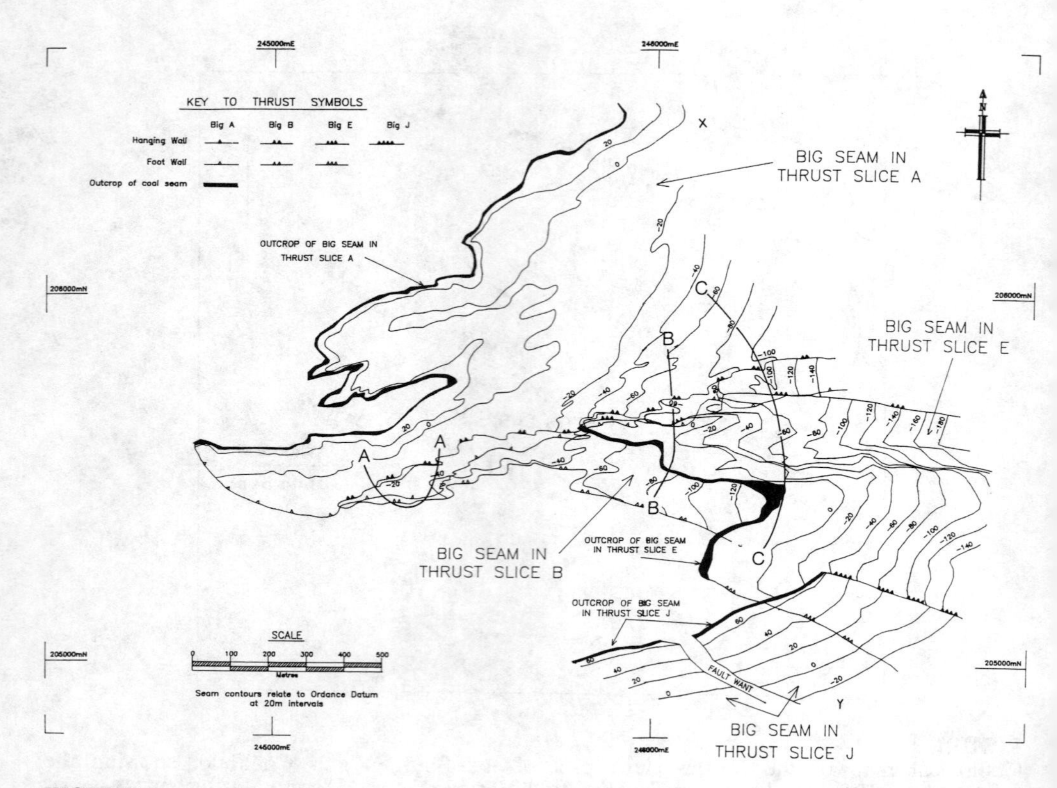

FIGURE 4
Site plan of Ffos Las Opencast Coal Site and the former adjoining Carway OCCS, showing the principal structures between the Llannon and Trimsaran Disturbances. The outcrop of the Big coal seam, present in four major thrust slices - Big A, Big B, Big E and Big J, is shown, with the hangingwall and footwall cut-offs of the Big seam in each of the thrust slices indicated. Contours on the Big seam are drawn at 20 m intervals, but to avoid confusion are not shown where overlain by a higher thrust slice. The location of section lines A (Fig. 11) & C (Fig. 13) are indicated, as are the end points (X + Y of the borhole section shown in Fig. 9.

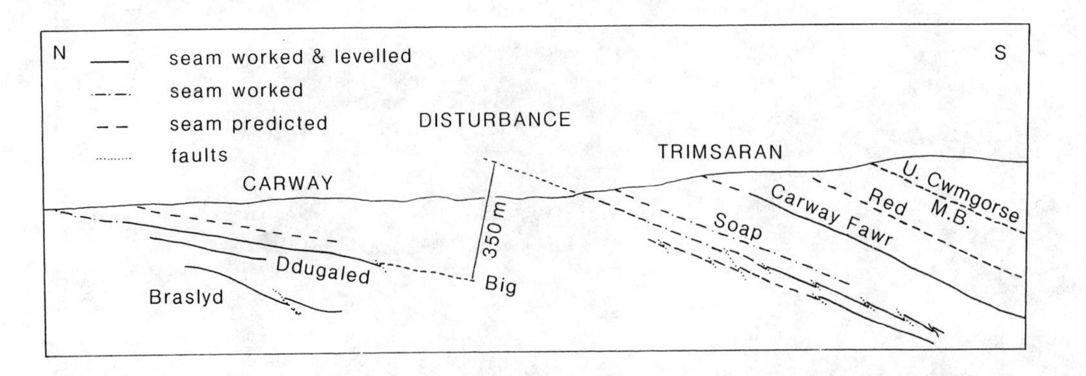

FIGURE 5
North-South cross-section between the old Carway and Trimsaran collieries to show the c.350m vertical throw of the Trimsaran Disturbance (after Archer 1968).

a

b

FIGURE 6
Examples of coal rashings bands, viewed from the west, within the seat earth of the Soap coal in the east of the Ffos Las OCCS.
A) shows classic S-C-C' fabrics indicating a 'top-to-the-north' sense of shear.
B) shows coal rashings deformed by an asymmetrical microfold indicating 'top-to-the-south' sense of shear.

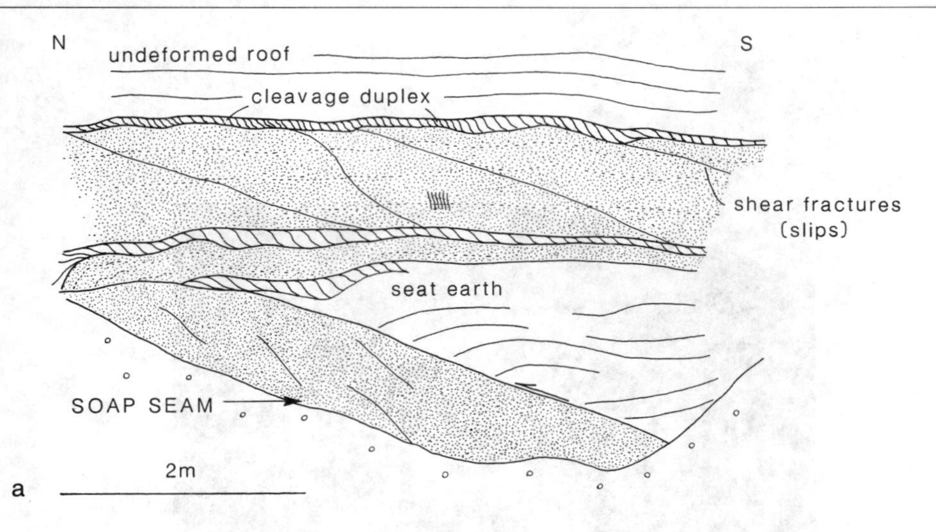

FIGURE 7
A Cleavage duplex development above, within and beneath the Soap coal viewed from the west in the centre of Ffos Las OCCS.
B detail of the intra-coal cleavage duplex shown in Fig. 7 a. Note the sigmoidal shape of the cleavage foliation that dips to the south, indicating a 'top-to-the-north' sense of shear.

FIGURE 8
Example of a typical thrust geometry in the coal-bearing sequences of Ffos Las OCCS, with long thrust flats developed in the seat earth floor of the Penny Pieces coal. The structural interpretation of these thrusts is given in the description of cross-section B and in Fig. 11.

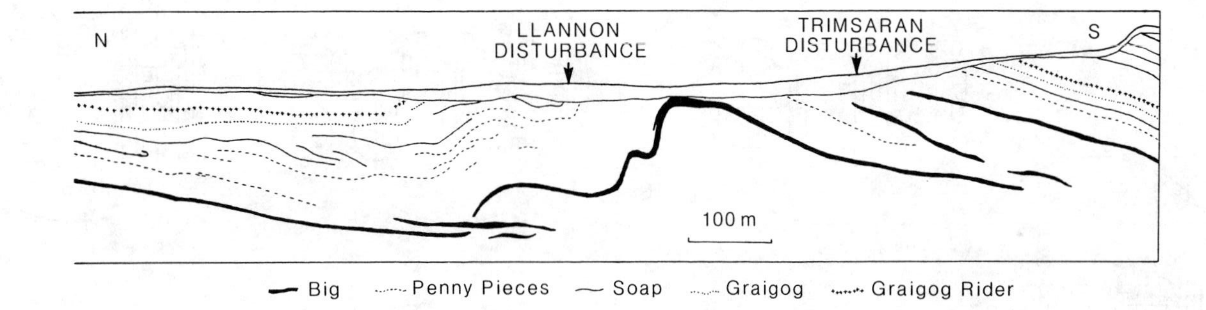

FIGURE 9
N-S cross-section through the Ffos Las site based on borehole data. Only coal seams are shown, although these are clearly repeated by thrusts, no interpretation is presented. The positions of the Llannon and Trimsaran Disturbances are indicated. The location of the section is shown in Fig. 4.

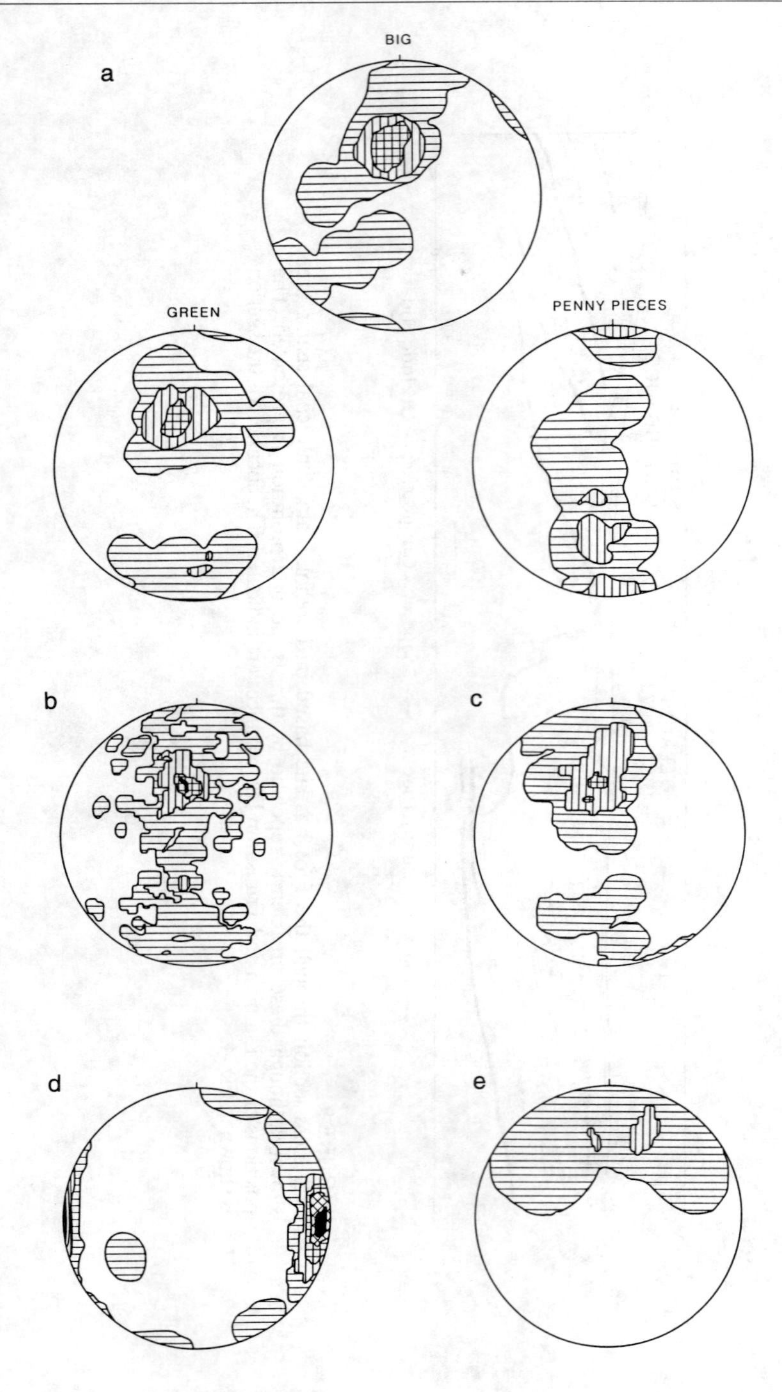
a
BIG
GREEN
PENNY PIECES
b
c
d
e

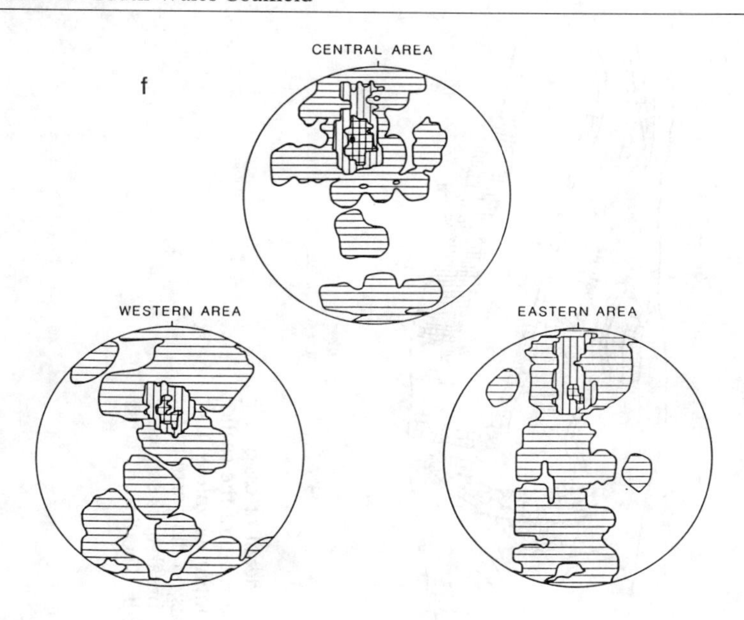

FIGURE 10
Stereographic analysis of structural data at Ffos Las.

A) Contoured stereoplots of poles to bedding planes from the western (n = 40), central (n = 76) and eastern (n = 76) subareas of the site. Axis construction gives fold axial plunges of 9½ to 102½ for the western subarea, 8½ to 082½ for the central subarea, and 9½ to 092½ for the eastern subarea.

B) Contoured stereoplot of poles to bedding planes (n = 268) for the whole site. Axis construction gives a fold axial plunge of 10½ to 091½.

C) Contoured stereoplot of the fold axes (n = 39) for the whole site. Point maximum: 9½ to 089½.

D) Contoured steroplot of poles to thrust planes (n = 38) for the whole site, showing a girdle distribution suggesting folding of thrusts. Point maximum gives preferred strike/dip of thrusts: 076½/44½S. Axis construction gives a fold axia plunge of 10½ to 086½.

E) Contoured stereoplot of poles to cleavage planes in cleavage duplexes for the whole site. Point maximum gives preferred strike/dip of 092½/54½S, suggesting a thrust transport towards 002½.

F) Contoured stereoplot of poles to bedding for individual seams over the whole site.

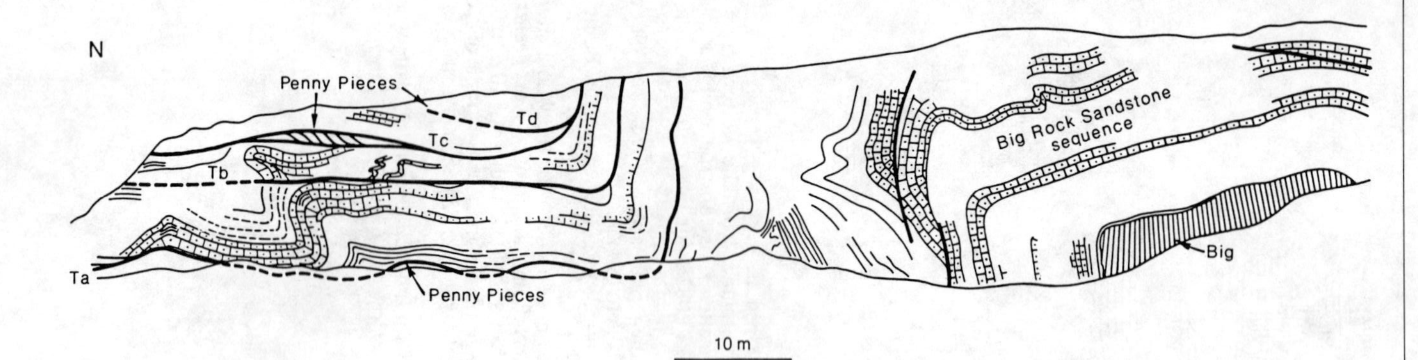

FIGURE 11
Cross-section B, constructed N-S across the centre of the site, showing detail of essentially layer-parallel thrusting and time relationships between structures. In the south of the section early thrusts have been folded by a later northward verging asymmetric antiform/synform pair, producing downward facing thrusts and associated hangingwall antiforms; in the north of the section, thrusts T_a, T_b, T_c, and T_d, shown also in Fig. 8, are developed in break-back sequence, each thrust cutting the hangingwall structure above the thrust beneath. The line of cross-section is shown in Fig. 4.

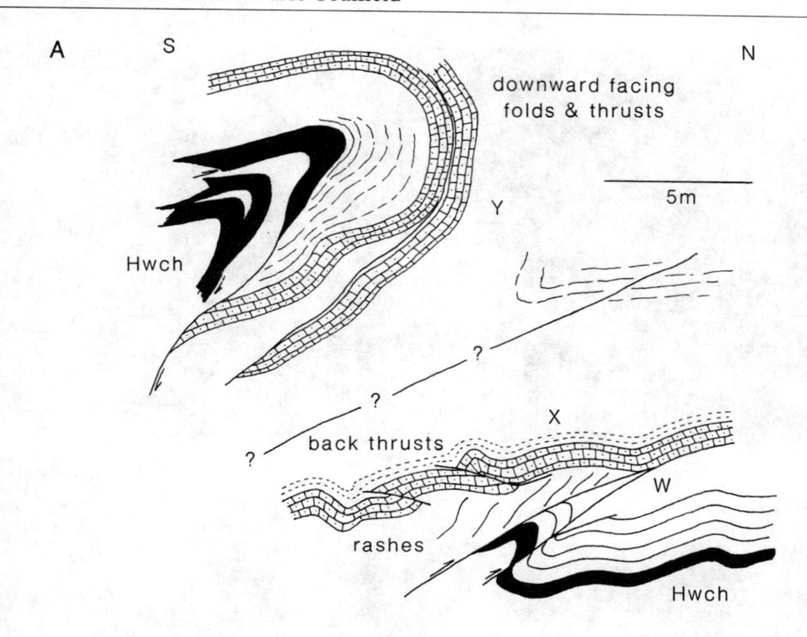
A
S
N
downward facing
folds & thrusts
5m
Y
Hwch
?
?
?
X
back thrusts
W
rashes
Hwch
Big Seam Sequence

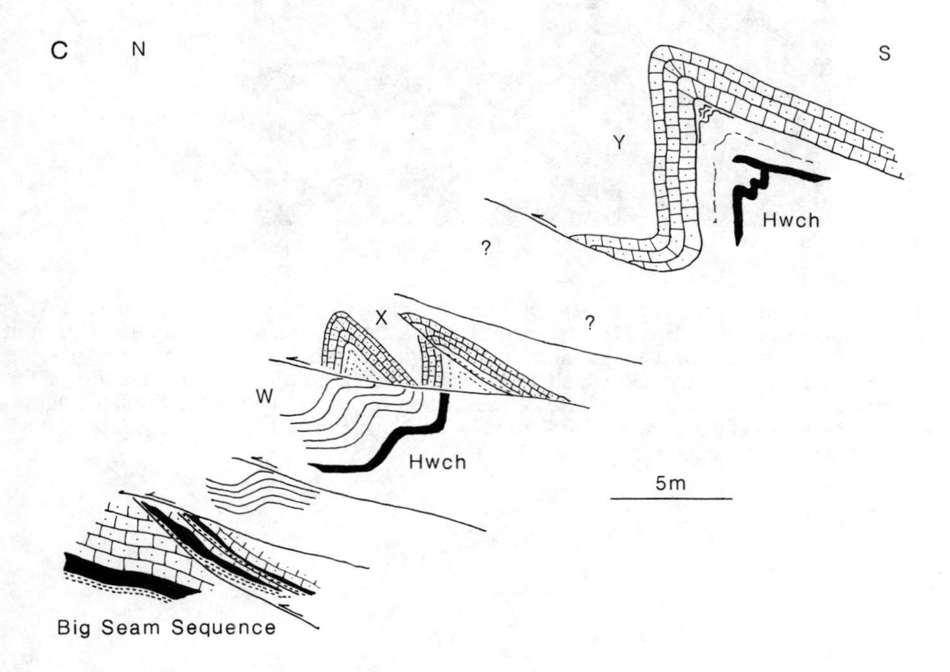
C
N
S
Y
Hwch
?
X
?
W
Hwch
5m
Big Seam Sequence

B

D

FIGURE 12
Sketch section and photographs of structures along section a in the SW corner of the site; A) + B) eastward facing section, C) + D) westward facing section. The line of sections is shown in Fig. 4. Note the presence of a major northward verging antiform/synform pair that can be traced across the two sections and which folds early bedding parallel thrusts that repeat the Hwch seam. The thrusts towards the base of the section tuncate structures in the hangingwall of the thrusts beneath and are interpreted as break-back structures. Note the presence of back thrusts in the east facing section (Fig.12A + 12B).

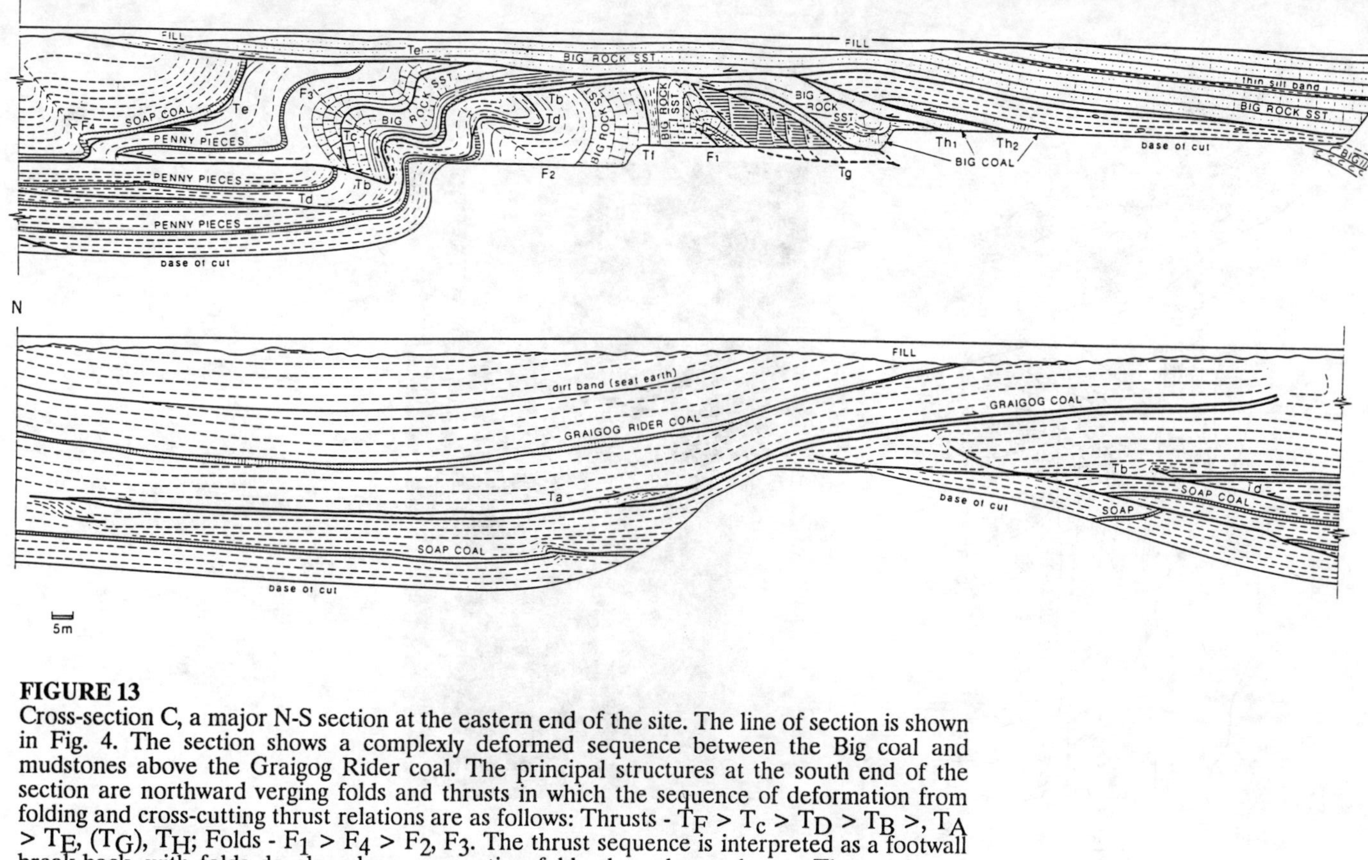

FIGURE 13
Cross-section C, a major N-S section at the eastern end of the site. The line of section is shown in Fig. 4. The section shows a complexly deformed sequence between the Big coal and mudstones above the Graigog Rider coal. The principal structures at the south end of the section are northward verging folds and thrusts in which the sequence of deformation from folding and cross-cutting thrust relations are as follows: Thrusts - $T_F > T_c > T_D > T_B >, T_A > T_E, (T_G), T_H$; Folds - $F_1 > F_4 > F_2, F_3$. The thrust sequence is interpreted as a footwall break-back, with folds developed as propagation folds above lower thrusts. The southward directed thrust in the north of the section beneath the Graigog coal and possibly also beneath the Soap coal are interpreted as a passive roof duplex. See text for detailed discussion.

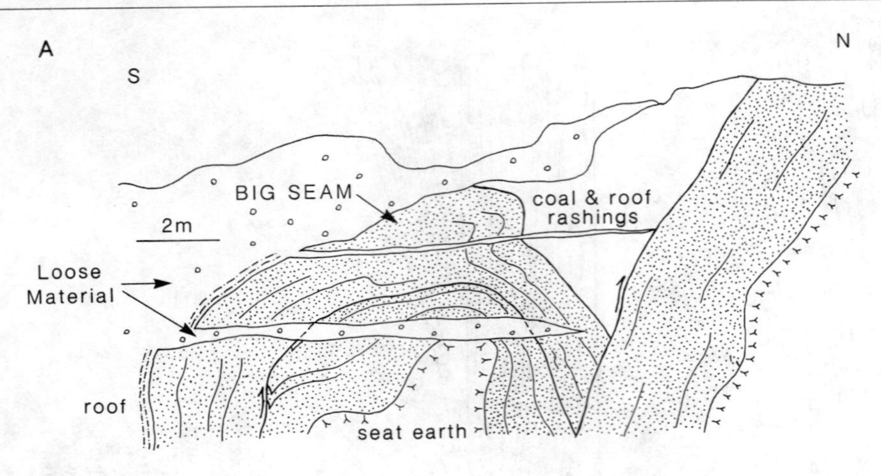
A
S
N
BIG SEAM
coal & roof
rashings
2m
Loose
Material
roof
seat earth

B

C

FIGURE 14
Sketch section and photographs of imbrication of the Big seam within Ffos Las OCCS, viewed from the east.
A) Sketch section and B) photograph of the imbricate structure with thrusts developed along the floor of the coal above the seat earth. The imbricate structure produces a thickening and thinning of the coal and is thought to be the explanation for seam thickness variations throughout the site, such as that seen in Fig. 9.
B) Detail of the vertical thrust repeating the Big seam in the south (right) of Fig.14A + B). Note strong footwall (synform), but only weak hangingwall deformation.

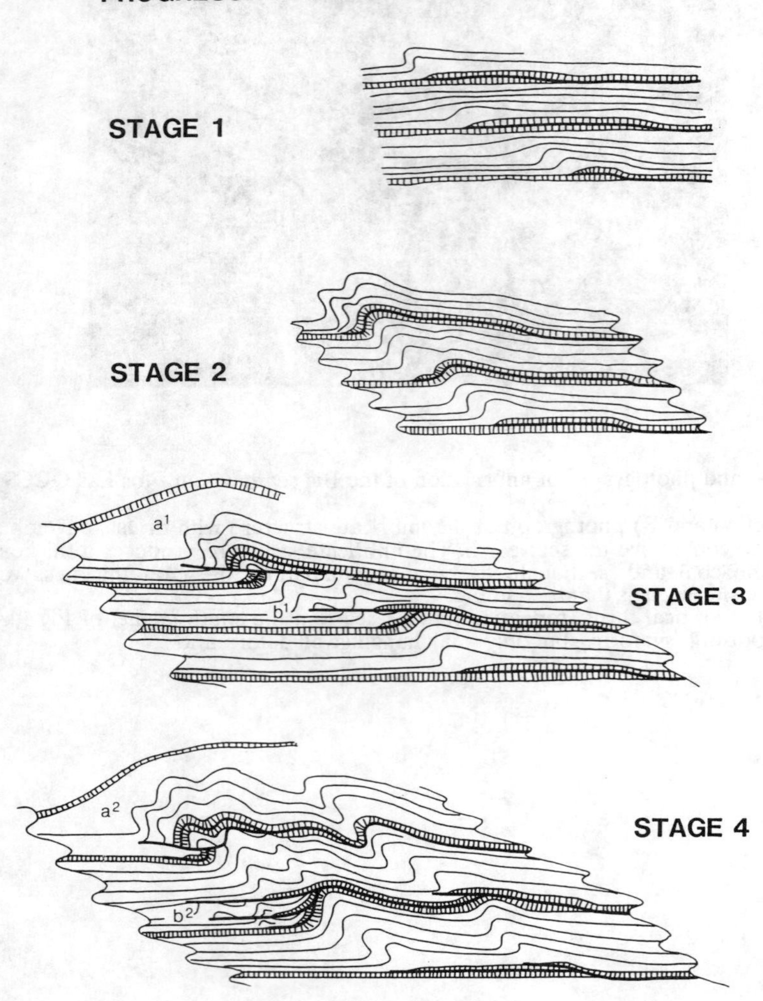

FIGURE 15
Model for the progressive, simultaneous, 'Easy-Slip' thrusting within the coal-bearing strata at Ffos Las OCCS. **Stage 1:** Thrusts develop simultaneously as flats along the floors of easy-slip coal seams, cutting up to the roof of the seams along short ramps; propagation folds grow at the thrust tips. **Stage 2:** Thrusts continue to propagate towards the orogenic foreland with amplification of the folds, until a lower propagation fold locks-up a higher thrust, producing downward facing cut-offs. **Stage 3:** Continued out-of-sequence movement on higher thrusts results in either foot wall (a^1) or hangingwall (b^1) break-back thrusting, note that thrusts locally cut down stratigraphy in the transport direction. **Stage 4:** Progressive out-of-sequence hangingwall break-back produces distinctive geometry (b^2), with the structure in a lower thrust slice being apparently unrelated to that in a higher thrust slice (of Figs. 8 + 11). Progressive footwall break-back (a^2) produces folded thrusts and thrusts cutting stratigraphy at a high angle to bedding (cf Fig. 13).

Economic Geology

Exploration for Bituminous Coal in NW Germany Research in Progress (Extended Abstract)

M.v.Sperber[1] / D.Schmitz[1] / Ä.Strack[2]

1 Deutsche Montan Technologie-Gesellschaft für Forschung und Prüfung GmbH,Institut für Lagerstätte und Vermessung, Hernestr.45, 4630 Bochum1, Germany

2 IBS-Institut für Baustoffuntersuchung und Sanierungsplanung GmbH, Am Torhaus 52, 6600 Saarbrücken, Germany

INTRODUCTION

More than 200 years of coal mining activity in an area of about 6,000 km^2 has exposed Upper Carboniferous Coal Measures of the Subvariscan foredeep to a depth of 1,500 m. The stratigraphy and structure of the paralic coal bearing strata are better documented than in any other coalmining area worldwide. This area therefore presents excellent prerequisites for investigating a section through the Variscan foredeep.

It is intended to give an impression of the scope and methods of exploration work in the Ruhr area and to encourage scientifically oriented geologists to participate in advanced investigations.

Due to the lack of geologists in the collieries' staff the geological evaluation and documentation has been delegated to the Montan Consulting GmbH, Essen, and to the Institute for Applied Geology of the Westfälische Berggewerkschaftskasse (WBK) in Bochum, which merged recently together with Bergbauforschung (Mining Research Center) into the newly established company Deutsche Montan Technologie - Gesellschaft für Forschung und Prüfung (DMT).

OBJECTIVES OF EXPLORATION-WORK AND REVIEW OF GEOLOGICAL FRAMEWORK

Due to the first 'oil crisis' in 1973 a large sized coal exploration program was initiated by the mining companies in the Ruhr district and later in the adjacent Erkelenz and Aachen areas (Fig. 1). It was requested to provide reliable estimates of mineable coal reserves of the future northern mining areas in order to support the estimation of coal production on a middle to long term basis.

The objectives are to delineate areas suitable for mining by exploring:-

- the structure and thickness of barren overburden.
- the structure of coal measures down to a depth of 1,500 m, including the orientation of the beds and fault planes and their displacements.
- identification of seams.
- assessment of extent, thickness and quality of mineable seams.

The geology of the coal bearing Upper Carboniferous is well known and well documented in many publication (Kukuk 1938, Fortschritte in der Geologie von NRW 1962-1965).
The general structure of the coal deposits in NW Germany can be summarized as follows:

- the surface of unconformity overlying the coal measures dips towards north with an increasing overburden thickness.
- overall thickness of the coal bearing strata approaches 3,000 m (Namurian C - Westphalian C).
- there are about 200 identifiable seams, about 50 of them can reach mineable thicknesses.
- the strata are folded and faulted; fold axes and thrust faults strike SW-NE, normal faults strike NW-SE.

In fact two exploration methods are well suited to achieve the required results. These are seismic reflection surveys and deep drillings. During all field operations, seismic surveying and drilling complement each other which helps to optimize seismic results and the positioning of exploration wells (Fig. 2).

Due to economic reasons the northern boundary of the exploration area is taken to be where the overburden thickness is in excess of 1.000 m.

EXPLORATION METHODS

(Fiebig et al. 1983, Klinger 1985, Sauer et al. 1985)

Seismic Survey

The first step in the chain of operations is the seismic survey. Linear seismology is used to give a two dimensional overview, following the traces of known fold-axes, with appropriate cross-lines. Line spacing is from several 100 m up to more than 2 km. The application of 'up to date' methods of seismic processing generates profiles show fairly clearly the unconformity between Upper Carboniferous rocks and younger overburden. Additionally, large structures such as the main anticlines and synclines, horsts and grabens and faults with greater than 30 m displacement can be recognized (Arnetzel 1983, Bornemann & Juch 1979, Dickel & Hell 1983).

On the other hand 3 - D seismology, which works with the aid of high frequency signals offers much better resolution possibilities. A network of shooting lines at right angles to each other and about 100 m apart is arranged on the surface for this purpose. The evaluation then supplies a dense network of underground points reproducing the underlying strata at distances of 12.5 to 25 m in grid form. All required lines of intersection, as well as horizontal planes, can be selected and combined to form a spatial image with the aid of the data system (Dickel & Lemcke 1981). Since 1971 more than 1,500 km of seismic reflection data have been obtained.

Drilling

The grid pattern for drilling is designed according to the results of the seismic evaluation. During the course of exploration the spacing between the boreholes narrows from about 3 km at the beginning to 1 km (or less) at the end. Since 1973 more than 700 deep boreholes have been drilled on a contract basis using truck mounted rigs. The wireline coring technique achieves an average progress of 30 - 50 m per day and a core recovery rate of 99 % in rock and about 96 % in coal has been achieved. The proposed drilling depth is about 1,500 m, although some have reached 2,000 m already. The overburden, comprised of barren strata, is normally penetrated by roller bits without recovering cores although drilling is continuously monitored by mud sampling. In, this case correlation and mapping of the different formations of overburden are achieved by means of geophysical borehole logs. In special cases, however, (e.g. exploring shaft locations), these rocks are cored as well. In contrast to the overburden, the Carboniferous rocks are cored continuously.

Geophysical borehole logging

All boreholes are geophysically logged, a task carried out by various international service companies using the most advanced technology available. With the help of combined efforts of research and developmental technology, a steady improvement of the results has been achieved.

A standard logging program includes the following measurements:

Overburden:
- natural radiation (gamma ray log);
- sound travel time and integration (sonic or acoustic log);
- resistivity.

Coal Measures:
- natural radiation;
- sound travel time;
- resistivity;
- dip and strike of strata interfaces and planes of separation (dip log);
- borehole diameter and deviation of borehole axis
- temperature
- gas entrainment in the drill mud.

The results are recorded on scales of 1:1,000, 1:200 and 1:40 for the coal measures and 1:1,000 and 1:200 for the barren measures.

Apart from the technical improvement of the geophysical logging tools, methods for a reliable geological interpretation of the logs have been worked out. Today the logs form indispensable data for lithostratigraphical correlation and determination of seam structure in case of core loss (Schmitz, D. 1983).

Core evaluation

The observations of lithological units, sedimentary features, fossils, seam structure, dip of strata, core recovery etc. are noted in a hand written record. The content is summarised in a columnar section on a scale of 1:200 in accordance with the DIN 22011 (1989).

A recording of all structural elements is executed graphically in a separate form, including continuous measurements of strata dip and relative azimuth, representation of fault planes, slickensides and joints with their spatial relationship to the stratification. In addition the cores have been photographed.

Sampling and testing

Coal samples are extracted from all seams thicker than 0.50 m for investigation of:

- general analysis (e. g. content of volatile matter, ash, sulphur etc.).
- technological properties (e.g. hardness, strength, cokability, ash fusion temperatur etc.).
- microscopic petrology (maceral analysis, rank determination, coke properties).

All analytical assessments and testing works are carried out in the coal producers' laboratories.

Additional investigations covering:

- microfauna and microflora.
- unusual fossils, plan associations.
- mineralogical assessments.

are carried out by the State Geological Survey (GLA) in Krefeld. Furthermore, thorough rock mechanical and hydrological investigation are undertaken in pilot boreholes. These are required to make shaft sinking safer and more economic.

Data presentation and further utilisation

The results of the investigation of each individual exploratory borehole are summarized and evaluated in a final report providing a permanent feedback for the ongoing exploration project. Within a certain planning area all final reports, together with the seismic findings establish the data for further investigation. Structural plans and cross-sections in adequate quantities form the most important basis for representing the geometry of the deposit. After the planning area is outlined by means of tectonic boundaries, the next planning step leads to preparation of plan views and seam plans with isolines of elevation, thickness and quality parameters. These plans form the basis for coal reserve calculations and mine layout.

GEOLOGICAL RESULTS

The creation and evaluation of new exposures in the northern Ruhr- and Erkelenz area have increased the existing knowledge and have in fact established a much more detailed picture of the stratigraphy and the structure of the hard coal deposits of NW-Germany.

Overburden

The area in question is covered by Upper Cretaceous strata with changing facies from west to east and south to north. The evaluation of the geophysical logs has delivered a detailed stratigraphic classification and makes continuous correlation of the overburden strata possible (Arnold & Wolanski 1964, Müller & Söne-Warnefeld 1985).

In terms of structure and stratigraphy the new findings confirm the difference between the western and the eastern part of the Ruhr district. In the western part rapidly changing thicknesses as of the Permian, Triassic and Jurassic sediments are observed. The presence or absence and the changing thickness of pre-Cretaceous formations is controlled by pre-, syn- and post-sedimentary movements of tectonic blocks along NW-SE trending normal and reverse faults (Bornemann 1980, Wolf 1985). In the eastern part of the Ruhr district the Permian, Triassic and Jurassic formations are missing due to pre-Cretaceous uplift and erosion. Both parts of the Ruhr district are covered by Upper Creataceous strata with changing facies from west to east and south to north. All Mesozoic formations are missing in the extreme western part and in the Erkelenz and Aachen areas, where Tertiary and Quaternary sediments lie directly on the Carboniferous.

Upper Carboniferous strata

The stratigraphic division of the coal measures which is used in the Ruhr district and adjacent areas were worked out at the beginning of this century and since then it has been improved continuously (Oberste-Brink 1930, Kukuk 1938, Fiebig 1971). It is based on paleantological and petrographical marker horizons (Fig. 3). Here, marine influenced bands and Kaolin-Kohlentonstein layers are of great importance. Many of them were found in the recently explored areas, establishing reference horizons for a reliable correlation. A few new Kaolin-Kohlentonstein layers have been discovered during the last 15 years. Now, more than 50 individual, distinguishable layers of Kaolin-Kohlentonstein layers are recorded. Here we have to mention the tremendous contribution of Dr. Burger in determining the occurence and nature of Kaolin-Kohlentonstein in Germany and worldwid (Burger 1982, Burger et al. 1984).

The increase of data resulting from drilling has brought a better understanding of the extent and development of marine influenced horizons. They can be differentiated in layers and zones of more or less distinctive marine character, it being shown that some layers with particular facies may be accepted as local marker horizons. Besides a typical flora and fauna and the specific properties of coal and rock they also show a characteristic shape in the gamma ray log response, which makes a more detailed stratigraphic correlation possible.

The results of thorough stratigraphic analysis are recorded in the stratigraphic-correlation maps, which give an image of the spatial development of seams and rock units. Strata from different districts can be stratigraphically connected, in spite of frequent facies changes combined with splitting and merging of seams. This was proved recently by correlating the Namurian C and Westphalian A of the Erkelenz Horst (north of Aachen) with the corresponding strata of the Ruhr district (Fig. 3; Strack 1988, Zeller 1985).
In addition to localized changes of facies in both space and time, regional trends also become evident. It is of major importance that the thicknesses of the strata and their coal content are determined. Coal content decreases from south to north in all stratigraphic units. In contrast to that, strata thicknesses incerease from Namurian C up to the Westphalian A from south to north, a tendency which is reversed in subsequent units (Fig. 4).

The progressively diminishing marine influence from the Namurian C to the dominantly terrestrial sedimentation of the Westphalian C has been confirmend by recording the flora and fauna in the pelitic rocks of the mining and exploration area (Fig. 5). A changing depositional environment becomes obvious by the grain size related rock composition (Fig. 6). Conditions of sedimentation are also reflected in the petrographic and chemical composition of coal (Strehlau 1990), displaying features which make identification easier. The petrological composition of the seams, however, exhibits a slower lateral change than the composition of the associated rocks. The reconstruction of the environmental conditions accomplished with the help of cores and geophysical logs is the subject of research carried out by working groups of the Universities of Bochum and Aachen (e.g. Conze 1984, David 1990, Steingrobe 1990).

The coalification of the seams, which was basically completed before folding, increases with the stratigraphic age. This simple relationship often varies considerably, depending on regional conditions. Small variations in the coalification gradient of the Ruhr district have been located in contrast to a remarkable deviation of gradients in the Erkelenz district.
The coal measures were heavily deformed by folding and faulting as mentioned above. The intensity of deformation decreases to the north-west. The exploration confirmed that though the amplitude of the major folds increases in this direction, its basic structure remains the same. This is also true of the thrust faults which run parallel to the fold axes. Wide, shallow

synclines and steep anticlines with complicated forms of compression are characteristic of the area of exploration (Fig. 7).

The specific forms of deformation which are dependent on different depth levels - Stockwerke -, have been worked out in the well known research programme called "Tiefentektonik" of the NW-German Coal District (Drozdzweski et al. 1980, 1985). Up and downwards plunging fold axes elevate towards the "Krefelder Gewölbe" in the west and the Beckumer cross-structure in the east. With special reference to the "Krefelder Gewölbe" (the translation for Gewölbe is 'vault', but the meaning is better expressed by 'axis culmination') the regional development of the strata thickness is of major interest. Thickness and facies of Upper Carboniferous sediments do not indicate any upland at the time of deposition.

The frequently quoted migration of the depocentre of the trough to the north during the Upper Carboniferous is in doubt because of the wide range of strata thicknesses from Namurian C to the upper Westphalian C which do not form an identifiable trend (Fig. 4). Extensive reconnaissance has confirmed the fact that the sedimentary environment was very uniform in time and space. It is worth considering why we find similar stratigraphic formations (referring to thickness, lithology, fauna and flora) again and again in similar forms in localities which are far apart.

A small part of the Variscan continental margin has been well exposed by coal mining and exploration activities. Even though a large number of questions concerning the environmental conditions of the coal measures has been raised, answers can possibly be found by emphasizing a less static way of looking at the geological situation. Reconstruction of the spatial movements of the presently known exposures in the northern hemisphere to fit its individual parts may better explain the facies and thickness analyses of the related sediments.

REFERENCES

Arnold, H. & Wolansky, D. (1964): Litho- und Biofazies der Oberkreide im südwestlichen Münsterland nach neuen Kernbohrungen. - Fortsch. Geol. Rheinld. u. Westf., 7, 421 - 478, Krefeld.

Arnetzel, H.H. (1983): Some aspects of reflection seismic coal field exploration. - Prakla Seismos sp. paper presented at the 'Workshop on Geophysical Exploration for Coal', April 1983, Nagpur (India).

Bornemann, O., & Juch, D. (1979): Tektonische Auswertung von seismischen Profilen und Bohrungen im Ruhrkarbon. - Z. dt. geol. Ges., 130: 77-91, Hannover.

Bornemann, O. (1980): Tiefentektonik der Lippe- und Lüdinghausener Hauptmulde. - In: Drodzewski et. al.: Beiträge zur Tiefentektonik des Ruhrkarbons. - 173-191, Krefeld.

Burger, K. (1982): Kohlentonsteine als Zeitmarken, ihre Verbreitung und ihre Bedeutung für die Exploration und Exploitation von Kohlenlagerstätten. - Z. dt. geol. Ges., 133: 201-255, Hannover.

Burger, K., Fiebig, H., & Stadler, G. (1984): Kaolin-Kohlentonsteine in den Explorationsräumen des Niederrheinisch - Westfälischen Steinkohlenreviers. - Fortsch. Geol. Rheinld. u. Westf., 32: 151- 169, Krefeld.

Conze, R. (1984): Sedimentologische Typisierung der feinklastischen Gestine des Ruhrkarbons. - Fortschr. Geol. Rheinld. u. Westf., 32, 187-230, Krefeld.

David, F. (1990): Sedimentologie und Beckenanalyse im Westfal C und D des nortwestdeutschen Beckens. - DGMK Forschungsbericht 384-3, 271 pp., Hamburg.

Dickel, U. & Heil, R. (1983): Fortschritte in der Übertageseismik zur Erkundung des Steinkohlengebirges. - Glückauf-Forsch. - H. 44: 219-225, Essen.

Dickel, U. & Lemcke, K.: Computer-Aided 3 D Seismic Interpretation for Coal Exploration in the Ruhr District, NW Germany. - In: Coal Exploration 3 (1981), Miller Freemann Publication, San Francisco.

DIN 22011 (1989) Rohstoffuntersuchungen im Steinkohlenbergbau. Bearbeitung von Untersuchungsbohrungen (Entworf). - Faberg, Berlin.

Drozdzewski, G., & Bornemann, O, & Kunz, E., & Wrede, V. (1980): Beiträge zur Tiefentektonik des Ruhrkarbons. - 192 pp., (Geol. L.-Amt Nordrh.-Westf.) Krefeld.

Drodzewski, G., Engel, H., Wolf, R., & Wrede, V. (1985): Beiträge zur Tiefentektonik westdeutscher Steinkohlenlagerstätten. - 236 pp., (Geol.-L-.Amt Nordrh.-Westf.) Krefeld.

Fiebig, H. (1969): Das Namur C und Westfal im Niederrheinisch-westfälischen Steinkohlengebiet. - C. R. 6. Congr. internat. Strat. Geol. Carbonif., Sheffield 1967, 1, 79-89, Maestricht.

Fiebig, H. (1972): Der Gesamtschichtenschnitt (overall-section) des niederrheinisch-westfälischen Steinkohlengebietes (Stand 1971). - In: Hedemann et. al.: Das Karbon in marin-paralischer Entwicklung. - C. R. 7. Congr. internat. Strat. Geol. Carbonif., Krefeld 1971, 1, 29-47, Krefeld.

Fiebig, H., Palm, H., & Pieper, B. (1985): New aspects in exploring the Upper Carboniferous of the Ruhr Area. - C. R. 10. Congr. Stratigr. Geol. Carbonif. Madrid 1983, 4, 125-134, Madrid.

Fortschritte in der Geologie von Rheinland und Westfalen.
Vol. 3 (1962): Das Karbon der subvariscischen Saumsenke. - Part 2: Das Steinkohlengebirge. Petrographie und Palaeontologie. - IX-XX, 423 - 865, Krefeld
Part 3: Das Steinkohlengebirge. Stratigraphie und Tektonik. - XXI - XXVII, 867 - 1282, Krefeld

Vol. 7 (1964): Die Kreide Westfalens. Ein Symposium. - 748 pp., Krefeld.

Vol. 13 (1966, 1967): Zur Geologie des nordwestdeutschen Steinkohlengebirges. (Part 1 und 2). - I - XXVI, 1 - 1444, Krefeld.

Vol. 18 (1971): Das höhere Oberkarbon von Westfalen und das Bramscher Massiv. - I-XII, 1 - 596, Krefeld.

Vol. 19 (1971): Die Karbon-Ablagerungen in der Bundesrepublik Deutschland. - I - VIII, 1 - 242, Krefeld.

Vol. 32 (1984): Nordwestdeutsches Oberkarbon (Part 1). - 1 - 339, Krefeld.

Vol. 33 (1985): Nordwestdeutsches Oberkarbon (Part 2). - 1 - 323, Krefeld.

Klinger, U. (1985): Entwicklungsstand der Explorationstechnik im Steinkohlenbergbau. - Z. Glückauf, 121, 13, 1018-1025, Essen.

Kukuk, P. (1938): Geologie des Niederrheinisch-westfälischen Steinkohlengebietes. - 706 pp., Springer Verlag, Berlin.

Müller, W., & Schöne-Warnefeld, G. (1985): Die Deckgebirgsgeologie der Reservefelder des Niederrheinisch-Westfälischen Steinkohlereviers. - In: Klein, J.: Handbuch des Gefrierschachtbaus im Bergbau. - Glückauf Betriebsbücher 31, p. 124-140, Glückaufverlag, Essen.

Oberste-Brink, K., & Bärtling, R. (1930): Gliederung des produktiven Karbons und einheitliche Flözbenennung im rheinisch-westfälischen Steinkohlenbecken. - Z. dt. geol. Ges., 82, 321-347, Berlin.

Sauer, A.F., Dickel, U., & Rack, P. (1985): Exploration und rißliche Darstellung von Steinkohlenlagerstätten als Grundlage bergbaulicher Planung. - Z. Glückauf, 121, 16, 1193 - 1199, Essen.

Schmitz, D. (1984): Das Erscheinungsbild von Kohlenflözen in geophysikalischen Bohrlochmnessungen. - Fortschr. Geol. Rheinld. u. Westf., 32: 231-241, Krefeld.

Steingrobe, B. (1990): Fazieseinheiten aus dem Aachen-Erkelenzer Oberkarbonvorkommen unter besonderer Berücksichtigung des Inde-Synclinoriums. - 325 pp., Diss. RWTH Aachen.

Strack, Ä. (1988): Stratigraphie in den Explorationsräumen des Steinkohlebergbaus. - Mitt. Westf. Berggewerkschaftskasse, 62, 210 pp. Bochum.

Strack, Ä. & Freudenberg, U. (1984): Schichtenmächtigkeiten und Kohleninhalte im Westfal des Niederrheinischen-Westfälischen Steinkohlenreviers. - Fortschr. Geol. Rheinld. u. Westf., 32, 243-256, Krefeld.

Strehlau, K. (1990): Facies and genesis of Carboniferous coal seams of Northwest Germany. - International Journal of Coal Geology, 15, 245-292, Elsevier Science Publ. B. V. Amsterdam.

Wolf, R. (1985): Tiefentektonik des linksniederrheinischen Steinkohlengebietes. 105-167, In: Drozdzewski, G., Engel, H., Wolf, R., Wrede, V.: Beiträge zur Tiefentektonik westdeutscher Steinkohlenlagerstätten. - 236 pp. Krefeld.

Zeller, M. (1985): Vorschlag eines Richtschichtenschnittes für das flözführende Oberkarbon (Westfal A und B) des Aachen - Erkelenzer Steinkohlenreviers. - Fortschr. Geol. Rheinld. und Westf., 33, 265-287, Krefeld.

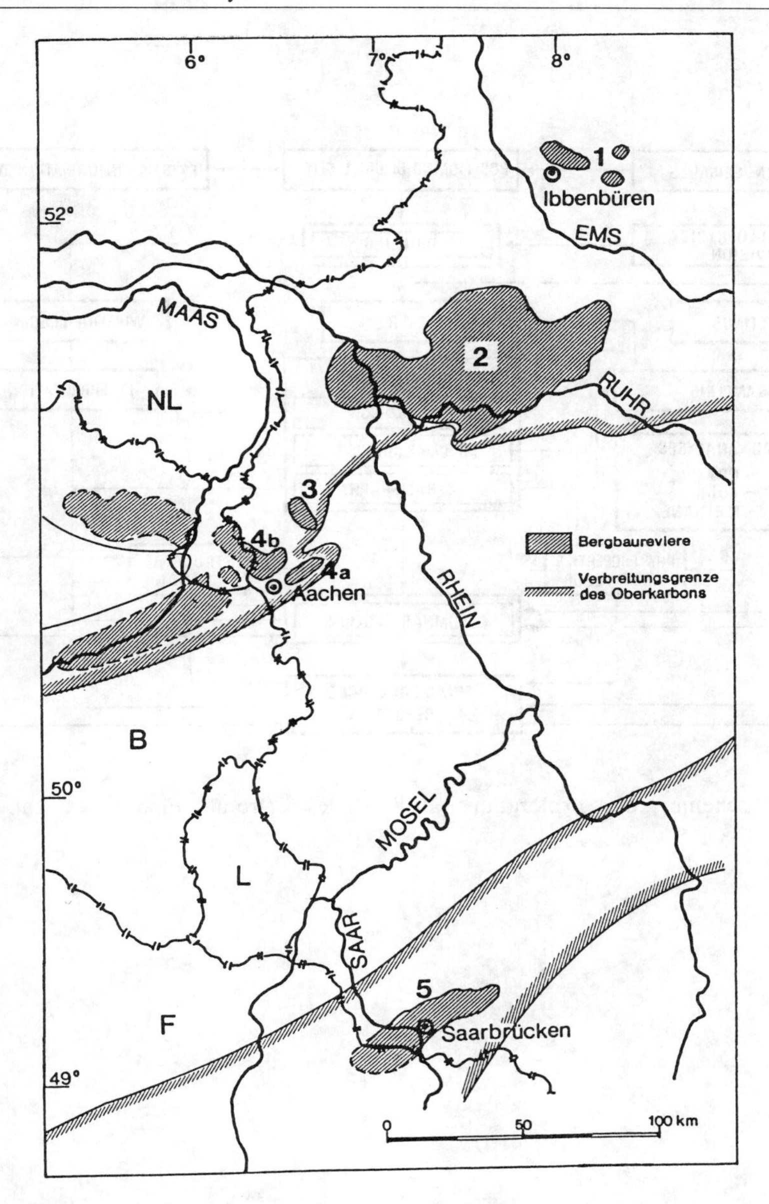

FIGURE 1
Location map of bituminous coal fields in West Germany. 1 = Ibbenbüren, 2 = Ruhr, 3 = Erkelenz, 4 = Aachen, a = Inde syncline, b = Wurm syncline, 5 = Saar.

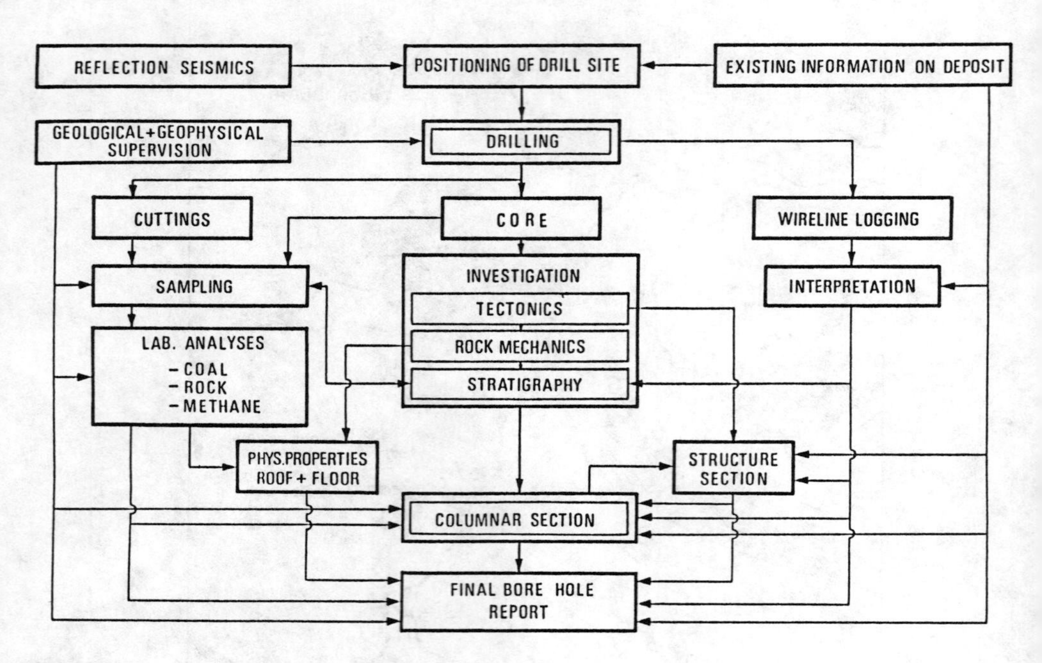

FIGURE 2
Evaluation scheme of exploration drill holes (from Fiebig et al. 1985).

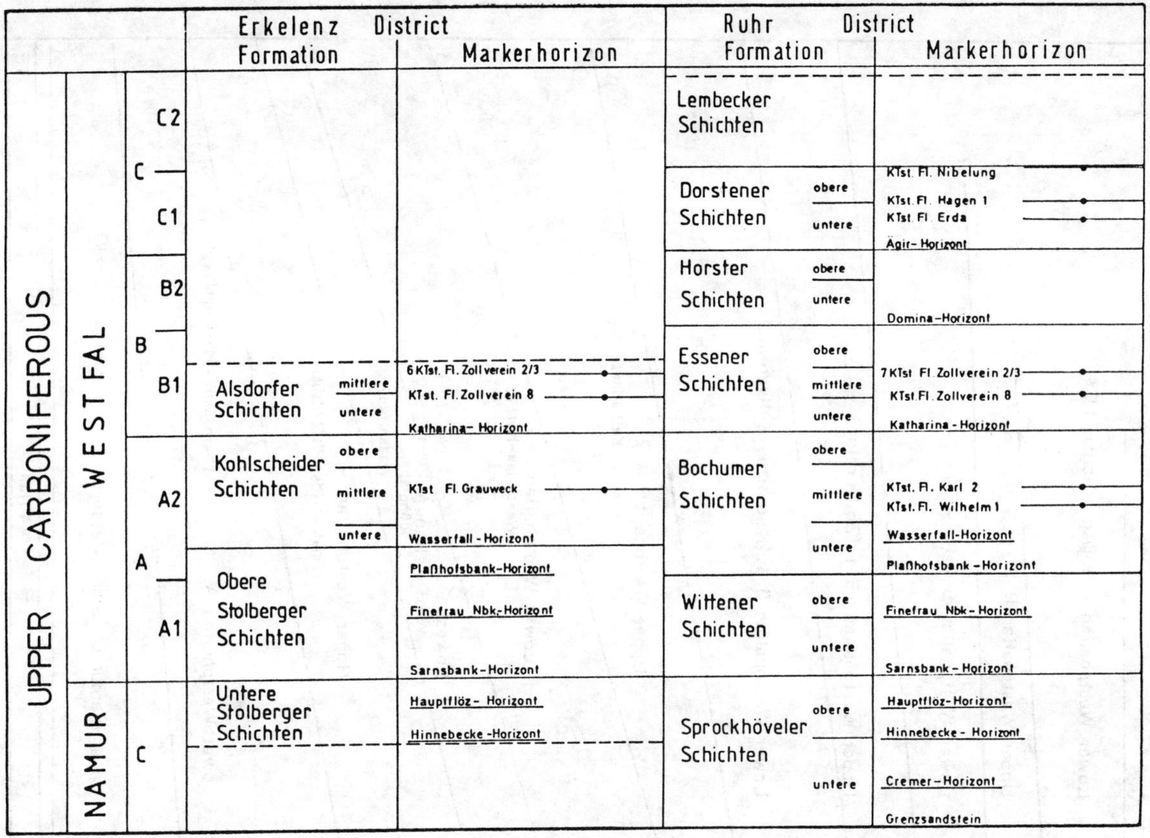

			Erkelenz District			Ruhr District		
			Formation		Markerhorizon	Formation		Markerhorizon
UPPER CARBONIFEROUS	WESTFAL	C2				Lembecker Schichten		
UPPER CARBONIFEROUS	WESTFAL	C1				Dorstener Schichten	obere untere	KTst. Fl. Nibelung KTst. Fl. Hagen 1 KTst. Fl. Erda Ägir-Horizont
UPPER CARBONIFEROUS	WESTFAL	B2				Horster Schichten	obere untere	Domina-Horizont
UPPER CARBONIFEROUS	WESTFAL	B1	Alsdorfer Schichten	mittlere untere	6 KTst. Fl. Zollverein 2/3 KTst. Fl. Zollverein 8 Katharina-Horizont	Essener Schichten	obere mittlere untere	7 KTst. Fl. Zollverein 2/3 KTst. Fl. Zollverein 8 Katharina-Horizont
UPPER CARBONIFEROUS	WESTFAL	A2	Kohlscheider Schichten	obere mittlere untere	KTst Fl. Grauweck Wasserfall-Horizont	Bochumer Schichten	obere mittlere untere	KTst. Fl. Karl 2 KTst. Fl. Wilhelm 1 Wasserfall-Horizont Plaßhofsbank-Horizont
UPPER CARBONIFEROUS	WESTFAL	A1	Obere Stolberger Schichten		Plaßhofsbank-Horizont Finefrau Nbk-Horizont Sarnsbank-Horizont	Wittener Schichten	obere untere	Finefrau Nbk-Horizont Sarnsbank-Horizont
UPPER CARBONIFEROUS	NAMUR	C	Untere Stolberger Schichten		Hauptflöz-Horizont Hinnebecke-Horizont	Sprockhöveler Schichten	obere untere	Hauptflöz-Horizont Hinnebecke-Horizont Cremer-Horizont Grenzsandstein

FIGURE 3
Simplified stratigraphic table of the coal bearing Upper Carboniferous strata in the districts of Erkelenz and Ruhr (Strack 1988).

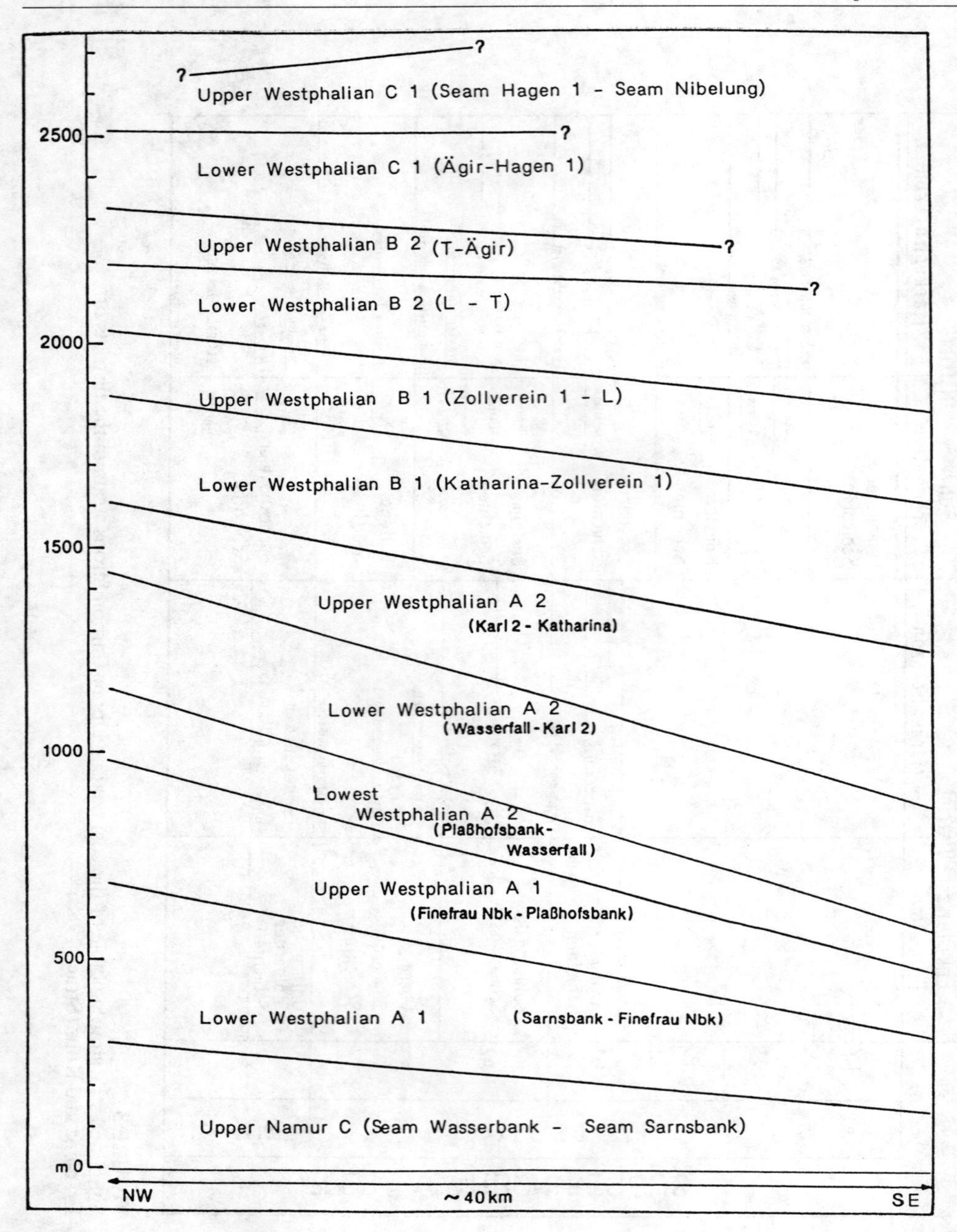

FIGURE 4
Development of formation thicknesses from Namurian C to Westphalian C1 (perpendicular to strike), showing that there is no clear trend of a migrating depocentre from south (Namurian C) to north (Westphalian C1), (Strack 1988).

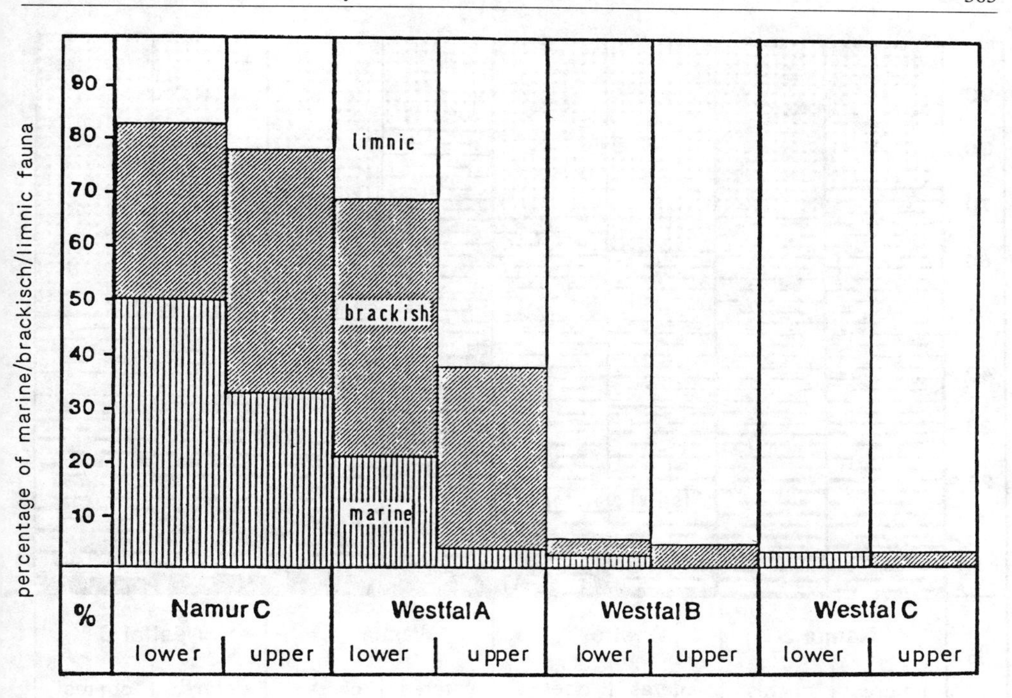

FIGURE 5
Changing of facies within the Upper Carboniferous coal measures in the northern Ruhr district, illustrating the decrease of the marine influence ascertained by fossil content in pelitic rocks. It demonstrates the transition from lower - to upper deltaic plain sedimentary environment (resulting from approx. 700 wells, with 30 fauna horizons per well in average), (Strack 1988).

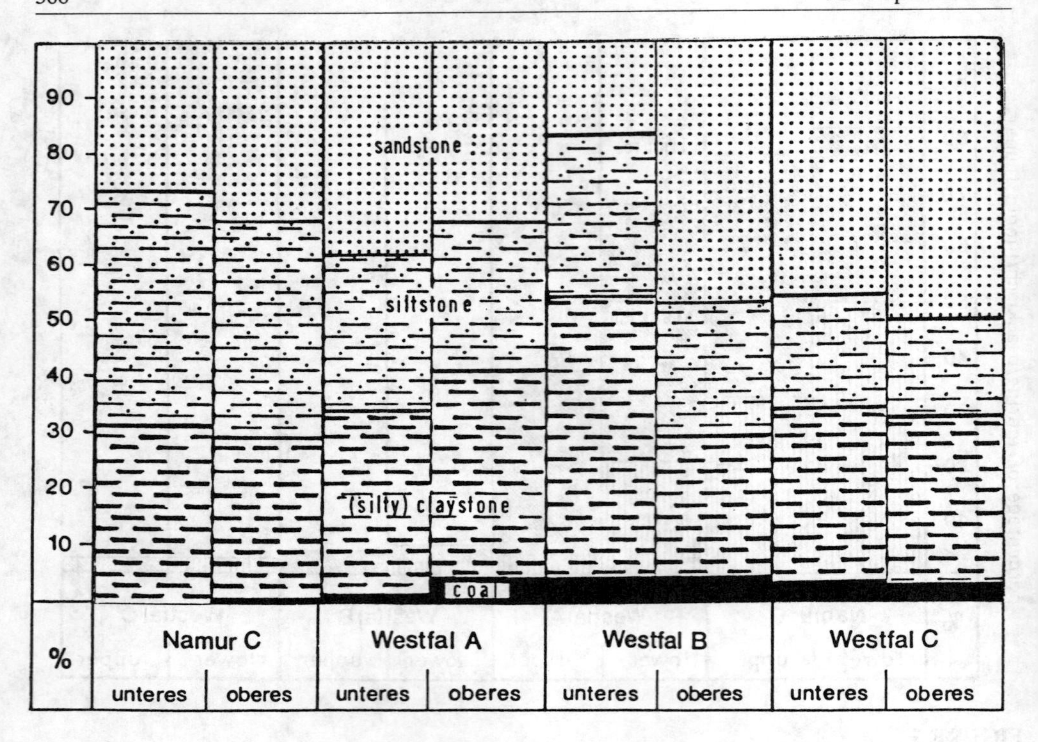

FIGURE 6
Grain size related composition of sediments of the Upper Carboniferous coal measures (evaluated from approx. 300 wells), (Strack 1988).

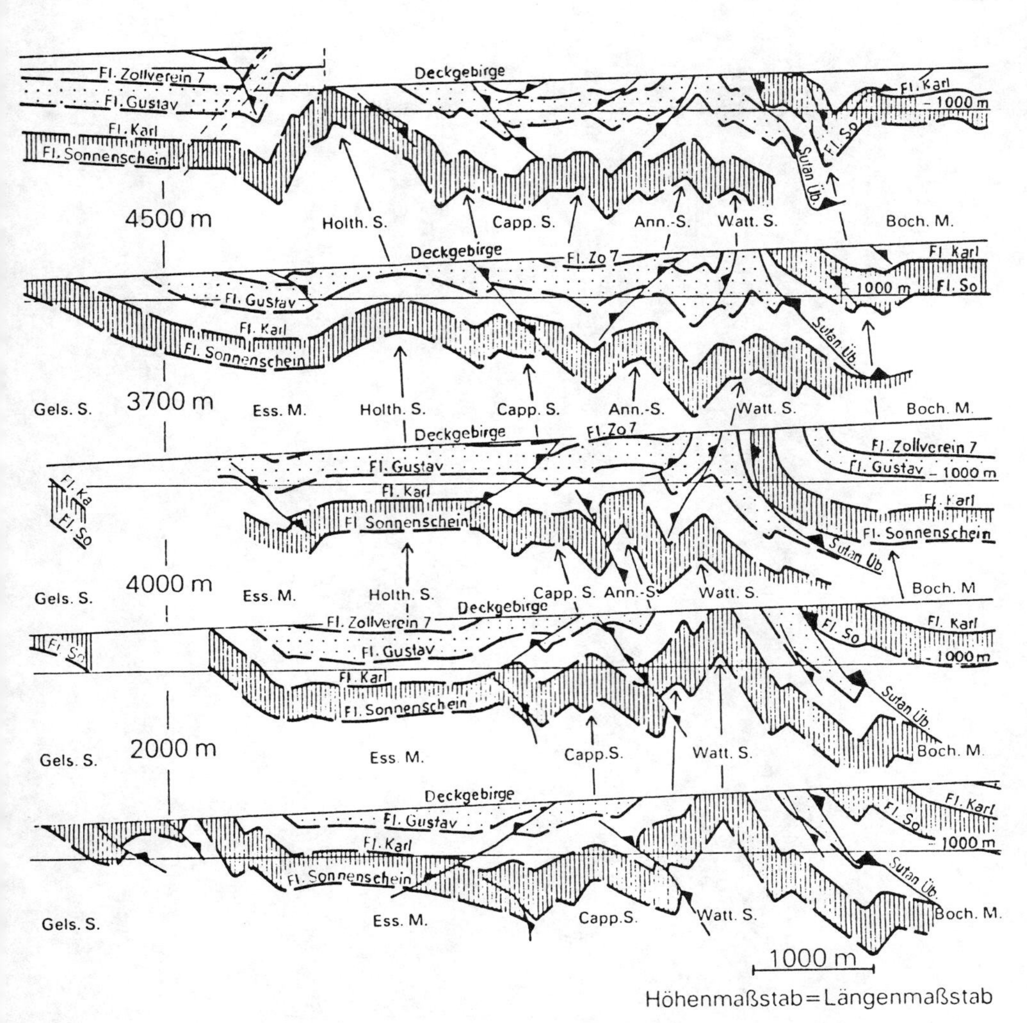

FIGURE 7
Series of cross sections in the area of exploration in the eastern part of the Essen synclinorium (Ruhr district) as an example for the style of folding and thrusting. (Kunz in: Drozdzweski et al. 1980).
S = Sattel = anticline, M = Mulde = syncline, Fl = Flöz = seam, Deckgebirge = overburden vertical scale = horizontal scale

The Origin of Kupferschiefer Mineralization in the Variscan Fold Belt of Southwestern Poland

S.Speczik
Geological Faculty, University of Warsaw, 02-089 Warsaw, Zwirki i Wigury 93, Poland

ABSTRACT

Fluid inclusion studies, vitrinite-liptinite rank measurements and organic geochemical studies were used to elucidate the tectonothermal history during formation of the Kupferschiefer mineralization in southwestern Poland.

A high-temperature geothermal gradient within the Rhenohercynian basement shows various heat anomalies that relate to paleohighs and tectonically deformed basement. These anomalies found within the southern part of the Fore-Sudetic Monocline correlate well with lower rank anomalies associated with Kupferschiefer Formation. The known base metal ocurrences of the Kupferschiefer type spatially relate to both types of paleogeothermal anomaly. Thus, it is envisaged, that one of the main controlling factors that lead to the formation of Kupferschiefer mineralization was the availability of thermal energy related to Variscan orogenesis.

INTRODUCTION

Kupferschiefer type ore bodies are characteristically tabular in shape, lack direct magmatic affinities and associated with intra-continental sedimentary facies. Such features are considered to reflect a syngenetic and syndiagenetic origin. Recently, plate tectonic concepts have caused a major reassessment of previously accepted genetic models, including the processes of ore deposit formation (e.g. Sawkins 1971, Sillitoe 1972, Petrascheck 1976, Bauman 1978, Sawkins and Burke 1980). Sawkins (1976), based on (Burek's (1973) geotectonic interpretations, connected the process of Kupferschiefer mineralization with intracontinental hot spot activity and rifting environments. In this concept the key role was played by oxygenated groundwaters heated by intracratonic hot spot activity.

Similar tectonic evolution of Rhenohercynian belt with bimodal volcanism, high heat flow, crust a thinning and rapid sedimentation into closed rift basin in the foreland of Hercynian orogeny is proposed by Jowett and Jarvis (1984). However, they locate the hydrothermal activity 15-20 Ma later after the last volcanism.

Tectonophysical phenomena are generally thought to be related to the geothermal gradient and its variability at the contact between the lithosphere and the upper mantle. As the correlation of major plate tectonic elements and the heat flow has been principally elucidated (Roy et al. 1972, Grimmer 1978), fluid inclusion studies, vitrinite and liptinite measurements as well as organic geochemical studies seem to be proper methods to search for a link between geotectonic processes and the Kupferschiefer mineralization in Central Europe. Organic matter is especially useful in such investigations. It records sensitively any thermal change in geological environments, and what is important, these processes are not affected by retrogresive alteration.

There is some evidence collected so far that points indirectly to the importance of the geotectonic influences on the Kupferschiefer mineralization. It has been already known that Kupferschiefer base metal occurences of an economic value are superimposed above a tectonically and thermally active transition zone between the Rhenohercynian and Saxothuringian belts of the European Variscides (Rentzsch 1981, Speczik et al. 1986). During most of thr Carboniferous, the Southern European continent was probably a rather thin, hot plate with a rigid crust. Due to Variscan subfluence (Weber 1978) the geothermal paleogradients of the Rhenohercynian belt are recognized as having the highest paleogeothermal gradient among European Variscides (Zwart 1962, Lorenc 1976). This area was probably thermally active also during Lower Permian, as indicated by Autunian volcanic activity. The present geothermal anomaly in the adjacent region is interpreted as being a deep Variscan thermal disturbance of the upper mantle (Majorowicz 1982). One of the major tectonic features in this area, the Dolsk deep fracture, is believed to be still active and producing remnant heat.

Thermal energy is a governing factor in the development of most of the ore-generating systems. Heat can stimulate and promote processes of base metal mobilization and preconcentration e.g. leaching of country rocks of their metal content by heated intraformational, connate or metamorphic waters, having an appreciable choride content. The energy (in any form) is also necessary for transport and metal emplacement in epigenetic type mineralizations. Elevated heat flow could be also a effective in producing highly reactive gas mixtures rich in base metals from any type of black shales or siltstones, in the way suggested by Walter and Buchanan (1969).

GENERAL GEOLOGY

The area under investigation roughly corresponds to the Fore-Sudetic Region of southwestern Poland (Fig. 1) and is spatially equivalent to the Rhenohercynian and partly the Saxothuringian belts of the Central European Variscides. The southern part of the Fore-Sudetic Region (Fore-Sudetic-Block) is separated from its northern part by the Middle Odra step-like fault structure. The geologic structure of the northern part of Fore-Sudetic region consists of two structural units: Precambrian and Paleozoic basement, folded and faulted and variously altered during diagenesis and metamorphism and its gently dipping (from 2½ to 5½) Permian, Mesozoic, and Cenozoic cover called the Fore-Sudetic Monocline. The main tectonic feature of the Fore-Sudetic Monocline is the Dolsk deep fracture that subdivides the Fore-Sudetic Monocline into southern and northern parts.

Precambrian and Early Paleozoic rocks crop out within Carboniferous strata in the cores of four anticlinal structures trending SE-NW. At least two of these structures: the Middle Odra Crystalline Zone and the Krotoszyn-Wolsztyn Uplift are thought to represent paleohighs during the Zechstein time. The Middle Odra Crystalline Zone spatially corresponds to the phyllite-schist Zone in Germany, which is thought to represent a transitional belt between theRhenohercynian and Saxothuringian. Precambrian and Early Paleozoic rocks consist of granites, granitic gneisses, gneisses, schists, hornfelses, and variously altered phyllites.

The most ubiquitous rocks of the Fore-Sudetic Monocline basement are Lower Carboniferous and to the north also Upper Carboniferous rocks. They consist mainly of graywackes, wackes, arenites and quartz arenites interbedded or intermixed with siltstones, shales and minor conglomerates. They are folded and faulted with the intensity of faulting and folding decreasing northwards. Carboniferous rocks are variously diagenetically altered (mature to hypermature stage), and in some locations even slightly metamorphosed. Various alteration processes differing in their kind and intensity have been recognized (e.g. albitization, choritization, hematitization, carbonatization and silicification). The intensity of these processes is higher in the southern part of the Fore-Sudetic Monocline basement and additionally increases towards paleohighs and Middle Odra Crystalline Zone. In these locations the entire Carboniferous sections were pervasively penetrated by an extensive hydrothermal alteration system. In other places the direction of fluids responsible for alteration is marked by a dense network of epigenetic veinlets that cut the Carboniferous formations (Speczik and Kozowski 1987).

In numerous drilling profiles a variable intensity alteration pattern has been observed. However, mostly the intensity of the alteration increases with the depth of the examined rocks. This may suggest an importance of the geothermal gradient and in some cases the horizontal or vertical flow of heat and mineralizing fluids along the microtectonic structures.The Rotliegendes foreland depressions were filled with clastic redbeds and saline lacustrine deposits, averaging from 600 to 800 m, and occasionally to 1600 m in thickness. The sedimentary Rotliegendes section is intruded by bimodal (basalt-rhyolite) igneous suites.

The Kupferschiefer Formation contains elements like Fe, Co, Ni, V, Cr, etc. and organic carbon in concentrations found commonly in other types of black shales. It is assumed that high concentrations of other base metals i.e. Cu, Pb, Zn, Ag were produced in epigenetic-type processes. This assumption is necessary when considering any genetic model based on geotectonic concepts. The Kupferschiefer is followed by a thick evaporitic Zechstein sequence and overlain by Mesosoic rocks. They are succeeded by Tertiary and Quaternary deposits, with thickness increasing in a northerly direction.

BASEMENT INVESTIGATIONS

Two methods, fluid inclusion studies and vitrinite rank measurements, were employed to estimate the paleogeothermal influence on Carboniferous rocks in the area of the Fore-Sudetic Monocline basement. Material gathered for more than 15 years is summarized in this chapter; part of this material has been presented earlier (Speczik 1979, 1985). The majority of these investigations has been published in polish, therefore some major conclusions are repeated. The study map of the paleogeothermal field of the Fore-Sudetic Monocline basement has been constructed from more than 200 fluid inclusion studies and 200 vitrinite measurements performed on material collected from 72 borings that pierced Carboniferous basement (Fig. 2).

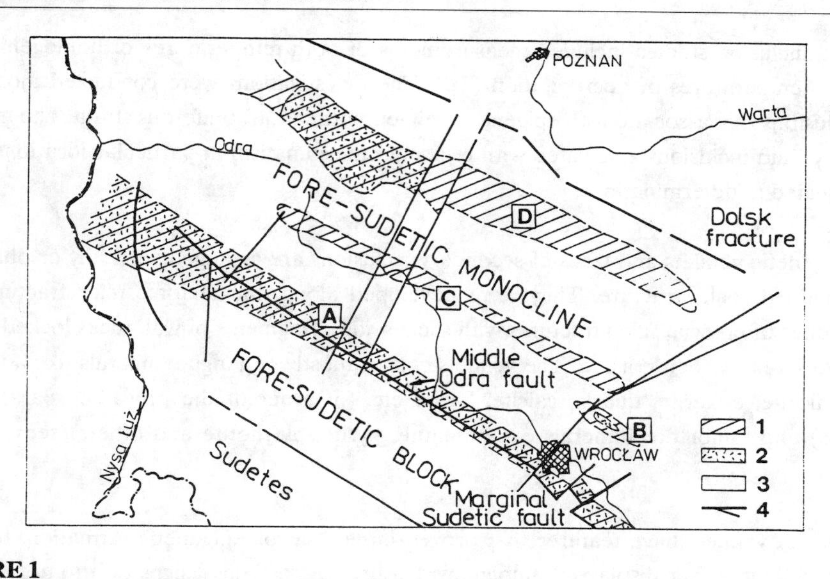

FIGURE 1
Geologic sketch map of the Fore-Sudetic Region.
1. Anticlinal structures in the basement of the Fore-Sudetic Monocline,
2. Precambrian and early Paleozoic rocks,
3. Lower and Upper Carboniferous (only in the area of the Fore-Sudetic Monocline),
4. Major faults. A - Middle Odra Cristalline Zone, B - Dobrzen Unit, C - Anticline Trzebnica-Bielawy, D - Krotoszyn-Wolsztyn Uplift.

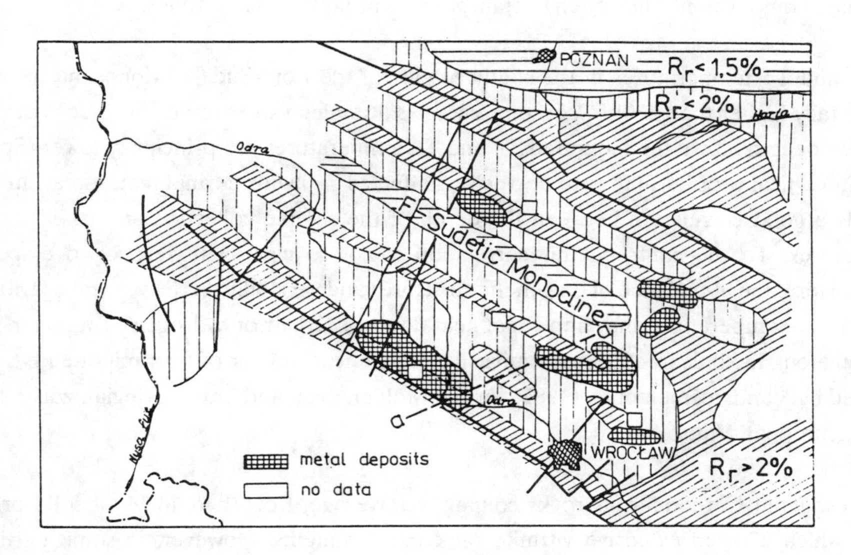

FIGURE 2
Map of equal paleogeothermal fields with marked positions of economic type mineralizations. Other explanations as in Fig. 1.

The fluid inclusion studies included measurements of both temperatures of homogenization (Th) and temperatures of decrepitation (Td). The investigations were conducted mostly on fluid inclusions that associate with epigenetic veinlets cutting Carboniferous strata. The trails of secondary fluid inclusions associated with regional metasomatism in particular locations were also subjected to determinations.

Both epigenetic veinlets and trails of secondary inclusions are arranged vertically or obliquely to the bedded rock structure. They were developed along thin tectonic microfractures, as healed zones of breccia, filled fractures with slickensides, fragments of wall rocks locked inside vein material have been noticed. The veinlets consist mostly of gangue minerals like adularia, albite, chlorite, ankerite, quartz, calcite, barite etc. (in order of the prevailing paragenetic sequence) and minor ore minerals e.g. hematite, pyrite, pyrrhotite and others (see Speczik 1988).

A number of veinlets have features that proved either dia- or epigenetic formation, but the majority of them either displaysed ambiguous features or they have signs of an intermediate origin. This may suggest acontinuous process of veinlet formation starting from diagenesis. Veinlets and signs of hydrothermal alteration cease at the contact between the Carboniferous and Lower Permian strata, indicating their Variscan (Sudetic or Asturian) age. Both primary inclusions of epigenetic veinlets as well as secondary inclusions of alteration zones have two- or multiphase composition. They often contain gaseous or liquid hydrocarbons.

The Th and Td temperatures in the southern part of the Fore-Sudetic Monocline basement are generally higher than 200½ C, in some samples attaining values up to 350½ C. North from the Dolsk deep fracture the measured Th and Td temperatures are principally lower (Speczik 1985). In vertical profiles that were pervasively affected by hydrothermal metasomatism there is mostly a good correlation between the depth of the sample collection and its Th and Td temperatures. To the contrary, this correlation is not so clear with respect to epigenetic veinlets. Here, the abundance of epigenetic structures and variability in the vertical distribution of Th and Td temperatures may imply that heat flow was largely of hydrogenic character. Heat migrated along tectonic fractures. A similar hydrogenetic character of heat migrated has been suggested by Van Breemen et al. (1982) for the Moldanubian and Saxothuringian zones of the Central European Variscides.

The investigated Carboniferous rocks contain, on average, from 1 to 14 % of solid organic matter, which allowed extensive vitrinite rank measurements. However, in some particular regions i.e. the Middle Odra Crystalline Zone and the Trzebnica-Bielawy Anticline, Carboniferous rocks (when present) to not contain organic material, which was presumably nearly entirely removed from the host rocks during extensive alteration processes. Elevations in

Carboniferous basement are generally recognized by a low content of vitrinite contrasted with an increased amount of inertinite grains. This might suggest oxidizing conditions during the observed epimetamorphic processes. The temperature and burial history are critical for interpretations of the change in vitrinite reflectance. The time factor has been neglected because of the long burial history of the Paleozoic Rhenohercynian belt in southwestern Poland.

The measured vitrinite rank values do not correlate with the present depth of the Carboniferous strata and with the geothermal gradient in the southern part of the Fore-Sudetic Monocline basement. Moreover, the vitrinite rank in most cases does not correspond to the maximum burial depth of the investigated rocks, calculated on the basis of the work by Majorowicz et al. (1984). The virtrinite rank is principally higher than expected, in majority of samples it exceeds 2 % Roil, reaching in some preparations up to 4-5% Roil. Several tens of km from the Dolsk deep fracture, the high geothermal paleogradients of the Fore-Sudetic Monocline are adjacent to areas of low paleogradients. Here, the vitrinite rank correlates well with the present depth of the sediments and the estimated Variscan paleogradients (Speczik 1988).

Inside the geothermal field of the Fore-Sudetic Monocline basement characterized by elevated paleogradients, additional positive anomalies in close proximity to paleohighs and highly tectonically disturbed regions have been encountered. High paleogeothermal field parameters of the Fore-Sudetic Monocline basement correspond with the findings of Teichmüller and Teichmüller (1979) for the Rhenohercynian belt of Germany. The results of the fluid inclusion studies and the vitrinite rank measurements correlate well within some classes (Speczik 1988). Additionally, paleogeothermal field elevations constructed in this manner correlate well with the positions of the Lubin-Sieroszowice deposits and other lower grade mineralizations found in this region. This may suggest that the availability of thermal energy during Carboniferous time might have an influence on the Kupferschiefer mineralization (Fig. 2).

KUPFERSCHIEFER INVESTIGATIONS

Vitrinite and liptinite rank measurements and organic geochemical studies were employed to examine directly the energetic factor of the mineralization processes with respect to Kupferschiefer formation. The vitrinite reflectance measurements remain the main method used; the other methods were conducted mainly for vitrinite rank determinations control.

The Kupferschiefer formation is mainly described as a thinly (0.2 - 1.2m) bedded bituminous marl or shale, formed in a euxinic and sapropelitic environment that in places is variously mineralized. This simplistic picture is not confirmed by detailed coal petrology studies of the organic matter (average content of about 5 %) associated with the Kupferschiefer. Variously bituminous and locally oxidized vitrinite is the dominating component of organic macerals found. Considering the entire Kupferschiefer basin, vitrodetrinite, composed of mostly structurless collinite with minor tellinite cell structures, dominates among the vitrinite group macerals. Vitrinite of more distal facies is variously cemented and impregnated with liptinite group macerals, predominantly along fissures anc cleats.

Liptinite (alginite) forms notable accumulations only in places that have high soluble organic material content, and correspond to some deeper shoals in relatively near-shore environments. In near-shore environments allochthonous inertinite dominates accompanied by lesser amounts of vitrinite. Vitrinite particles in near-shore environments and in areas of ore-grade mineralisation reveal symptoms of various intensity oxidation. The oxidation indicies, occurring as patchy textures with differences in refractive index and colour, or as oxidation rims at edges and along fissures, could be connected with two genetic processes: oxidation by mineralizing solutions related to epigenetic Rote Fäule facies, and oxidation related to weathering or redeposition. Based on the environmental and sedimentological parameters at least in more distal facies and in deposits that relate to small paleohighs, the secondary oxidation by epigenetic solutions has been assumed.

This inference is supported by the two types of vitrinite reflectograms that have been obtained. The first type, with a very regular distribution pattern, is characteristic for areas of low vitrinite reflectance that in most cases have no associated ore grade mineralization. The second type, with an irregular bimodal distribution pattern, reflects variable oxidation of vitrinite (vitroinertinite) and is characteristic of those regions of the Kupferschiefer formation that bear economic type mineralizations (Fig. 3). The process of epigenetic vitrinite oxidation (related also to some extent with the temperature of solutions) elevates the rank of vitrinite. The latter observation does not fully agree with the prevailing statements that pore fluid chemistry has no effect on coal rank. It is, however, plausible that in deposits that are situated in near-shore environments both processes i.e. epigenetic oxidation and oxidation due to weathering were responsible for changes in the vitrinite rank.

In contradistinction to the Carboniferous basement the random (mean) vitrinite reflectance of the Kupferschiefer formation was found to be generally low. More than 80 % of results fall in a narrow range between 0.5 to 0.9 % Roil. There is a good correlation between the depth of sample collection and the random vitrinite reflectance (Fig. 4), except for samples with very high content of organic carbon (espectially liptinite group marcerals). They have relatively lower reflectance what is connected with extremly high contents of reducing agents. The samples that reveal pronounced symptoms of oxidation and were collected close to paleohighs

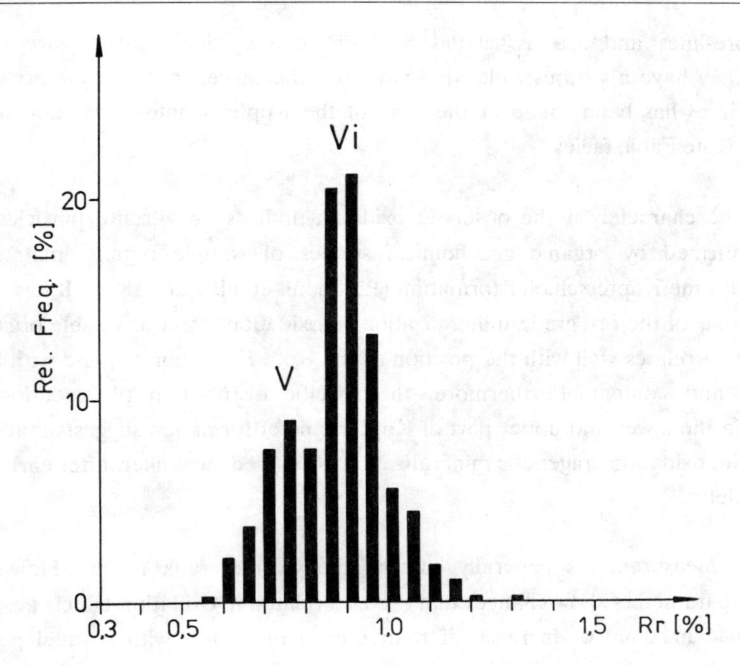

FIGURE 3
Vitrinite reflectogram of a Kupferschiefer sample. Borehole G-1, depth 815.5m. V-vitrinite, Vi-vitroinertinite.

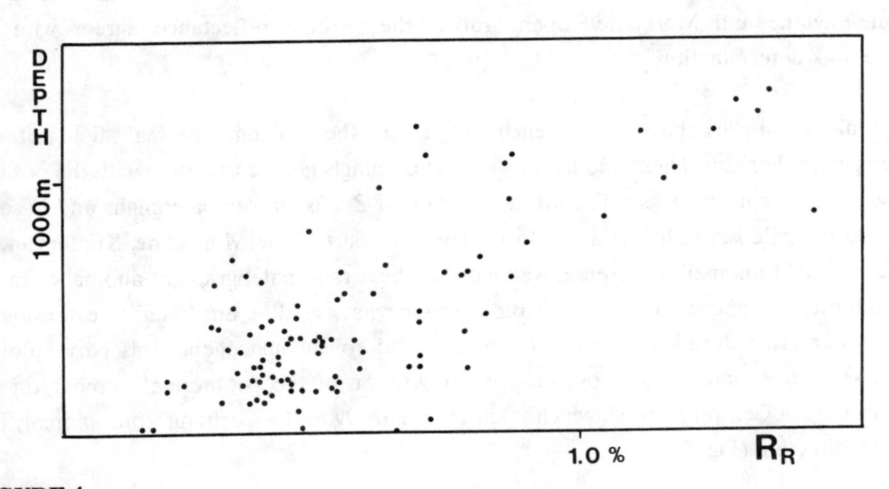

FIGURE 4
A correlation between the depth of sample collection and random reflectance. Samples from Sieroszowice-Rudna mines.

or paleoshore-lines and are associated with ore grade mineralizations are the second exception. They have a variously elevated vitrinite reflectance. In ore grade areas the most oxidized vitrinite has been found at the base of the Kupferschiefer formation or near the contact with Rote Fäule facies.

The epigenetic character of the observed oxidation indices on vitrinite particles has been recently confirmed by organic geochemical studies of soluble organic material that is associated with the Kupferschiefer formation (Püttmann et. al. 1987, 1899). It has been found that in the areas of the ore grade mineralization an oxidation effect of soluble organic matter appears that correlates well with the position of the Rote Fäule horizon and with base metal composition and content. Furthermore, the specific distribution of oxidation-indicating compounds in the lower and upper part of Kupferschiefer formation suggests that oxidation - associated with oxidation epigenetic mineralization - occurred most likely after early diagenesis of Kupferschiefer.

Fluorometric measurements generally confirm the vitrinite rank results. However, some fluorescence parameters show changes that are also related to oxidation effects i.e. green shift of fluorescence maximal or increase of fluorescence intensities with parallel petrographic changes of associated vitrinite. To the contrary, organic geochemical chemofossil parameters like Pristane/Phytane, Moretane/Hopane ratios, CPI, give no information about maturity of organic matter in the areas of the ore grade mineralization. In those places the elaborated chemofossils were nearly equally affected by secondary oxidation. In areas of low (primary-syngenetic) mineralization that lack oxidation indices calculated on the basis of Pristane/Phytane and Moretane/Hopane ratios, the vitrinite reflectance agrees with the vitrinite rank determinations.

Low rank anomalies have been encountered in the paleogeothermal field of the Kupferschiefer horizon. These anomalies show astonishingly good correlation with the position of base metal occurrences as well with the location of Zechstein sea paleohighs and in some cases also the paleoshore line (Fig. 5). In the southern part of the Monocline, Kupferschiefer geothermal field anomalies correlate well with the high rank paleogradient anomalies in the Carboniferous basement. In a northerly direction, in what could be attributed to extension an of the lower crust during the formation of the Mid-Polish aulacogen, this correlation is unclear. This is supported also by the present position of the geothermal anomaly in the southern part of Central Europe, which is shifted relatively to the Carboniferous anomaly in a northerly direction (Fig. 6).

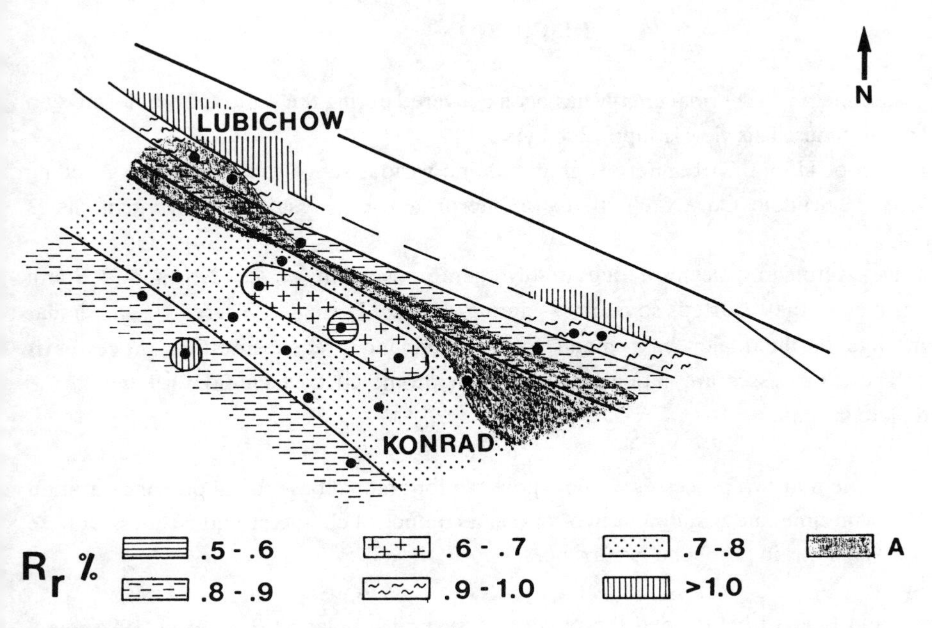

FIGURE 5
Map of vitrinite rank in the area of Konrad mine. A - position of ore grade mineralization. Black dots mark the location of investigated samples.

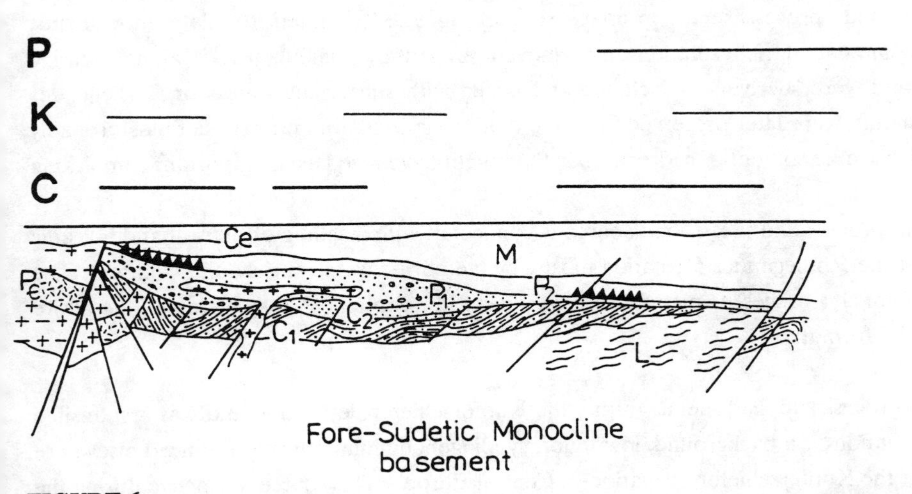

FIGURE 6
Schematic cross-section through the northern part of the Fore-Sudetic Region with positions of Carboniferous (C), Kupferschiefer (K), and present (P) geothermal anomalies. P - Precambrian, L - Lower Carboniferous, C1, C2 - Lower and Upper Carboniferous, P1, P2 - Lower and Upper Permian, M - Mesozoic, Ce - Cenozoic. Black triangles mark positions of ore grade mineralizations.

DISCUSSION

It is suggested that most thermal energy has been delivered during the Late Carboniferous and Early Permian times. This view is supported by:

1) along chain of Upper Carboniferous granitoid and various subvolcanic intrusions trending parallely to the Middle Odra fault, that are interpreted as a result of convergent plates movement;

2) Extensive Autunian volcanism due to divergent movement of the North and South European plates that started sometimes during Lower Permian. Therefore, a similar (Variscan) age of the major base metal mobilization and preconcentration processes is assumed. These processes are thougt to have been promoted by heat flow related to the Variscan plate motion.

It is conceiveable that two processes were responsible for gradual base metal preconcentration in Late Variscan time and resulted in two-source, a composite character and richness of base metal mineralizations in particular occurrences.

Subfluence and later rifting divided the plates into several subplates (Behr et al. 1984): as a consequence of deviations in particular sub-plate movements, favourable tectonic and energetic conditions were established in certain regions. These regions, now recognizable as paleohighs and deep fractures, correspond credibly to major zones of subfluence and rifting. Intraformational processes (the primary source suggested) related to plate movements, evolved saline metal rich solutions that moved toward the paleohighs or along tectonic fractures and were successively included and mixed with subsurface waters of Rotliegendes age. The metals were later preserved in a kind of underground aquifers. These fluids probably had an appreciable content of hydrocarbons that might have acted as an important complexing agent.

The second process and the second source is connected with leaching of detrital and volcanic material of the Rotliegendes formation mostly by meteoritic water enriched in chlorides. The effectivness of this process is manifested by the very low content of base metals in the entire Rotliegendes formation.

The latter process supplied metals during the Kupferschiefer deposition and was responsible for the formation of a background, low tenor, syndiagenetic mineralization noticed elsewhere throughout the Kupferschiefer formation of Central Europe. This source was active during the deposition of the Kupferschiefer bed. The mixing of sea water with mineralized fluids resulted in a uniform composition of the syndiagenetic mineralization.

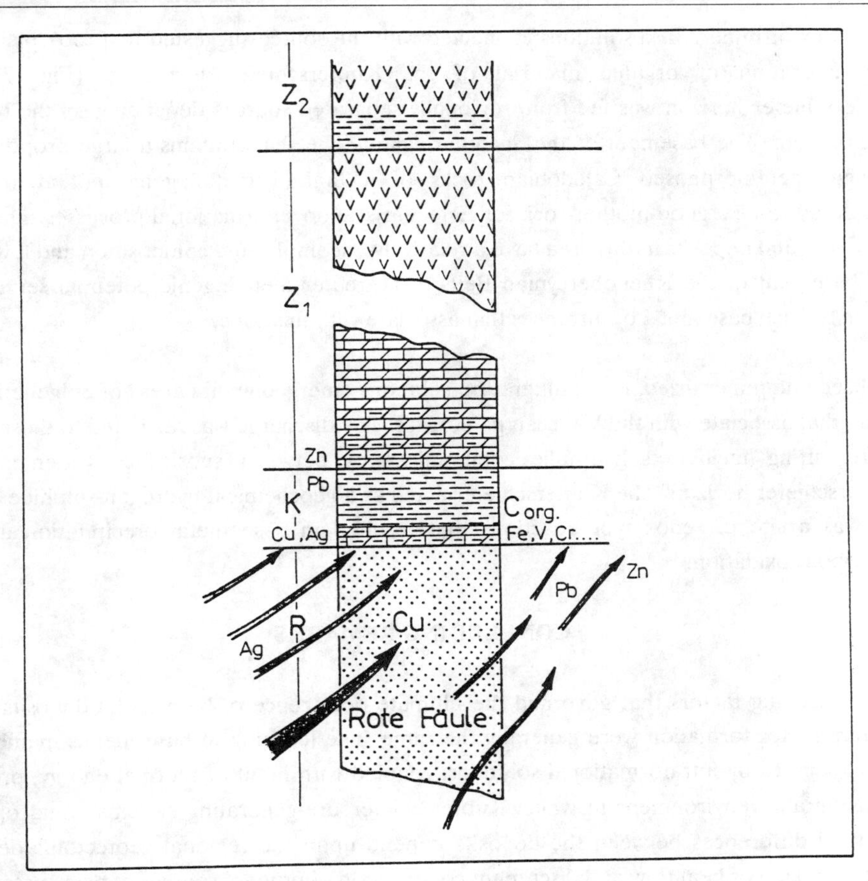

FIGURE 7
Generalized lithological profile in the Lubin-Sieroszowice area. Already diagenetically altered, Kupferschiefer formation with elevated content of Fe, V, Cr, Corg etc., invaded by epigenetic mineralized fluids that relate to the Rote Fäule facies. R - Rotliegendes sandstones and conglomerates, K - Kupferschiefer marls/shales, Z1, Z2 - Zechstein evaporitic cyclothems.

The influx of mineralized solutions connected with the source suggested first took place after the maturation of organic material of the Kupferschiefer formation (Fig. 7). The Kupferschiefer horizon was fed from regionally separated sources depending on the regional development. The basement of the Sacothuringian Zone that contains a large proportion of strongly metamorphosed Caledonian rocks of a limited metallogenic potential is not considered to be a good mother-rock for Late Variscan intraformational processes. Therefore the base metal deposits in this area have a relatively uncomplicated composition and low tenor. The basement of the Rhenohercynian Belts of a greater metallogenic potential served as a source for rich base metal occurrences that associate with this zone.

Oxidized and mineralized, low temperature solutions emerged within areas of epigenetic Rote Fäule, that associate with the Variscan paleohighs. The discharge was restricted to those places where rifting facilitated hydraulic communications between subsurface water and the Kupferschiefer horizon. The Kupferschiefer acted as a geochemical hydrogen-sulphide barrier and was a site of redox type reactions that resulted in base metal precipitation and hydrocarbons oxidation.

CONCLUDING REMARKS

The controlling factors that governed the ultimate occurrence of base metal deposits in the Kupferschiefer formation were generally the same. The leaching of base metals mainly from crustal sources by intraformational solutions, coupled with input of thermal energy, provided the favourable environment in which Kupferschiefer ore-generating systems could operate. The final differences between the deposits depend upon the regional geotectonic development, direction of heat flow and discrepant composition of primary sources of base metals.

Acknowledgements

Sincere thanks are expressed to Dr. W. Püttmann (RWTH Aachen, FRG) for kind cooperation and for his interesting results of organic geochemical studies, that to some extent support the genetic concept presented here.

REFERENCES

Baumann, L., 1978: Zur Bedeutung der Plattentektonik für die Metallogenie-Mineralogie. Z. Geol. Wiss. 6, 1357-1377.

Behr, J.J., W. Engel,W. Franke, P. Giese and K. Weber, 1984: The Variscan Belt in Central Europe: Main structures, geodynamic implications, open questions. Tectonophysics 109, 15-40.

Burek, P.J., 1973: Structural deductions of the initial age of the Atlanctic Rift Systems. In: Tarling, D.H. and S.K. Runcorn (eds) Implications of Continental Drift to the Earth Sciences. Academic Press 2, 815-830.

Grimmer, A.,. 1978: Wärmefluß und seine Beziehungen zu plattentektonischen Elementen. Z. Geol. Wiss. 11, 1329-1337.

Jowett, C. E. and Jarvis G.T., 1984: Formation of foreland rifts. Sediment. geol. 40, 51-72.

Lorenz, V., 1976: Formation of Hercynian subplates, possible causes and consequences. Nature 262, 374-377.

Majorowicz, J., 1982: The ambiguities in tectonic interpretation of geothermal field patterns in the platformic areas in Poland. Przegl. Geo. 2, 86-94.

Majorowicz, J., S. Marek and J. Znosko, 1984: Paleogeothermal gradients by vitrinite reflectance data and their relation to the present geothermal gradient patterns of the Polish Lowland. Tectonophysics 103, 141-156.

Petrascheck, W.E., 1976: Mineral zoning and plate tectonics in the Alpine-Mediterranean area. Geol.Soc. Canada Spec. Pap. 14, 353-359.

Püttmann, W., H.W. Hageman, C. Merz and S. Speczik, 1988: Influence of organic material on mineralization processes in the Permian Kupferschiefer Formation, Poland. Org. Geochem. 13, 357-363.

Püttmann, W., C. Merz and S. Speczik, 1989: The secondary oxidation of organic material and its influence on Kupferschiefer mineralization of southwest Poland. Appl. Geoch. 4, 151-161.

Rentzsch, J., 1981: Mineralogical-geochemical prospection methods in Central European Copper Belt. Erzmetall 34, 492-495.

Roy, R.F., D.D. Blackwell and E.R. Decker, 1972: Continental heat flow. In: Robertson, E.C. (ed.). The nature of Solid Earth. McGraw-Hill, New York.

Sawkins, F.J., 1976: Metal deposits related to intracontinental hotspot and rifting environments. J. Geol. 84, 653-671.

Sawkins, F.J. and K. Burke, 1980: Extensional Tectonics and Mid-paleozoic massive sulfide occurrences in Europe. Geol. Rundschau 69, 349-360.

Sillitoe, R.H., 1972: A plate tectonic model for the origin of Porphyry Copper deposits. Econ Geol. 67, 184-197.

Speczik, S., 1979: Ore mineralization in the basement Carboniferous rocks of the Fore-Sudetic Monocline (SW Poland). Geol. Sudetica 14, 77-122 (In Polish).

Speczik, S., 1985: Metallogeny of pre-Zechstein basement of the Fore-Sudetic Monocline (SW Poland). Geol. Sudetica, 20, 37-100. (In Polish).

Speczik, S., and Kozlowski, A., 1987: Fluid inclusion study of epigenetic veinlets from the Carboniferous rocks of the Fore-Sudetic Monocline (southwest Poland). Chem. Geology, 61, 287-298.

Speczik, S., C. Skowronek, G. Friedrich, R. Diedel,C. Schumacher and F. P. Schmidt, 1986: The environment of generation of some base metal occurrences in Central Europe. Acta. Geol. Polon. 36, 1-34.

Speczik,S., 1988: Relation of Permian Base Metal Occurrences to the Variscan paleogeothermal Field of the Fore-Sudetic Monocline (Southwestern Poland). In: Friedrich, G. and P.M. Herzig (eds.). Base Metal Sulfide Deposits. Springer Verlag Berlin-Heidelberg, 12-24.

Teichmüller, M. and R. Teichmüller, 1979: Diagenesis of coal (Coalification). In: Larsen, G. and G. Chilingar (eds.) Diagenesis in sediments and sedimentary rocks. Elsevier. Amsterdam 207-246.

Van Breemen, O., M. Aftalion, D.R. Bowes, A. Dudek, Z. Misar, P. Povondra and S. Vrana, 1982: Geochronological studies of the Bohemian massif, Czechoslovakia, and their significance in the evolution of Central Europe. Trans. R.Soc. Edinb. Earth Sci. 73, 89-108.

Walker, A.L. and A.S. Buchanan, 1969: The production of hydrothermal fluids from sedimentary sequences. Econ. Geol. 64, 919-922.

Weber, K., 1978: Das Bewegungsbild im Rhenoherzynikum - Abbild einer varistischen Subfluenz. Z.dt. Geol. Ges. 129, 249-281.

Zwart, H.J. 1967: The duality of orogenic belts. Geol. Mijnbow 46, 263-309.

Post-Variscan Evolution

The Post - Variscan Development of the Rhenish Massif

W.Meyer
Geologisches Institut der Universität, Nußallee 8, 5300 Bonn1, Germany

ABSTRACT

A survey of the post Carboniferous history of the Rhenish Massif is presented, with emphasis on the uplift processes which started in the late Cretacecus and still continue. They are connected with several phases of continental volcanism.

HISTORY OF UPLIFT AND DENUDATION

After the Variscan deformation and metamorphism, which took place between 330 and 300 Ma (Ahrendt et. al. 1983), the Rhenish Massif was strongly denuded. During the Lower Permian (Rotliegendes) the denudation nearly achieved the present level, although the surface of the massif was still weakly undulated parallel to the NE-SW trending Variscan folds. Thus several troughs developed, which were filled with debris under fluvial conditions. During the Lower Triassic (Buntsandstein) the area was nearly peneplained, but at some points the Variscan fold pattern was preserved in the form of minor swells. In the western part of the Rhenish Massif the Eifel North-South Depression developed as a small subsiding basin and provided a connection between the North German - Netherland Basins in the north and the sedimentary basins of eastern France and southern Germany during Triassic and Lower Jurassic times. Most parts of the massif outside the Eifel Depression remained unconvered during the Triassic and Jurassic. The northern and northwestern margins of the massif were affected by some marine ingressions during the Upper Cretaceous. This gives evidence that the massif was a peneplain only slightly above sea level at the end of the Mesozoic. Since Jurassic times warm-age weathering has taken place in the whole Rhenish Massif, leading mainly to a decomposition of the slates down to more than 20 m below the surface.

A distinct change in the structural history of the massif was caused by younger uplift. The first uplift movements took place in the northwestern part (Aachen-Maestricht region) at the end of the Cretaceous and in the early Tertiary. In spite of these movements the surface of the massif did not rise much above sea level, so a short marine transgression during Middle and Upper Oligocene covered large areas of the western and central Rhenish Massif (Sonne 1982). The main uplift started at the end of Oligocene and continued with changing intensities up to the present.

VOLCANISM AND MINERALIZATION

The early phase of uplift is connected with strong volcanism in the Hocheifel area, which, after some Cretaceous precursors, had a maximum at the end of the Eocene and the beginning of the Oligocene (42 - 34 Ma, Lippolt 1983). Most volcanoes had an alkali basaltic composition; in the centre of the volcanic field there occur fractionation products, e.g. hawaiites, mugearites, benmoreites, trachytes (Huckenholz 1983). The volcanoes are distributed in an elongated field parallel to the Eifel North-South Depression. This is the last feature connected with the Eifel Depression.

With the beginning of the main uplift at the end of the Oligocene a NW-SE fault system developed, which led to the subsidence of a large triangular shaped basin, the Lower Rhine Embayment. In the centre of the rising Rhenish Massif the small Neuwied Basin subsided synchronously. It is located at the intersection of two graben-like troughs, the NE-SW trending Mosel trough and the NW-SE trending Rhine trough. The down-faulting in each of these troughs amounts to 150 m. The combined displacements give the maximum subsidence of the Neuwied Basin, which is in fact 300 m (Meyer 1988).

The Cenozoic uplift of the Rhenish Massif is connected with a hydrothermal Pb-Zn-Cu mineralisation which seems to have affected nearly the whole massif. The most important deposits are in the Aachen-Stolberg region (replacement in carbonates) and near Mechernich and Maubach in the northern Eifel Depression (dissemination in Triassic sandstones). Beside that, there are widespread veins, e.g. in the western Eifel region (Bleialf, Rescheid) and the northern Sauerland area (Brilon). The exact age of this mineralisation is still in discussion; according to Large et al. (1983) and Schaeffer (1984) a maximum probably could have occurred in late Mesozoic or early Tertiary times.

This second phase of strong uplifting and rifting was also accompanied by volcanism. In the southern continuation of the Lower Rhine Basin there developed a volcanic field (now cut by the Rhine Valley). Most of the K/Ar ages have values about 25 Ma (Lippolt 1983), at the end of the Oligocene. The elongation of this field is parallel to the NW trending rift structures of the Lower Rhine Basin and the Rhine Trough; volcanic dykes and chains of volcanoes show the same orientation. Another volcanic centre is located farther east in the Westerwald area. The average composition is that of alkali basalts. Due to fractionation intermediate to acidic lavas and pyroclastics occur in some centres (Siebengebirge, near Bonn, and southwestern Westerwald).

CONTROL OF RECENT GEOMORPHOLOGY

Because of the increasing height differences between the Massif and the subsiding Lower Rhine Basin, river systems developed. The clay material of the decomposed Palaeozoic rocks has been denudated and transported into small grabens or into the large basin, where thick clay deposits are intercalated with lignite seams or marine sands. The Rhine has crossed the Rhenish Massif since the Middle Miocene; the Mosel valley is probably older. Despite the uplift since the Oligocene, both valleys remained rather broad and had gentle slopes, even at the beginning of the Quaternary.

In the Rhenish Massif a period of strong uplift started during the late Quaternary. The rivers were forced to cut narrow valleys with steep flanks into the uprising plateau. Thus the present landscape developed. This phase of strong uplift, which is still active, began before 0.5 Ma, but the exact date of the beginning of these processes is still unknown; estimated ages range from 400,000 to 800,000 years b.p. The Quaternary uplift has been accompanied by volcanic activity, concentrated in two fields: one extended in NW-direction in the western Eifel between Bad Bertrich and Ormont; the other is located in the eastern Eifel near the Rhine and the Neuwied Basin. The western field has mainly produced basaltic material. The eastern volcanic area, besides basalt volcanoes, contains three complexes with phonolithic and trachytic material. Here violent eruptions of tuff and pumice led to the collapse of huge calderas. The three intermediate caldera complexes are the Rieden, the Wehr and the Laach complex; the latter with Lake Laach.

REFERENCES

Ahrendt, H., Clauer, N., Hunziker, J.C. & Weber, K. (1983): Migration of Folding and Metamorphism in the Rheinisches Schiefergebirge Deduced from K-Ar and Rb-Sr Age Determination. - In: Martin, H. & Eder, (eds.): Intracontinental Fold Belts, Berlin, Heidelberg (Springer) 323-338.

Fuchs, K., v. Gehlen, K., Mälzer, H., Murawski, H., Semmel, A. (eds.) (1983): Plateau Uplift, Berlin, Heidelberg (Springer), 411 p.

Huckenholz, H.-G. (1983): Tertiary Volcanism of the Hocheifel Area. - In: Fuchs, K. et al. (eds.): Plateau Uplift, Berlin, Heidelberg (Springer) 121-128.

Large, D., Schaeffer, R. & Höhndorf, A. (1983): Lead Isotope Data from Selected Galena Occurrences in the North Eifel and North Sauerland, Germany. - Mineral. Deposita 18, 235-243.

Lippolt, H.J. (1983): Distribution of Volcanic Activity in Space and Time. - In: Fuchs, K. et al. (eds.): Plateau Uplift, Berlin, Heidelberg (Springer), 112-120.

Meyer, W. (1988); Geologie der Eifel. - 2nd. ed., Stuttgart (Schweizerbart) 615 p.

Meyer, W., Albers, H. J., Berners, H.P., v. Gehlen, K., Glatthaar, D., Löhnertz, W., Pfeffer, K.H., Schnütgen, A., Wienecke, K. & Zakosek, H. (1983): Pre-Quaternary Uplift in the Central Part of the Rhenish Massif. - In: Fuchs, K. et al. (ed.): Plateau Uplift, Berlin, Heidelberg (Springer), 39-46.

Quitzow, H.W. (1974): Das Rheintal und seine Entstehung. Bestandsaufnahme und Versuch einer Synthese. - Centenaire de la Soc. Géol. de Belgique. - L' Evolution Quaternaire des Bassins Fluviaux: 53-104.

Schaeffer, R. (1984): Die postvariszische Mineralisation im nordöstlichen Rheinischen Schiefergebirge. - Braunschw. geol.-paläont. Diss. 3, 206 p.

Sonne, V. (1982): Waren Teile des Rheinischen Schiefergebirges im Tertiär vom Meer überflutet? Mainzer geowiss. Mitt. 11, 217-219.